Marcel Schneider

The Imperial Qín Dynasty

Welten Ostasiens / Worlds of East Asia / Mondes de l'Extrême Orient

Im Auftrag der Schweizerischen Asiengesellschaft –
On behalf of the Swiss Asia Society –
Au nom de la Société Suisse-Asie

Volume 37

Marcel Schneider

The Imperial Qín Dynasty

Elements of Governance as Reflected in the *Lǐyē* 里耶 Manuscripts

DE GRUYTER

The open access publication of this book has been published with the support of the Swiss National Science Foundation.

ISBN 978-3-11-079183-9
e-ISBN (PDF) 978-3-11-079185-3
e-ISBN (EPUB) 978-3-11-079195-2
ISSN 1660-9131
DOI https://doi.org/10.1515/9783110791853

Library of Congress Control Number: 2024940483

Bibliographic information published by the Deutsche Nationalbibliothek
The Deutsche Nationalbibliothek lists this publication in the Deutsche Nationalbibliografie; detailed bibliographic data are available on the internet at http://dnb.dnb.de.

Cover image: Marcel Schneider
Typesetting: Integra Software Services Pvt. Ltd.
Printing and binding: CPI books GmbH, Leck

MIX
Papier | Fördert gute Waldnutzung
FSC® C083411

www.degruyter.com
Questions about General Product Safety Regulation:
productsafety@degruyterbrill.com

Contents

Acknowledgements

There are guiding figures whose wisdom and unwavering support light the way through the complexities of research and discovery. To Prof. Dr. Wolfgang Behr and Prof. Dr. Hans B. Thomsen, I extend my deepest gratitude. Your guidance, patience and encouragement have been the driving forces behind this journey. I am particularly indebted to Prof. Dr. Chén Wěi and the Lǐyē Museum of Qín Slips, whose willingness to share insights and engage in extensive interviews made my research both enlightening and enjoyable. To my family and friends, especially my wife Xu Zelin and my two little whirlwinds of joy, Célia and Marlon – thank you for keeping me grounded throughout the roller coaster of academia and life. Your love, laughter and gentle reminders to live in the moment have been invaluable, flipping my perspective in the most delightful and unexpected ways. My heartfelt thanks also go to the Institute of Asian and Oriental Studies at the University of Zurich, De Gruyter and the Swiss National Science Foundation. Your support has been essential in bringing this work to a wider audience. Special thanks to Xú Dá, copy editor Simon Pinney, and to Maxim Karagodin, Katja Lehming and Kathleen Prüfer from De Gruyter for their assistance in shaping this book.

https://doi.org/10.1515/9783110791853-203

Periodization

The following list comprises an overview of the historical dynasties referred to in this book.[1]

Shāng 商 / Yīn 殷	**ca. 1500–1045 BC**
Western Zhōu 周	**ca. 1045–771 BC**
Eastern Zhōu 周	**770–221 BC**
Springs and Autumns period	770–476 BC
Warring States period	453–221 BC
Qín 秦 Dynasty	**221–206 BC**
Hàn 漢 Dynasty	**206 BC–220 CE**
Western Hàn 漢	206 BC–9 CE
Xīn 新 Dynasty (Wáng Mǎng 王莽 *interregnum*)	9–23 CE
Eastern Hàn 漢	25–220 CE

1 This periodization table is based on Barbieri-Low and Yates 2015: Section 1.2. For more on the periodization of Ancient China see, for example, Higham 1996; Keightley 1999; Lǐ Xuéqín 2002b; Shaughnessy 2009.

https://doi.org/10.1515/9783110791853-204

Abbreviations

Shuìhǔdì	Shuìhǔdì Qín tomb bamboo manuscripts (*Shuìhǔdì Qín mù zhú jiǎn* 睡虎地秦墓竹簡), excavated in 1975 at modern-day Chéngguān Shuìhǔdì 城關睡虎地 in Yúnmèng 雲夢 County, Húběi Province. The tomb is dated 217 BC.[2]
"Qín lǜ shíbā zhǒng"	Documents titled "Eighteen Categories of Qín Statutes" (*Qín lǜ shíbā zhǒng* 秦律十八種) in the *Shuìhǔdì* texts.
Yuèlù	Yuèlù Academy Qín manuscript collection (*Yuèlù shūyuàn cáng Qín jiǎn* 岳麓書院藏秦簡), purchased in 2007 by the Yuèlù Academy 岳麓書院 at the Hong Kong antique market. The texts span both the pre-imperial and imperial Qín periods (late 3rd century BC).[3]
"Wéi yù děng zhuàng"	Documents titled "Four Types of Documents for Trying Criminal Cases and Other (Procedures)" (*Wéi yù děng zhuàng sì zhǒng* 爲獄等狀四種) in the *Yuèlù* texts. These texts date back to the Qín pre-imperial period and contain criminal cases from 246 to 222 BC.[4]
Zhāngjiāshān	Zhāngjiāshān Hàn bamboo texts (*Zhāngjiāshān Hàn jiǎn* 張家山漢簡), excavated in 1983 and 1984 from a Western Hàn dynasty tomb of an imperial official who was active between 202 and 194 BC. The tomb is dated 186 BC and located near modern-day Jīngzhōu 荊州, Húběi Province.[5]
"Èrnián lǜlìng"	Documents titled "Statutes and Ordinances of the Second Year" (*Èrnián lǜlìng* 二年律令) in the *Zhāngjiāshān* texts.
"Zòuyànshū"	Documents titled "Book of Doubtful Cases Reported to the Higher Authorities" (*Zòuyànshū* 奏讞書) in the *Zhāngjiāshān* texts. These documents mainly contain legal cases from 200 to 195 BC that were forwarded to a higher authority to be adjudicated due to their ambivalence or difficulty.[6]
Shǐjì	The *Records of the Grand Historian* (*Shǐjì* 史記) is a history compiled by Sīmǎ Tán 司馬談 (165–110 BC) and his son Sīmǎ Qiān 司馬遷 (135–86 BC) under the auspices of Emperor Wǔ 武 of Hàn (141–87 BC).[7]

2 Gāo Mǐn 2000.
3 Zhū Hànmín and Chén Sōngzhǎng 2010.
4 Lau and Staack 2016: 11–12; 310.
5 *Jīngzhōu dìqū bówùguǎn* 1985.
6 Lau 1999: 38.
7 Qiān Sīmǎ 1979.

https://doi.org/10.1515/9783110791853-205

1 Introduction

Excavated manuscripts written on bamboo, wood and silk dating from the fifth to the late second century BC have become important sources of evidence for all fields of scholarly research on Early China. The sheer amount of recently excavated material has revealed invaluable information on a wide range of topics, proving itself essential for discerning and, more often than not, reassessing conventional perspectives on the Qín dynasty as reported in transmitted texts. For all its fame and notoriety, the imperial Qín period remains one of the most heavily misconceived periods in the history of China.

Two of the first political opponents to fiercely question the achievements of the Qín and ignite contentious debate regarding the alleged misuse of power by the Qín Emperors were Xiàng Yǔ 項羽 (r. 206–202 BC), an insurgent warlord from the former state of Chǔ 楚, who annihilated the imperial capital in 206 BC, and Liú Bāng 劉邦 (r. 202–195 BC), the latter founder of the Hàn dynasty, subsequently known as Emperor Gāozǔ 高祖 of Hàn.[8] These figures spearheaded large scale military insurgencies in order to shape public opinion and satisfy a population that, according to the sources accredited to them both, had been systematically deprived of its cultural heritage, basic rights and liberties.[9] Aside from certain nuanced assessments, literature and material culture began to form the basis of a prescriptive historical narrative, irrevocably setting the tone, pace and direction of how the short-lived and much-vilified Qín dynasty would be portrayed ever since.[10]

Against this backdrop, a proliferation of Qín-related anecdotes, historically flattened and contextually modified, transformed the image of the empire from that of a unifying power to one of governmental fiasco. Ironically, the subsequent rise of the Hàn dynasty was strengthened by the very use of coercive mechanisms and strategic reforms.[11] These included, among other things, administrative, economic, legal and military adjustments in newly occupied territories that enhanced the ability to harness resources from conquered populations.[12] The Qín administration had already possessed efficient means of resource extraction, which, being too valuable to be abandoned, were readily appropriated by the Hàn Emperors *eiusdem generis* for their ascent to political hegemony. After the

8 See Tián Rénlóng 1992.

9 See Wáng Xiānqiān 1985: 12.328.

10 See Dull 1983.

11 On the logic of domination in Early China, see Hui 2005: 10, 29, 33.

12 For a discussion on the balance between offensive and defensive measures in political theory, see Glaser and Kaufmann 1998; Hopf 1991; van Evera 1999.

https://doi.org/10.1515/9783110791853-001

Qín collapsed, the burden was on the Hàn sovereigns to restore traditional values, promote scholarly activity, retain imperial power and create an inclusive, sustainable model of rulership.[13]

Accusations of pure violence or coercion regarding the efforts of the Qín to abolish the order formerly belonging to the Kings of the 'Three Dynasties' (*sān dài* 三代) should immediately raise eyebrows. The timeline of the Xià-Shāng-Zhōu chronology, a 1996 multi-disciplinary project commissioned by the People's Republic of China, and the millennia-spanning sense of cultural continuity it attributes to the vast territories of the Three Dynasties are highly problematic and cast doubts on the methods applied by the research teams.[14] Furthermore, mobilization for economic production and warfare based exclusively on violence and coercion would certainly not have sufficed to sustain the new social structure envisioned by the political architects of the Qín.

For instance, a strong communal belief in social status and mobility, embedded within a system of meritocracy, shared ancestry, ritual practices, or conceptions of legal subjectivity among large segments of the ruling class and wider population, was essential in establishing a foundation for the creation and expression of political legitimacy. As not only transmitted sources but also excavated material corroborate, the Qín dynasty was neither a "legalist totalitarian regime" that completely abandoned state rituals, nor did it sharply distinguish between proponents of other traditionalist or progressive ideas.[15] Recent excavations have already brought to light preliminary discoveries relating to *Lǎozǐ* 老子, *Mòzǐ* 墨子, *Guǎnzǐ* 管子 and the *Annals of the Warring States* (*Zhànguó cè* 戰國策).[16]

The unanimous rejection of doctrinal pluralism is not what these sources confirm, and the inherited miscellany of ideological teachings continued to directly impact the power of the sovereign. Intellectual diversity may have been encouraged by the numerous scholars of the Warring States who sought political refuge and economic opportunity at the newly established court.[17] However, there was no strict hierarchical order to this network of doctrines, only a vague notion of complementary interests that should be understood as an extension of

13 See Kern 2000: 155–163; Lewis 2007: 51; Pines 2004.

14 For more information on the Xià-Shāng-Zhōu Chronology Project (*Xià Shāng Zhōu Duàndài Gōngchéng* 夏商周断代工程) commissioned by the People's Republic of China, see Keightley 1999; Lǐ Xuéqín 2002b; Shaughnessy 2009.

15 Hulsewé 1985; Kanaya 1960: 242–243; Kern 2000: 188.

16 Loewe 1977: 102, 115.

17 Fields 1989: 23.

the intellectual developments of the late Warring States and early imperial periods.[18]

It cannot be ruled out that tyrannical measures were employed to suppress clandestine activity within the court, as well as political criticism fuelled by fear, resentment, mistrust, cultural isolation or rejection. After all, the Qín reforms had profound consequences for both the structure of government and the lives of individual people. Nonetheless, many of these reforms were not fully implemented within the limited time span of the empire and took many decades or centuries to be completed. To name one example, the military and bureaucratic effort of the Qín to unify the writing system seems absent in excavated material and appears only once in the *Shǐjì*.[19]

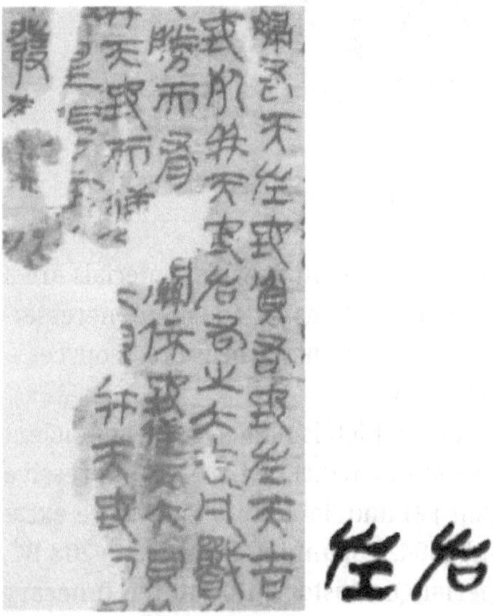

Fig. 1.1–1.3: Section of the Western Hàn dynasty *Mǎwángduī* 馬王堆 manuscript (Fig. 1.1) in which the character '*zuǒ* 左' appears in both clerical script (Fig. 1.2) and small seal script (Fig. 1.3).[20]

One reason for the slow implementation of the writing reform may have been the still prevailing orthographic 'irregularities' (*bù zhèng* 不正) during the Qín and Early Hàn periods (Figs. 1.1–1.3).[21] Excavated records from a Western Hàn dynasty tomb at Zhāngjiāshān 張家山 (186 BC), formulating punishments for officials who

18 Galambos 2004; Hulsewé 1985; Shelach and Pines 2006.

19 Galambos 2004: 181–182, 192.

20 Galambos 2006: 44.

21 Traditional scholarship mainly relies on the transmitted texts of the "Little School" (*Xiǎoxué* 小學), the "Treatise on Literature" (*Yìwénzhì* 藝文志) in the *Book of Hàn* (*Hànshū* 漢書), and *Explaining the Unit Characters and Analyzing the Compound Characters* (*Shuōwén jiězì* 說文解字) postface did not practice the 'regular' (*zhèng* 正) script, lend credibility to the presumption that orthographic standards were not fully adopted, let alone fully centralized by Lǐ Sī's 李斯 (ca. 280–208 BC; served 246–208 BC) reforms.[22] Epigraphic evidence on stone and bronze highlight the gradual changes in character variability over hundreds of years (Fig. 1.4).

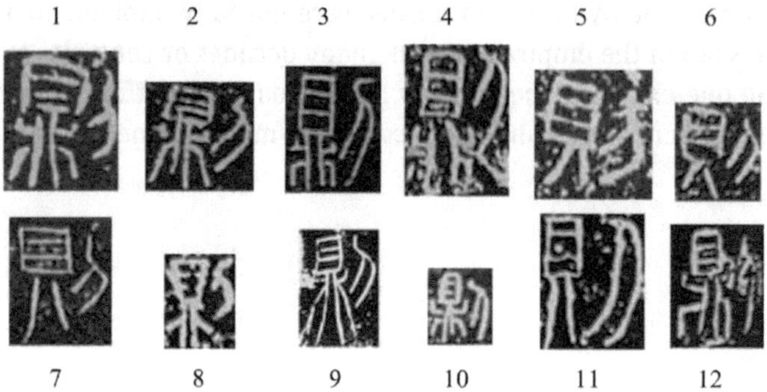

Fig. 1.4: Graphic variants of the character '*zé* 則' on Qín dynasty edict plates.[23]

Written artifacts, stone stelae and information inscribed on durable materials are a treasure trove for researchers studying ancient civilizations. Although not necessarily free from tampering or reworking, these texts – unlike transmitted sources – are available to us in their original, preserved form.[24]

One of these sources is the *Lǐyē* 里耶 corpus, which is named after an ancient township in present-day Lóngshān 龍山, Húnán Province, where an impressive collection of over 37,000 wooden 'strips' (*jiǎn* 簡) and 'tablets' (*dú* 牘) were excavated from wells and other sites in 2002 and 2004.[25] Spanning from 222 to 208 BC, the manuscripts cover the majority of imperial Qín history. Due to the funerary significance of sites at which other text corpora were found, Mark Edward Lewis has linked a substantial collection of early legal writings to ritual practices "remi-

for details on the Qín writing reforms. This has created misconceptions about the structural consistency of Warring States characters (Galambos 2006: 31).

22 Lǐ Xuéqín 2002a. Lǐ Sī 李斯 was the closest advisor to the first two Emperors and the chief political architect of the Qín dynasty.

23 Wáng Huī 1990 in Galambos 2006: 37.

24 On the question of the oral and written nature of early philosophical texts or the long prehistory of writing materials used in the production of Western Zhōu bronze inscriptions, see Shaughnessy 2015; Škrabal 2019. The same may hold true for the preceding Shāng oracle bone inscriptions, which could be considered secondary or tertiary sources (Keightley 2001).

25 Chén Wěi 2012; Gāo Yīzhì 2013: 239–244

niscent of the Zhōu bronzes," arguing that these either protected "the deceased in the afterlife" or equipped "the tomb with all the materials needed to continue the deceased's mode of living in the world beyond."[26] Insofar as the texts themselves support the idea that their presence at burial sites could have served as an official marker of social status or professional expertise, they blur the line between ritual practice and political authority. The *Lǐyē* manuscripts, on the other hand, were actual working documents that grant modern scholars a unique opportunity to redefine our knowledge of the logistical, social, legal, political and ritual responsibilities of the first Chinese empire.

This book is concerned with overall governance at the southern fringes of the Qín dynasty, as reflected in excavated manuscripts from Lǐyē well J1. By analyzing new textual evidence, it aims to revise our understanding of the empire, suggesting that the concentration of authority in imperial times was more moderate than previously assumed and largely counterbalanced by relatively autonomous regional administrations. It also suggests that there was a strong aim toward both unification and historical preservation, as seen in the dynasty's own sense of internal diversification, intellectual openness and adoption of the standards set by earlier rulers. Not only do these findings shed new light on the Qín "legalist" tradition, they also draw attention to customs adopted from earlier periods that were partially institutionalized under the rule of the First Emperor and continued in subsequent periods.

For the dynasty to succeed, it had to quell popular resistance, prevent political disintegration and ensure the effectiveness of its institutions. To achieve this, strategies were employed, and alliances formed to win over various societal groups, such as the advisors of the court, government officials of the imperial units, soldiers on the battlefields and commoners from the 'four quarters' (*sì fāng* 四方). Due to the hostile environment from which the empire rapidly emerged, any endeavor to enact a radically new, all-inclusive change in governance would have been short-lived. Neither a complete departure from previous traditions nor strict adherence to them would have provided a viable solution.[27]

This book has particularly benefitted from the analysis of the *Lǐyē* texts transcribed and published in the first volume of *Lǐyē Qín jiǎndú jiàoshì* 里耶秦簡牘校釋 by Chén Wěi 陳偉.[28] Unless stated otherwise, I will be drawing mainly upon my own translations of text passages from layers 5, 6, 8 – and in some cases – 9, 12, 14, 15 and 16 of Lǐyē well J1. Altogether, the artifacts from these layers span

26 Lewis 2007: 227, 229. My own *pīnyīn* transcription.
27 Kiser and Cài Yǒng 2003: 512–513.
28 Chén Wěi 2012.

from 222 to 209 BC.[29] Other materials, such as the *Yuèlù* and *Shuìhǔdì* texts, or objects of material culture from the Late Warring States or Early Hàn periods, will be taken into consideration where analogies or the potential for complementary approaches exist.

Transmitted sources seem far removed from actual day-to-day practices, but they too are useful for cross-comparisons. Comments accompanying paleographic evidence will be examined through secondary sources to identify both paradigmatic occurrences and prototypical patterns of governance in Qiānlíng County (*Qiānlíng xiàn* 遷陵縣) under the administrative sovereignty of Dòngtíng Commandery (*Dòngtíng jùn* 洞庭郡). Examples will cover the postal system, the lunar calendar and timekeeping, administrative geography, social mobility, land ownership, taxation, written reports on trade with *corvée* labor, criminal proceedings and the allocation or redistribution of essential resources.

Chapter two will provide a summary of the Lǐyē archaeological reports, examining the physical properties and quantitative data of the manuscripts, along with observations on the selected terms '*shǒu* 手', '*fā* 發' and '*xíng* 行'. Chapter three deals with calendar and timekeeping systems, which ensured the timely cultivation of crops, the upkeep of administration, the allocation of goods to where they were most needed and the collection, archiving and retrieval of information. A thorough understanding of these activities also allows us to date damaged manuscripts or manuscript fragments and to reconstruct distances between locations. Chapter four discusses approaches to the formation of the first empire, including its territorial fragmentation into smaller units called 'commanderies' (*jùn* 郡) and 'counties' (*xiàn* 縣). As will be demonstrated, this organizational structure had already been introduced in the Chǔ 楚 (740–223 BC) and Qín kingdoms (337–221 BC), and later provided the groundwork for the division of government into 'Bureaus' (*cáo* 曹) and 'Offices' (*guān* 官), even in the early stages of the empire. Chapter five explores the lowest tier of Qín society, which included workers, convicts and conscript soldiers, all crucial to meeting the empire's primary needs. The Qin dynasty's systems of human trade, *corvée* labor and military conscription were maintained by civil servants and other individuals who were appointed to supervise and control members of the lower classes as well as those sentenced to compulsory work. The last chapter focuses on the generation, storing and redistribution of resources, including detailed grain tax calculations for households at the 'district' (*xiāng* 鄉) level.

The ability to gather, archive and retrieve oral and written information from territories under the empire's control, along with the capacity to exchange informa-

29 The collection includes 35 documents or document fragments excavated from layer 5, 40 from layer 6 and, the majority of remaining documents, 2,552 from layer 8.

tion between lower and higher levels of government, was crucial for building trust between institutions, upholding the rule of law and enabling sound leadership.[30] Conscious efforts were made by the administration in Dòngtíng and other commanderies to sustain 'statutes and ordinances' (*lǜlìng* 律令) that guaranteed the right to legal recourse. Numerous excerpts from the 'statutory provisions' (*lǜ* 律) in the *Lǐyē* corpus demonstrate that judicial institutions were accountable to the general populace, meaning they had to publicly justify the use of coercion or re-evaluate conflicts in relation to the law upon request.

The foundational work carried out by the excavation team at the Húnán Provincial Research Institute of Cultural Relics and Archaeology (*Húnán shěng wénwù kǎogǔ yánjiūsuǒ* 湖南省文物考古研究所), together with contributions made by Chén Wěi and other scholars of Early China in combining and reconstructing manuscript texts from multiple, sometimes heavily damaged fragments, has become indispensable for a meticulous analysis of the *Lǐyē* archive. Nonetheless, the results of their labor should remain open to debate.

For instance, some characters that appear to have been discontinued after the fall of the Qín empire in 206 BC are not found in dictionaries and are therefore susceptible to ambiguous interpretation. There is also limited consensus on text punctuation and the sequence of manuscripts or fragments. Even though matching texts are archived by number and text row or column, it is difficult to isolate which part of the text is attributed to which document, thus complicating the identification of specific officials at the administrative level.

The material used for these writings – whether bamboo or wood – conveys additional information through factors such as length, width, shape, cut-off corrections, empty spaces between characters, blackened regions, side notches, and more. These elements can significantly deepen our understanding of early textual culture, particularly of how messages were transmitted and received. Transcriptions, of course, tell us nothing about the handwriting or positioning of texts within the actual documents. In archaeological practice, researchers might be missing the precise excavation locations for documents from the various layers of well J1, failing to identify any noticeable clusters in particular areas of each layer, or lacking records of blank documents. Such information, if available, could be readily used to determine the physical age of the archived material via modern dating methods.

Despite these shortcomings, the *Lǐyē* corpus suggests an alternative understanding of the Qín dynasty that is both important and unique. For this reason, it should be treated as new epigraphic evidence that could help address unan-

30 For a discussion of the differences between an archive and library in the Former Hàn dynasty, see Fölster 2018.

swered questions and verify or challenge existing records in complementary sources. Each individual source, on its own, can be deceptive and does not provide a representative image of an entire polity. It is a fallacy to assume that knowledge specific to one region allows for comprehensive conclusions about others or the interactions between different locations. Undoubtedly, Qín society was far more multifaceted than any single material could convey. While the *Lǐyē* documents cannot capture the empire in all its nuance and entirety, they do offer a fascinating glimpse into the many microstructures and complexities present at one of its fringes. Whether or not these were remotely characteristic of the surrounding areas at the time, they still provide a detailed perspective on the brief yet historic emergence of the imperial regime.[31] This is a partial commentary on the ongoing excavation, restoration and transcription efforts conducted mainly by Chinese research teams and scholars. As more material is uncovered and further research published, I look forward to additional contributions to the field.

Technical Conventions

Corresponding texts are combined from the lowest to the highest archival number with a plus sign ("+") in between them. Text rows are assigned Roman numerals and the text columns alphabetical letters. Graphs that are not clearly decipherable are put in lenticular brackets ("【】"), and unrecognizable characters are replaced with squares ("□"). A series of dots (". . .") indicates a damaged document in which characters are present yet undecipherable. Black dots in the mid-column ("•") represent the irregularly deployed black dots on the original manuscripts. These are often used to separate textual units, quote sources in technical texts or hint at summations in statistical reports. A square with a diagonal line through it ("☐") refers to a missing part of a document which is unknown in length.[32]

31 This book is based on my dissertation submitted to the University of Zurich in November 2021. It does not account for studies published after its submission, or shortly before.
32 For a discussion of punctuation in early Chinese texts, see, for example, Guǎn Xīhuá 2002.

2 The *Lǐyē* 里耶 Manuscripts

2.1 General Background

The *Lǐyē* 里耶 corpus is named after an ancient township in present-day Lóngshān 龍山, Húnán Province, where an impressive collection of over 37,000 wooden 'strips' (*jiǎn* 簡) and 'tablets' (*dú* 牘) were excavated from wells and other sites in 2002 and 2004 (Fig. 2.1).[33] Of the 37,000 mainly wooden documents, over 18,000 are blank, while over 17,000 bear an approximate 100,000 graphs, making it by far the largest Qín manuscript corpus ever discovered.[34] The manuscripts include archival records linked to the Qín administration of Qiānlíng County (*Qiānlíng xiàn* 遷陵縣) in Dòngtíng Commandery (*Dòngtíng jùn* 洞庭郡), shortly before the unification of the empire in the 25th year of King Zhèng 政 of Qín (222 BC), and encompass most of the dynasty's history, extending up until the second year of Emperor Èrshì 二世 of Qín (208 BC).[35]

What makes the discovery of the corpus so fascinating is the relatively well-preserved physical condition of certain texts. Due to partial publication, these texts have not yet undergone philological analysis or systematic translation into modern languages. Unlike comparable Qín manuscripts – those discovered in tombs at Fàngmǎtān 放馬灘, Yúnmèng Lónggǎng 雲夢壟崗, Wángjiātái 王家臺 and Shuìhǔdì 睡虎地, or others acquired obscurely from the Hong Kong antique market, like the *Yuèlù* archives, the *Lǐyē* manuscripts were excavated *in situ* and seem to have been written and stored for everyday use.[36] An analysis of the cache is underway to ad-

---

33 Chén Wěi 2012; Gāo Yīzhì 2013: 239–244.

34 *Húnán shěng wénwù kǎogǔ yánjiūsuǒ* 2003; *Húnán shěng wénwù kǎogǔ yánjiūsuǒ* 2007; *Lǐyē gǔchéng Qín jiǎn yǔ Qín wénhuà yánjiū* 2009; Lǐ Xuéqín 2003; Mǎ Yí 2007; Shěn Sòngjīn 2003; Wáng Huànlín 2007; Yóu Yìfēi 2013. The number of documents in comparison to the *Lǐyē* archive is as follows: *Yúnmèng Lónggǎng* 雲夢壟崗, 293 Warring States bamboo strips and one wooden tablet; *Yuèlù*, 2,000 Qín bamboo strips from the state of Chǔ; *Fàngmǎtān* 放馬灘, 460 Qín bamboo strips; *Wángjiātái* 王家臺, 814 Qín bamboo strips; *Shuìhǔdì* 睡虎地, 1,155 Qín bamboo strips.

35 The first two lines of *Lǐyē* manuscript 8-757 state: "(What is known) today (as) Qiānlíng became a county in the 25th year (of King Zhèng) [222 BC] (*jīn Qiānlíng niànwǔ nián wéi xiàn* 今遷陵廿五年爲縣)". King Zhèng 政 of Qín, who became August Emperor of Qín (Qín Shǐhuáng 秦始皇) in 221 BC, will be referred to as the First Emperor in this book. "Emperor Èrshì 二世 of Qín," literally meaning "Second Generation Emperor of Qín," will be referred to as the Second Emperor.

36 For a summary of Qín-related epigraphic sources, see Lǐ Xuéqín 2003; Wáng Huī and Chéng Xuéhuá 1999. For a general overview of bamboo and silk manuscript discoveries in the 20th century, see Pīn Yǔqiān and Duàn Shū'ān 2006. The average span of document preservation for archives was roughly 13 years in the Early Hàn dynasty (Wáng Guìhǎi 1999: 227–232) and 10 years

https://doi.org/10.1515/9783110791853-002

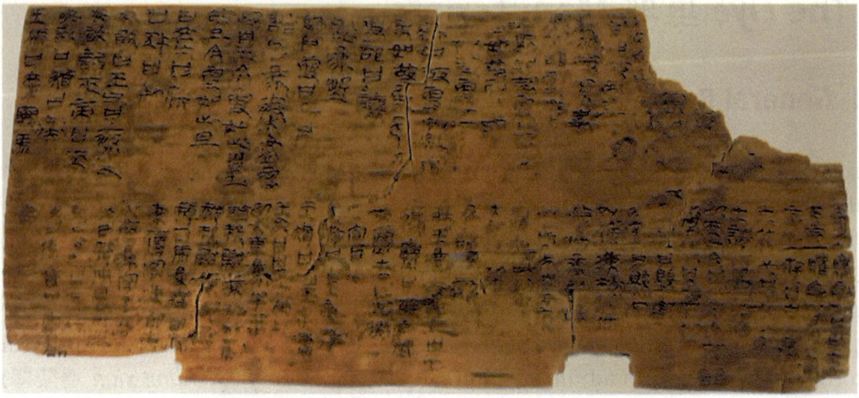

Fig. 2.1: Polychrome photograph of *Lǐyē* wooden 'tablet' (*dú* 牘) 8–461. Ink on wood. Dimensions: 12.5 x 27.4 x 0.6 cm.[37]

dress historical gaps and inaccuracies, and provide supplementary information to researchers in addition to the very rare imperial Qín objects currently available.

Dòngtíng Commandery was situated at the periphery of the empire in the former state of Chǔ 楚. This region borders the Yí 夷 River to the north, Dòngtíng Lake (*Dòngtíng hú* 洞庭湖) and the Zī 資 River to the east, and encloses the region where the Yuán 沅 River originates in the west. In the south lies Tánchéng 鐔成 County (Fig. 2.2).[38] Qiānlíng County is located in the northwestern area of Dòngtíng Commandery and is connected with other parts of the region via the Yǒu 酉, Yuán 沅 and Lǐ 澧 River valleys.

The archeological site of ancient Lǐyē is situated at the southernmost tip of Lóngshān County on the western side of the Yǒu 酉 River valley and covers approximately 5,500 square meters.[39] At an average altitude of 254.5 meters above sea level, its climate is relatively temperate, with adequate rainfall that allows for the cultivation of millet, rice and other grains. The region was logistically accessible and open to trade with other territories through overland transportation routes and waterways. The first two lines of manuscript 8-167+8-194+8-472+8-1011 state:

in the Wú Kingdom (Sòng Shǎohuá 2008: 261). A possible reason for this may have been the danger of uncontrollable fire outbreaks (Ma Tsang Wing 2020a: 545–546).

37 *Húnán shěng wénwù kǎogǔ yánjiūsuǒ* 2012a: 14.

38 Guō Tāo 2017; Yóu Yìfēi 2015.

39 *Húnán shěng wénwù kǎogǔ yánjiūsuǒ* 2007.

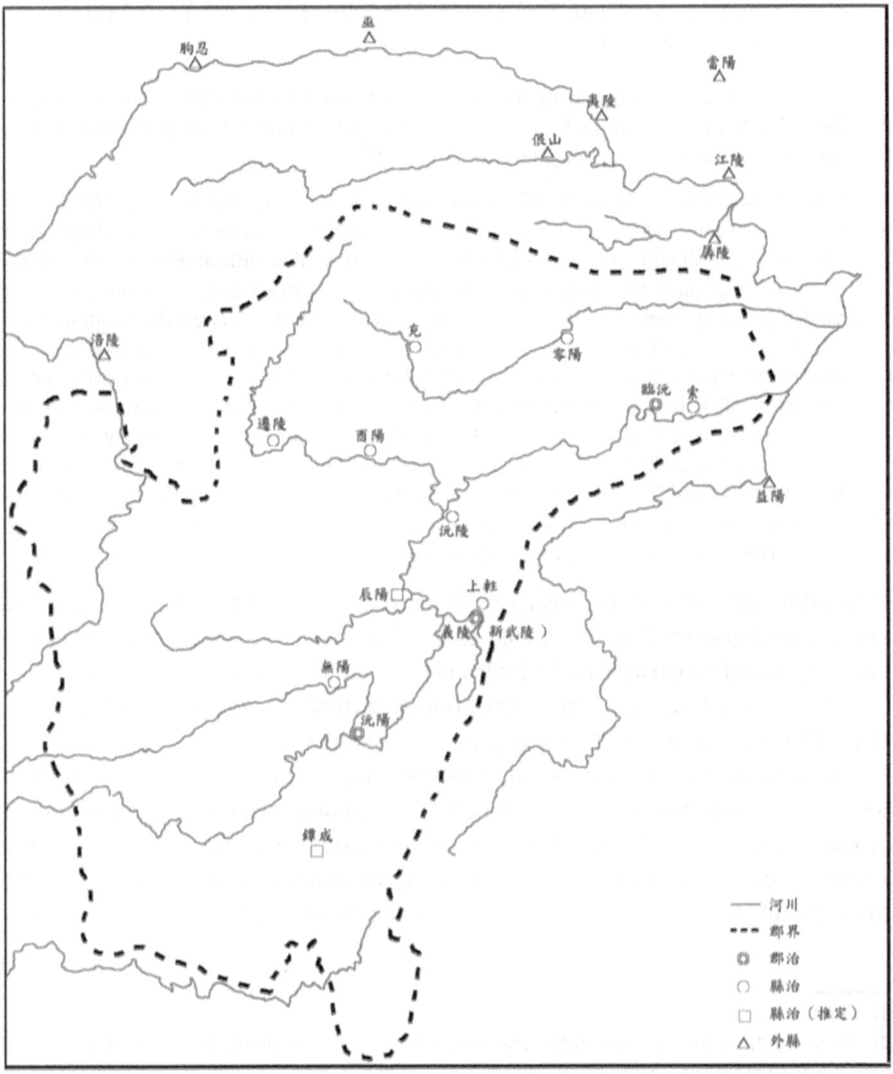

Fig. 2.2: Reconstructed map of Dòngtíng Commandery.[40]

40 Tán Qíxiāng 1982 in Yóu Yìfēi 2015: 42.

8-167+8-194+8-472+8-1011 I–II: 校長寬以遷陵船徙卒史┛【酉陽, 酉陽】□□【船】□元(沅)陵, 寬以船屬酉陽校長徐. [. . .][41]

Section Commander Kuān uses the boats of Qiānlíng to move Commandery Secretaries┛【to Yǒuyáng, Yǒuyáng】□□【boat】□ Yuánlíng (County). Kuān provides a boat to Xú, Section Commander of Yǒuyáng (County). [. . .][42]

Comments: Chén Wěi and Gāo Mǐn argue that an official with the title '*xiàozhǎng* 校長' could have been in charge of a local 'post' (*tíng* 亭) in the Qín dynasty.[43] According to their view, this title should be interchangeable with 'Commanding Official of a Post' (*tíngzhǎng* 亭長). I do not share this opinion. The criminal cases on strips 1–2, 4, 25 and 68 of the "Wéi yù děng zhuàng", and strip 100 of a Hàn dynasty segment in the "Zòuyànshū" both mention '*xiàozhǎng* 校長' and '*tíngzhǎng* 亭長' separately.[44] Information in the *Lǐyē* corpus is scarce on this matter. There is no mentioning of the official title '*tíngzhǎng*'. Perhaps, manuscript 8-149+8-489 (column B, line IV) provides additional information, given that it closely aligns with the hierarchical ordering of strips 471–472 of the "Èrnián lǜlìng". In contrast to the "Wéi yù děng zhuàng" and "Zòuyànshū", which both assign '*xiàozhǎng*' and '*tíngzhǎng*' to authorities at the local level, the "Èrnián lǜlìng" and *Lǐyē* texts associate 'posts' (*tíng* 亭) with '*xiàozhǎng*' – officials at the county level. Yuánlíng 沅陵 and Yǒuyáng 酉陽 were counties in Dòngtíng Commandery, and Yǒuyáng was a county adjacent to Qiānlíng.[45]

Paleolithic and Neolithic findings, revealing traces of prehistoric habitation in the area, were discovered by provincial excavation teams in 1996.[46] In preparation for the planned construction of a hydro electric energy plant, ongoing archaeological efforts, which began in 2002, have revealed further evidence of Shāng, Zhōu, Chǔ (of the Late Warring States), Qín and Hàn settlements.[47]

Two main sites have been documented in excavation reports: the ancient township of Lǐyē, dating back to the Late Warring States in the Chǔ Commandery of Qiánzhōng 黔中, which, after the Qín annexation, was assigned to Qiānlíng County under the administrative sovereignty of Dòngtíng Commandery; and the township of Wèijiāzhài 魏家寨, founded in the Western Hàn dynasty.[48] Chǔ set-

41 Chén Wěi 2012: 101.

42 All translations are my own unless stated otherwise. For a guide on reading and translating early Chinese texts, see Gassmann and Behr 2005.

43 Chén Wěi 2012: 90–91; Gāo Mǐn 1971.

44 See Lau and Staack 2016: 311–312.

45 Yóu Yìfēi 2015.

46 *Húnán shěng wénwù kǎogǔ yánjiūsuǒ* 2007.

47 Fán Guódòng 2009; *Guójiā wénwù jú* 2003: 62–69; *Húnán shěng wénwù kǎogǔ yánjiūsuǒ* 2003, 2007; *Húnán shěng wénwù kǎogǔ yánjiū suǒ, Xiāngxī Tújiā Miáo zú zìzhìzhōu wénwù chǔ* 2003; *Riya Shinkan kōdokukai* 2004.

48 Evidence that Qiānlíng was part of, or in communication with, Dòngtíng Commandery appears in 36 *Lǐyē* documents. "*Qiānlíng Dòngtíng* 遷陵洞庭": 8-181+8-1676, 8-188, 8-189, 8-230, 8-305, 8-333, 8-372+8-1337, 8-443, 8-507, 8-513, 8-515, 8-524, 8-553, 8-556, 8-828, 8-947, 8-1116, 8-1127+8-2397,

tlers had already fortified the township of Lǐyē with a protective moat and wall (Fig. 2.3–2.5). It measured 235 meters in length along the western flank (168 meters extant) and approximately 150 meters to the north (85 meters extant) and south (97 meters extant). Its width varied between 5.4 and 11 meters, with the western partition being the widest at 7 to 11 meters. This section also features remnants of a town gate built during the Hàn dynasty. According to the excavation report, Qín wooden strips were also discovered beneath the wall structure in 2006, but they are still under preliminary investigation. A section of the eastern wall and parts of the town facing the riverbank are missing, perhaps owing to erosion caused by the Yǒu River, whose course was gradually widened in the 20th century.

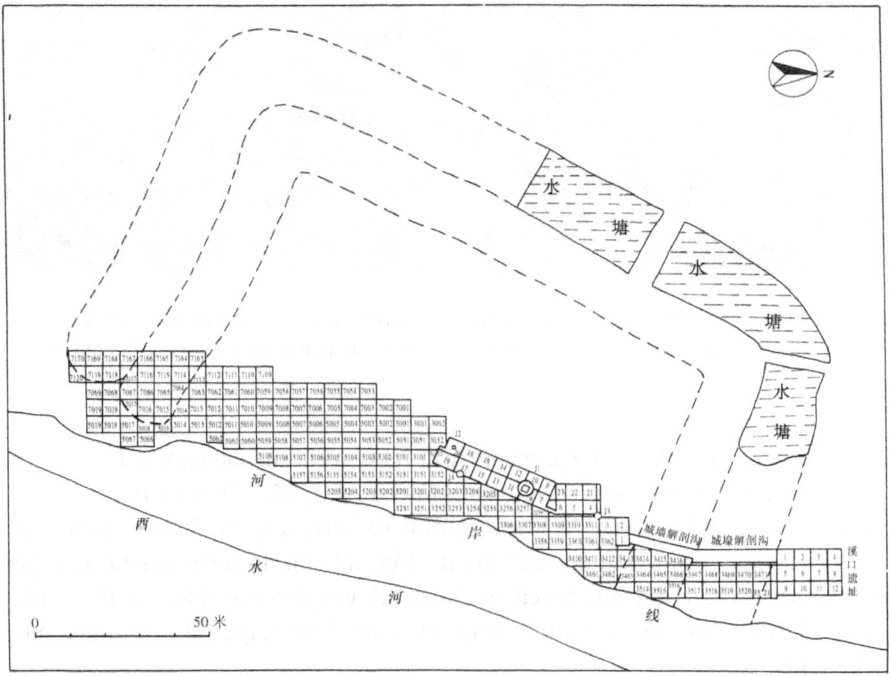

Fig. 2.3: Archeological map of the ancient township of Lǐyē, with the Yǒu 酉 River in the east and the remains of the surrounding moat and township walls.[49]

8-1244, 8-1253, 8-1382, 8-1497, 8-1573, 8-1594, 8-1637, 8-1653, 8-1682, 8-1684, 8-1884, 8-1935, 8-1948; *"Qiānlíng Dòngtíng* 遷陵洞庭*"*: 8-695, 8-976, 8-1826; *"Qiānlíng Dòngtíng jùn* 遷陵洞庭郡*"*: 8-1149; *"Qiānlíng Dòngtíng jùn* 遷陵洞庭郡*"*: 8-469.

49 *Húnán shěng wénwù kǎogǔ yánjiūsuǒ* 2007: 12.

Fig. 2.4: Photograph of the archeological site of Lǐyē, with the reconstructed remains of the surrounding moat and township walls, as seen from the main gate in the northwest. Orientation: north.[50]

In adjacent areas outside the archeological remains of the ancient Lǐyē township, a total of 236 Warring States tombs at Màichá 麥茶, 255 Western Hàn tombs at Qīngshuǐpíng 清水坪, and 70 Western and Eastern Hàn tombs at Dàbǎn 大板 were excavated. They contained an array of burial objects, including bronze, pottery vessels, *bì*-discs (*bì* 璧) as well as coins and weapons that adorned the resting place of distinguished individuals (Fig. 2.6–2.10).[51] No imperial Qín tombs have been found to date.

Several wells were discovered by archeologists alongside the river within the township (Fig. 2.12–2.13). One of these, designated 'J1' and located in survey area T9, measures 3.5 to 5.5 meters in diameter and reaches a depth of 19.4 meters. Starting from layer 5, approximately 3.8 meters below the surface, 13 of the total

50 My photograph taken on May 14, 2019 during a field trip to Lǐyē.

51 For more information on sumptuary laws intended to regulate social hierarchies in early funerary practices, see Brashier 2011; Gāo Chóngwén 2011; Wu Hung 2010.

Fig. 2.5: Photograph of the archeological site of Lǐyē, with the reconstructed remains of the surrounding township walls (on the right) and the modern wall (in the middle) that protects the archeological site from the Yǒu 酉 River, as seen from the main gate in the northwest. Orientation: southeast.[52]

18 layers contained fragmentary Chǔ documents written on bamboo, as well as one document written on a rectangular wooden tablet (in layer 5), a few iron and bronze weapons (in layers 12, 13 and 17) and the entire cache of presently retrieved Qín wooden manuscripts (in layers 5 to 17).[53] Most Qín manuscripts were recovered from layers 6A, 6B, 8A, 9C, 10C, 12 and 15 (Fig. 2.11).

Based on current evidence, Qín officials in Dòngtíng Commandery notably seem to have preferred wood over bamboo for official writing purposes.[54] Compared to other manuscripts discovered in tombs, which are almost exclusively written on bamboo, with the exception of maps depicted on wood, the *Lǐyē* corpus exhibits particularities in terms of materiality, dimensions and content.[55] It does

52 My photograph taken on May 14, 2019 during a field trip to Lǐyē.

53 Chén Wěi 2009.

54 For a discussion on the use of bamboo and wood as writing carriers, refer to Liú Ruì 2003.

55 See Hsu Mei-Ling 1993 concerning wooden maps found at Fàngmǎtān 放馬灘.

Fig. 2.6–2.10: Hàn dynasty objects recovered from tombs in the surrounding area of Lǐyē. From left to right: bronze *hú*-kettle (*tóng hú* 銅壺), M331:1, dimensions unknown; *bì*-discs with colored glaze (*liúli bì* 琉璃璧), M339:1, dimensions unknown; decorated bronze sword (*tóng jiàn* 銅劍), M334:1, dimensions unknown; bronze *dǐng*-vessel (*tóng dǐng* 銅鼎) with inscription, M84:5, dimensions unknown; iron coins (*wǔ tiě qián* 五鐵錢), M99, dimensions unknown.[56]

56 *Húnán shěng wénwù kǎogǔ yánjiūsuǒ* 2007: color print appendixes 47, 48, 57 and 65.

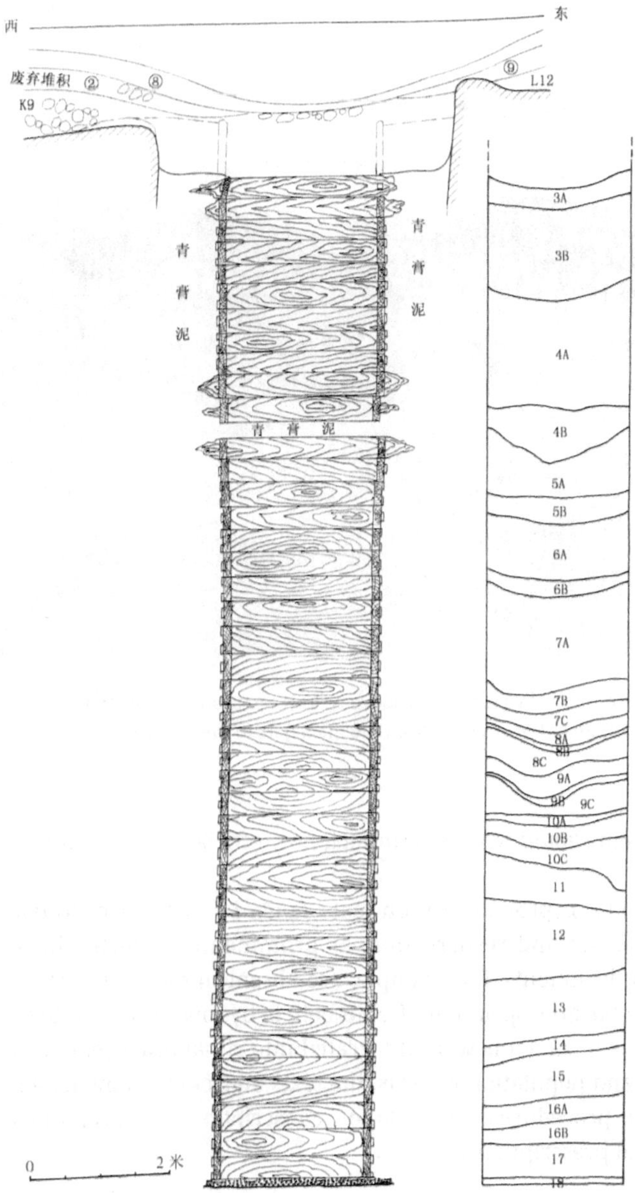

Fig. 2.11: Sketch of well J1 in survey area T9. Profile view of layers 3A–18.[57]

57 *Húnán shěng wénwù kǎogǔ yánjiūsuǒ* 2007: 43.

Fig. 2.12: Photograph of modern protective structures above wells J1 (excavated), J2 (intact) and J3 (intact) within the archeological site of Lǐyē, as seen from the main gate in the northwest. Orientation: east.[58]

not reveal how the system *should* work but, in fact, how officials and administrative relations *did* work.[59]

'Day books' (*rì shū* 日書), 'prognostication texts' (*zhān bǔ* 占卜) or narratives concerned with the afterlife and resurrection, as prominently reflected in the *Shuìhǔdì* or *Fàngmǎtān* documents, for example, are absent in the *Lǐyē* corpus.[60] There is no sign of popular appropriation of elite beliefs by low-ranking officials or any indication that the documents were integrated into ritual practice. Beyond household registration and population records, the texts address land ownership, logistics, fiscal and legal procedures, *corvée* labor, conscription, warfare, official terminology and medical prescriptions.

58 My photograph taken on May 14, 2019 during a field trip to Lǐyē.

59 Yates 2012–2013: 299.

60 For more information on the *Shuìhǔdì* or *Fàngmǎtān* documents, see Lewis 2007: 236; Pines 2004: 38–39.

Fig. 2.13: Photograph of excavated well J1, with modern protective beams made of wood.[61]

As it appears, Qiānlíng County, particularly the township of Lǐyē, was of pronounced administrative and military importance. The Qín administration sent a wide array of people to the region, including soldiers, state officials, workers and others.[62] It may come as unexpected that such weight was given to Qiānlíng, yet one should not underestimate the significance of remote areas in the creation of an imperial identity and the legitimization of the dynasty's rule over vast lands and peoples.[63] Qiānlíng County is not mentioned as an administrative unit under the Qín dynasty in

61 My photograph taken on May 14, 2019 during a field trip to Lǐyē.

62 On forced migration and resettlement in the Qín dynasty, see Barbieri-Low 2019. On the role and practice of local officials in the Late Warring States and early imperial governments, see Sou 2013.

63 Dòngtíng Commandery was much larger and better connected than previously assumed by Charles Sanft and others (*Húnán shěng wénwù kǎogǔ yánjiūsuǒ* 2007: 235; Sanft 2014a: 92–93). Even border regions and settlements in the very south of the Qín and Hàn territories, such as Lǐngnán 嶺南, were probably not "loosely connected to the imperial structure" or "sparse to nonexistent," as described by Mark E. Lewis (Lewis 2007: 11, 151). Rather, they seem to have been characterized by dynamic transportation networks and policies, combining both local and imperial influences (Brindley 2015; Demandt 2020; Yao 2016). Comprehensive studies on interaction, exchange and migration processes along the Eurasian trade routes further reveal the very complex political, geo-

transmitted or excavated historiography and appears for the first time in a *Lǐyē* entry that was recorded one year before the proclamation of the empire.[64]

> 8-757 I–II: [. . .] 今遷陵┘ 廿五年爲縣 [. . .][65]
>
> [. . .] (What is known) today (as) Qiānlíng ┘ became a county in the 25th year (of King Zhèng) [. . .]
>
> Comment: The 25th year of King Zhèng converts to 222 BC.[66]

After the military conquest of the state of Chǔ by the Qín in 223 BC, Qiānlíng County did not become an isolated part of the empire but was instead actively integrated through frequent communication, economic exchange and shared access to resources. Manuscript 52 from layer 16 is a meticulous record of the distances between Qiānlíng and other administrative units:

> 16-52 I–VII: 鄢到銷百八十四里┘ 銷到江陵二百卌里┘ 江陵到孱陵百一十里┘ 孱陵到索二百九十五里┘ 索到臨沅六十里┘ 臨沅到遷陵九百一十里┘ 千四百卌里.[67]
>
> (From) Yān to Xiāo (is) 184 miles┘ (From) Xiāo to Jiānglíng (is) 240 miles┘ (From) Jiānglíng to Chánlíng (is) 110 miles┘ (From) Chánlíng to Suō (is) 295 miles┘ (From) Suō to Línyuán (is) 60 miles┘ (From) Línyuán to Qiānlíng (is) 910 miles┘ 1,440 miles.
>
> Comments: One 'mile' (*lǐ* 里) was a measure of distance in the Qín and Hàn dynasties equivalent to approximately 415.80 meters.[68] As stated by Mǎ Yí, Yóu Yìfēi and Xiè Kūn, Yān, Chánlíng and Suō were counties in the Qín empire, while Xiāo and Jiānglíng were place names.[69] Yàn Chāngguì discusses imperial Qín communication routes not in terms of the hierarchical position of administrative units but by considering topographical features and the course of rivers.[70] When a document had to be transported by boat either downstream or upstream, Yàn Chāngguì advocates the view that this was referred to as '*xià* 下' or '*shàng* 上', respectively. He reaches the same conclusion as Mǎ Yí but adds that Suō was an important strategic hub situated in the north of Lǐyē, where the main 'communication pathways' (called

---

graphical and social flow of goods, information and people at the fringes of the empire and beyond in Early China (i.e. Brosseder 2015; Christian 2000; Di Cosmo 2015).

64 Chén Wěi 2011a, 2011b; Yóu Yìfēi 2015.

65 Chén Wěi 2012: 217.

66 Chén Wěi 2012: 217.

67 Transcription taken from Mǎ Yí 2005.

68 Hulsewé 1961: 206–207.

69 Mǎ Yí 2005; Yóu Yìfēi 2015; Xiè Kūn 2015.

70 Yàn Chāngguì 2013.

'*dào* 道') of Dòngtíng Commandery converged, making it a crucial intersection for the transportation of documents. Suǒ County (*Suǒ xiàn* 索縣) also appears in the "Èrnián lǜlìng".[71]

In order to incorporate newly conquered territories, the Qín sought to replace the decentralized system of government (*fēnfēng zhì* 分封制) with a system based on the parceling of land into 'commanderies' (*jùn* 郡), 'counties' (*xiàn* 縣), 'marches' (*dào* 道) and 'districts' (*xiāng* 鄉), which were further subdivided into 'settlements' (*yì* 邑), 'townships' (*lǐ* 里), 'sections' (*xiào* 校), 'subsections' (*máo* 髳), 'posts' (*tíng* 亭) and 'postal units' (*yóu* 郵).[72]

Officials at Qiānlíng County level supervised at least three 'districts' (xiāng 鄉): Èrchūn 貳春, Qǐlíng 啓陵 and Dū 都. Whenever individuals or 'households' (*hù* 戶) migrated from one district to another, it was the responsibility of the 'District Incumbent' (*xiāngshǒu* 鄉守) to register and forward the matter to the Qiānlíng County authorities "in accordance with the statutes and ordinances" (*rú lǜlìng* 如律令).[73] *Lǐyē* manuscript 8-142, although heavily damaged, contains records of the counting and reallocation of people from one administrative unit to another:

8-142 I–III, rI: 二月辛未[8], 都鄉守舍徒薄(簿)☐┘ 受倉隸妾三人, 司空城☐┘ 凡六人. 捕羽, 宜、委、□☐ └二月辛未[8]旦, 佐初□☐[74]

In month 2 on a *xīnwèi*, Shè, Incumbent of Dū District, (compiles) the register of convicts ☐┘ receive three (female) Bondservants from the Granary, (from) the Commissioner of Public Works. Wall ☐┘ In total, there are six people. Catch bird feathers, Yí, Wěi, □☐ └Month 2 in the morning of a *xīnwèi*, Assistant Chū □☐

Comments: The number in the brackets denotes the position of the day in the 'sexagenary cycle' (*gānzhī* 干支). '*Xīnwèi* 辛未' is in position 8 of the 60-day cycle. For a discussion of calendar and timekeeping systems, see chapter three.

71 Yàn Chāngguì 2006.

72 For a discussion of 'townships' (*lǐ* 里) and 'posts' (*tíng* 亭) in the Qín dynasty, see Lau and Staack 2016; Zhāng Jīnguāng 2004: 587–602. 'Marches' (*dào* 道) were regions similar to counties in size, but inhabited by native tribes (Yates 2012–2013: 306). The term 'settlement' (*yì* 邑) appears on manuscript 8-1236+8-1791, the term 'subsection' (*máo* 髳) on manuscript 8-149+8-489, and the term 'postal unit' (*yóu* 郵) on manuscript 8-1432. For an overview of the construction of territories in the Qín empire, see Tong 2021.

73 One way to interpret 'statutes and ordinances' (*lǜlìng* 律令) is as components of a legal system consisting of 'penal codes' (*lǜ* 律) for the punishment of criminals (and potentially civil codes for non-criminals), and 'administrative codes' (*lìng* 令) for the functioning of administration (Zhāng Zhāoyáng 2008).

74 Chén Wěi 2012: 82.

The term '*shǒu* 守' generally stood for anyone holding office in the *Lǐyē* texts, hence the official title 'Incumbent' is a fitting translation, as suggested by Lau and Staack.[75]

The phrase 'register of convicts' (*tú báo (bù)* 徒溥(簿)) is often preceded by the verb 'to make' or 'to compile' (*zuò* 作).[76] In the *Lǐyē* documents, '*tú(lì)* 徒(隸)' refers to individuals with a criminal record or 'convicts' who were compensated with basic necessities and a fixed state income to repay fines, redemption fees or debts.[77] For more information on *corvée* labor, see chapter five.

Bird feathers were used for a multitude of purposes, including weaponry (e.g. arrows), currency, writing utensils, fans, dance performances, clothing decoration, hangings, jewelry, boxes, hats, carriages, architecture, etc. and were therefore required in large quantities.[78]

A disregard for the rules or failure to uphold deadlines typically resulted in fines or punishment. Strip 184 from the "Statutes Concerning the Delivery of Documents" (*Xíng shū lǜ* 行書律) in the *Shuìhǔdì* explicitly states:

> *Shuìhǔdì*, "Statutes Concerning the Delivery of Documents" (*Xíng shū lǜ* 行書律), strip 184: 行命書及書署急者, 輒行之; 不急者, 日觱(畢), 勿敢留. 留者以律論之.[79]
>
> When forwarding (royal) commands as well as documents marked "urgent", these are to be forwarded immediately. Those that are not urgent are to be fully dealt within a day; one must not venture to withhold them. Cases of withholding are to be condemned according to the statutes.[80]

Lǐyē manuscript 8-137 recounts the failure of a 'servant' (*pú* 僕) to forward reports on domestic animals to higher authorities, which ultimately exposed the wrongdoings of his superior and resulted in the punishment of the Official in Charge of the Documents:

> 8-137 I–III, rI–II: ☒□朔戊午[55], 遷陵丞遷告畜官僕足, 令┛ ☒□毋書史, 畜官課有未上, 書到亟日 ┛ ☒□守府事已, 復視官事如故, 而子弗 ┗☒事, 以其故不上, 且致劾論子, 它承┛ ☒就手.[81]
>
> ☒□ new moon on a *wùwǔ*, Prefect Qiān of Qiānlíng reports to Servant Zú from the Office of Domestic Animals, orders┛ ☒□ not the Official in Charge of the Documents, the evaluations

75 See, Lau and Staack 2016: 86, footnote 486.

76 Examples can be found on manuscripts 8-199+8-688, 8-285, 8-681, 8-686+8-973, 8-787, 8-815, 8-1069+8-1434+8-1520, 8-1143+8-1631, 8-1207+8-1255+8-1323, 8-1278+8-1757, 8-1340, 8-1425, 8-1559 and 8-1742+8-1956.

77 See Cáo Lǚnìng 2015; Gāo Héng 1983; Lǐ Lì 2009.

78 For a study on the documented history of featherwork in Early and Medieval China, see Milburn 2020. For an overview of the trade, taxation and use of feathers during the Qín dynasty, based mainly on readings of the *Lǐyē* manuscripts, see Wáng Zǐjīn 2016.

79 Chén Zhìguó 2008.

80 Translation is taken from Hulsewé 1985: 85, with minor modifications.

81 Chén Wěi 2012: 77.

of the Office of Domestic Animals exist (but have) not yet been sent to the higher authorities. The documents are urgently due┘ ☒□ The matters of the Governor have ended. (He) repeatedly inspected the matters of the office as he did in the past, but he did not └☒ matters, for this reason did not declare to the higher authorities, and moreover exposed the crimes of his master and brought about his judgment, it received┘ ☒ By the hand of Jiù.

The bureaucratic institutions of the imperial administration aimed to "panoptimize" the fundamental elements of human life across different levels of government. To gain an overview of disposable assets, the provincial, prefectural and local administrations needed to collect extensive personal and institutional data, which was then shared among authorities to facilitate a rapid and accurate flow of information. Literate individuals, both before and during the imperial period, generated enormous amounts of written records in labor-intensive scripts. What set the Qín dynasty apart from its predecessors was not merely the creation of administrative records, but its mastery of document archiving and exchange.[82] All relevant information had to be promptly catalogued, accessed and processed at the appropriate time and place, a task as challenging for imperial officials as procuring writing materials and maintaining meticulous records.[83]

2.2 Physical Properties of the Manuscripts

Layer 5 of well J1 revealed several heavily damaged Chǔ characters on bamboo fragments.[84] Judging from the photographs printed in *Lǐyē Qín jiǎn (yī)* and the 2012 transcriptions provided by Chén Wěi, a large percentage of the wooden Qín texts from layers 5 to 17 were either lost or have suffered irreparable damage due to physical exposure, natural decomposition or fire.[85] The majority of strips and

82 On the strategic military importance of document storage and archiving in early imperial times, see Péng Hào, Chén Wěi and Kudô Motoo 2007.

83 Assuming it took a skilled craftsman five minutes to carve and polish each wooden document. For the estimated 100,000 document fragments in the *Lǐyē* corpus, the manufacturing process would have required 500,000 minutes, equaling 694.4 days (500,000 minutes / 720 minutes) or almost two years of continuous work. This is assuming 12-hour workdays without breaks or days off, and includes the time needed to fell and transport the necessary trees to the workshop. Around four to five years might be a more realistic estimate. However, considering that some documents consist of multiple fragments, a total production time of around three years seems plausible.

84 Yates 2012–2013: 292.

85 The first volume of *Lǐyē Qín jiǎn* published by the Cultural Relics Publishing House (*Wénwù chūbǎnshè* 文物出版社) in January 2012 contains high-resolution, polychrome photographs with

tablets are made of 'fir wood' (*shānmù* 杉木), while others are made of either 'pine wood' (*sōngmù* 松木) or tree species of unknown origin.[86] Since radiocarbon or mass-spectrometry dating have yet to be applied to the wood or ink, the age and physical authenticity of the documents remain unverified.

In terms of dimensions, the documents average 23 centimetres in length, equivalent to one Qín 'foot' (*chǐ* 尺), with widths varying from 1.4 to 5 centimeters. Narrow 'strips' (*jiǎn* 簡) usually contain one or two lines, whereas 'tablets' (*dú* 牘) are wider and, depending on the content, feature columns consisting of 2 to 12 lines, with up to 32 graphs per line or a maximum of approximately 370 graphs per side (Fig. 2.14). Given that the texts often continue on the *verso*, or sides of certain multi-angled documents, one strip or tablet often forms a self-contained unit. No overwritten graphs seem to exist. Whenever a mistake needed correction, the part of the document containing the error was cut off with a so-called 'scratch knife' (*xuē* 削) or 'document knife' (*shū dāo* 書刀) and stored separately (Fig. 2.15).[87] Some wooden documents were bound with twine (now decayed) and likely stored in labeled boxes, none of which have survived in Lǐyē.[88]

Robin Yates speculates on the basis of Early Hàn statutes found at Zhāngjiāshān 張家山 that the nondisclosure of information was ensured through the careful storage of documents in labeled and sealed boxes.[89]

> *Zhāngjiāshān*, "Statutes on Households" (*Hù lǜ* 戶律), strips 331–333: 民宅園戶籍、年細籍、田比地籍、田合籍、田租籍, 謹副上縣廷, 皆以篋若匣匱盛, 緘閉, 以令若丞、┘ 官嗇夫印封. 獨別為府, 封府戶. 節(即)有當治為者, 令史、吏主者完封奏(湊)令若丞印. 嗇夫發, 即襍治為.┘ 其【事】已, 輒復緘閉封臧(藏). 不從律者: 罰金各四兩. 其或為詐(詐)偽, 有增減也, 而弗能得: 贖耐.[90]
>
> As for the registers of the people's houses, gardens and households, the detailed registers of ages, the land registers with the neighboring fields, the registers of the names of the fields, and the registers of the field taxes, copies are to be conscientiously forwarded up to the

transcriptions of the 2,627 bamboo and wooden strips and tablets excavated from layers 5, 6 and 8 of well J1.

86 This and the following information is taken from *Húnán shěng wénwù kǎogǔ yánjiūsuǒ* 2007: 179–180.

87 For a drawing of a 'scratch knife' (*xuē* 削) and other writing utensils, such as a bamboo writing brush, or an ink stone and rub stone found at Fènghuángshān 鳳凰山, tomb M168 (dated 167 BC), see Selbitschka 2018: 454, 456.

88 Labeled boxes made of thread and bamboo, containing written documents, were excavated from tomb M1 at Mǎwángduī 馬王堆 (Zhāng Jīn 2016).

89 For a discussion of sealing procedures, see Chén Zhìguó 2008.

90 Barbieri-Low and Yates 2015: 798.

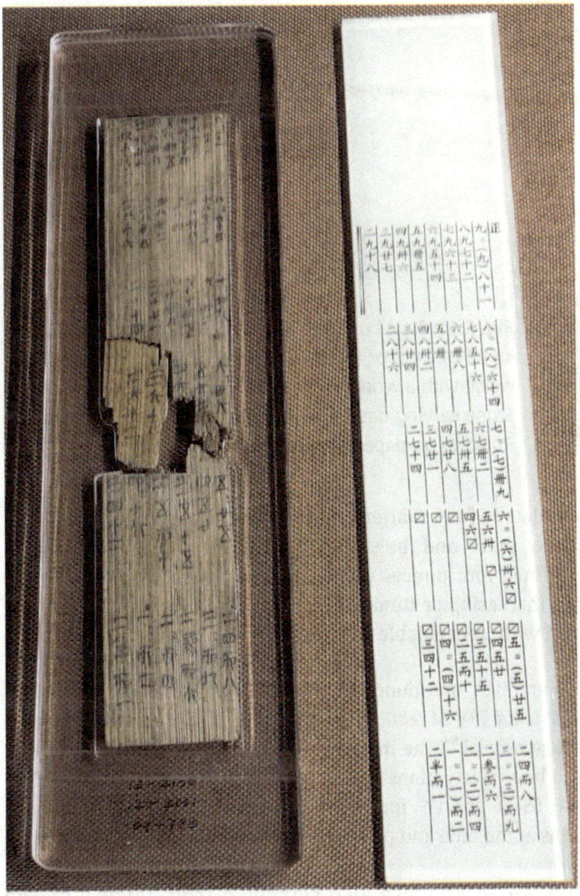

Fig. 2.14: Photograph of an original *Lǐyē* tablet consisting of three different fragments enclosed in a customized protective casing, with a corresponding transcription in modern Chinese characters. On display at the Lǐyē Museum of Qín Slips.[91]

County Court, and in every case are to be held in a trunk or a coffer or an armoire and closed and sealed, using the seal of the Magistrate, or the Assistant, or the Bailiff of the (relevant) office. They are to be placed by themselves in a document repository and the door of the repository is to be sealed. When it is necessary to put some of them in order or create (new ones), the Magistrate's scribe and the official in charge are to check to see if the seals are whole and to match (the impressions) with the seals of the Magistrate or the Assistant, and the Bailiff is to open (them). Then they are to put them in order or create (new ones)

91 My photograph, taken at the Lǐyē Museum of Qín Slips (*Lǐyē Qín jiǎn bówùguǎn* 里耶秦簡博物館) on May 14, 2019.

Fig. 2.15: Detail of a kneeling scribe carrying a 'scratch knife' (*xuē* 削) or 'document knife' (*shū dāo* 書刀) at his waist, with a writing utensil tucked behind his right ear. Stone carving dating back to the late Eastern Hàn dynasty. Rubbing on paper. Dimensions unknown.[92]

together. When they have completed 【the matter】, they are to store and immediately close and seal (the container) once more, and the storehouse is to be sealed. Those who do not obey the statutes are to be fined four ounces of gold each. Should anyone make false and fraudulent (entries), either increasing or diminishing them and (the officials) are not able to catch them, (punish them with) redeemable shaving.[93]

Comments: In the *Yuèlù* bamboo archives, one 'ounce' (*liǎng* 兩) of gold was equivalent to 1/16 of a 'catty' (*jīn* 斤), three 'maces' (*chuí* 錘), 24 '*zhū* 銖' or 576 coins, which was an estimated three-month salary for an average official.[94] The market value of gold set the value of other tradable goods but was prone to large fluctuations depending on the time and location of exchange. We learn from strips 46, 55–56 and 119, found in tomb 247 at Zhāngjiāshān, that one ounce of gold was worth 315 coins at the time and place specified in the texts.[95]

Twenty-two so-called 'labels' (*sì pái* 笥牌) were also retrieved from layer 8 by the team of archeologists.[96] On average, the labels are 10 to 13 centimeters long and

92 Zēng Zhāoyù, Jiǎng Bǎogēng and Lí Zhōngyì 1956: Plate 28, number 6.
93 Translation is taken from Yates 2012–2013: 327–328, with minor modifications. My *pīnyīn* transcriptions.
94 My estimation is based on the calculation in Yú Zhènbō 2013 that evaluates the value of gold on the basis of recovered *Yuèlù* documents. For a discussion of coins in Early China, see Thierry 2017.
95 Péng Hào 2001.
96 This corresponds to a negligible fraction (approx. 0.01329%) of all *Lǐyē* strips and tablets containing text. For a study on *Lǐyē* labels, see Zhāng Jīn 2016. For a study on bamboo boxes as a document storage, see Zhōu Bèi and Gěng Xiāngxīn 2011: 61–64, 72.

5 to 7 centimeters wide.[97] All labels share the following characteristics: 1) they are rectangular in shape with two rounded corners at the top, 2) the top section is stained with a line of black ink (ca. 1–3 centimeters in width), 3) except for one label, they contain two or four punctures in, or just below, the black line, possibly serving as fixing devices for twine, and 4) the exact date is recorded in the first line on the right half of the label (Fig. 2.16, 2.17). In addition, certain labels mention the official responsible for the box and/or its contents. A contents list and the destination to which the documents were to be sent are also provided. Scholars speculate that these labels were produced not only for archiving purposes but also for transportation.

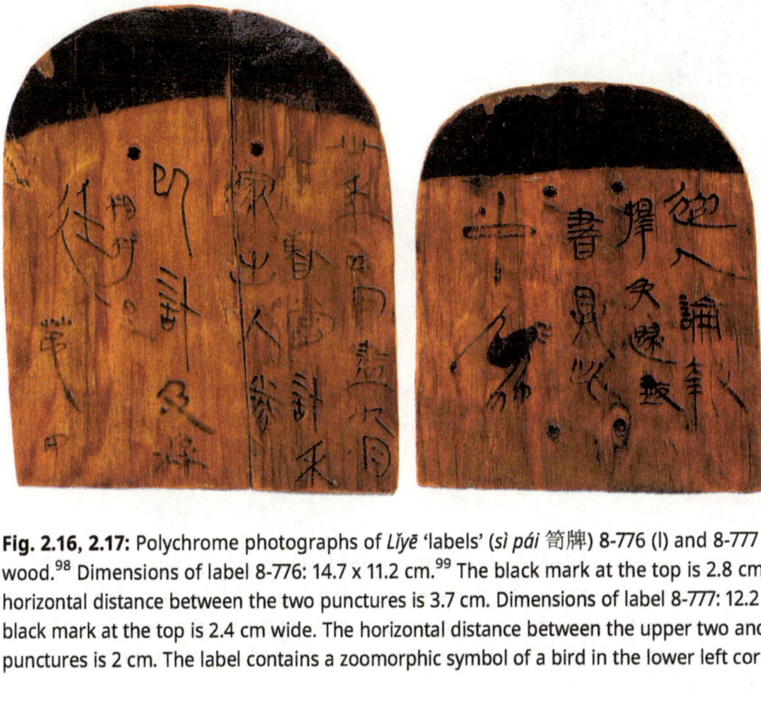

Fig. 2.16, 2.17: Polychrome photographs of *Lǐyē* 'labels' (*sì pái* 笥牌) 8-776 (l) and 8-777 (r). Ink on wood.[98] Dimensions of label 8-776: 14.7 x 11.2 cm.[99] The black mark at the top is 2.8 cm wide and the horizontal distance between the two punctures is 3.7 cm. Dimensions of label 8-777: 12.2 x 10.2 cm. The black mark at the top is 2.4 cm wide. The horizontal distance between the upper two and lower two punctures is 2 cm. The label contains a zoomorphic symbol of a bird in the lower left corner.

The calligraphic and structural characteristics of each label vary substantially. Despite many labels being manufactured after the implementation of the Qín

97 The longest label is 8-1536, with a length of 16.7 cm. The shortest label is 8-285, with a length of 7.5 cm (Zhāng Jīn 2016).

98 *Húnán shěng wénwù kǎogǔ yánjiūsuǒ* 2012a: 13.

99 These and following dimensions are taken from Zhāng Jīn 2016.

writing reform, 'seal script' (*zhuàn shū* 篆書) remained in common use.[100] Only six labels, or 27.3 percent of the total, make use of the typical 'small seal script' (*xiǎo zhuàn* 小篆), which is rare for Qín manuscripts unearthed in other areas.[101] Each label was written by a different individual. Label 8-1201 even resembles the embellished 'large seal script' (*dà zhuàn* 大篆), as seen in Late Zhōu dynasty inscriptions found on jade, clay, stone and bronze (Fig. 2.19, 2.20–2.22).[102] Interestingly, label 9-2309 seems to depict a bronze bell in the upper left corner (Fig. 2.18). Whether this

Fig. 2.18, 2.19: Photographs of *Lǐyē* 'labels' (*sì pái* 笥牌) 9-2309 (l) and 8-1201 (r). Ink on wood.[103] Dimensions of label 8-1201: 10.5 x 12.3 cm.[104] The black mark at the top is 2.5 cm wide. The horizontal distance between the upper two punctures is 5.8 cm, and between the lower two, 5.9 cm. The space between the upper left and lower left puncture is 1.7 cm, and between the upper right and lower right, 2 cm.

100 Wáng Huànlín 2016. For an in-depth analysis of the orthography and the lengthy process of script unification in early Chinese writing, see Galambos 2004, 2006.

101 *Húnán shěng wénwù kǎogǔ yánjiūsuǒ* 2007. The six labels are: 8-94, 8-214, 8-419+8-612, 8-1201, 8-1536 and 8-1776.

102 *Húnán shěng wénwù kǎogǔ yánjiūsuǒ* 2017: 248.

103 *Húnán shěng wénwù kǎogǔ yánjiūsuǒ* 2012a: 153; *Húnán shěng wénwù kǎogǔ yánjiūsuǒ* 2017: 248.

104 These and the following dimensions are taken from Zhāng Jīn 2016.

symbol and the depiction of the bird or 'hook' (*gōu* 鉤) on labels 8-777 and 8-1201 served as embellishments or conveyed important information remains open to debate until more evidence comes to light.

Fig. 2.20–2.22: Examples of the character '*dāng* 當' in inscriptions on bronze and clay from the Warring States period. Detail of a bronze spearhead found at Wǔdāng 武當 (*Wǔdāng máo* 武當矛) (l); detail of a clay stamp seal collection found at Shízhōng 十鐘 (m) (*Shízhōng yìn jǔ* 十鐘印舉); and, for comparative purposes, detail of *Lǐyē* label 8-1201 (r).[105]

Both the spacing and linear arrangement of the characters on the labels are generously laid out and show a relatively high degree of consistency, which may be attributed to their peculiar form. Labels written in archaic or Chǔ script might have facilitated the transportation of documents to regions where the Qín writing reform had not yet been introduced, and a mutually comprehensible set of characters was still required for communication. Messengers traveling hundreds of kilometres through vast, often contested territories, such as those belonging to the former Chǔ and other states, faced constant threats of attack by insurgent groups. Although highly speculative, it cannot be ruled out that these labels were purposefully designed to disguise the content stored in the boxes.

Documents designated by the research teams as 'postal document strips' (*yóu shū jiǎn* 郵書簡) stand out due to their irregular shape. Unlike the labels, these strips seem to have been incorporated into one text or bundle of texts, possibly serving as 'cover pages' (*jiǎn* 檢) or subsections of documents meant for dispatch to different regions. Based on his readings of these sources and related material, Yóu Yìfēi concludes that Dòngtíng Commandery comprised at least 14 counties.[106] Unfortunately, the physical characteristics of these strips have not been systematically documented or referenced in transcriptions and, as a result, have yet to receive proper scholarly attention. Classification is further impeded by the dissimilar for-

105 Gāo Míng 1983: 843.
106 Yóu Yìfēi 2015.

mats of archaeological document numbers used by Chén Wěi and the editors of *Lǐyē Qín jiǎn (yī)*.

As the name suggests, postal document strips – recognizable by the pointed tip at the bottom – may represent the early development of an empire-wide postal or courier system.[107] Whereas some strips have their upper section stained with blank ink, others seem to have been intentionally left blank. The vertical-linear arrangement of characters is consistent, and the spacing is generous throughout the collection. A systematic study of these strips could reveal how they were categorized, archived and read, potentially offering insights into the various stages of their production, the status of one or more documents within the corpus, and whether correlations exist between the length, content or linguistic features of strips belonging to the same text (Fig. 2.23, 2.24).

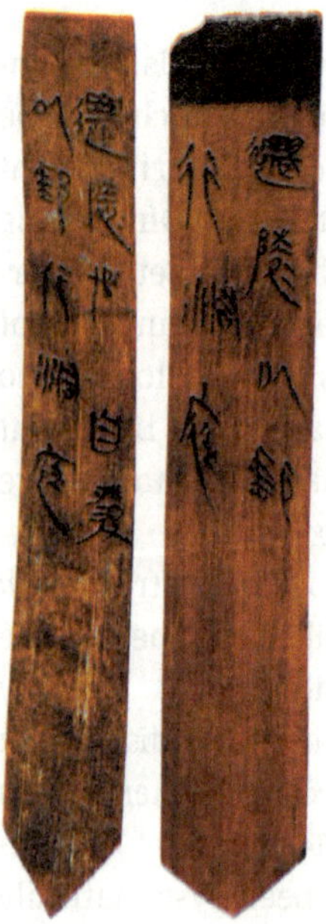

Fig. 2.23, 2.24: Polychrome photographs of *Lǐyē* 'postal document strips' (*yóu shū jiǎn* 郵書簡) 9-46 (l) and 9-24 (r). Ink on wood. Dimensions unknown.[108] The top section of strip 9-46 is not stained with black ink and reads: "Sent by the Prefect of Qiānlíng himself. Use the imperial postal system to deliver to Dòngtíng" (*Qiānlíng chéng zì fā yǐ yóu xíng Dòngtíng* 遷陵丞自發以郵行洞庭). The top section of the strip on the right is stained with blank ink and reads: "(From) Qiānlíng. Use the postal system to deliver to Dòngtíng" (*Qiānlíng yǐ yóu xíng Dòngtíng* 遷陵以郵行洞庭).

107 For a comparative study on the shape of various *Lǐyē* manuscripts, see Yáo Lěi 2015. For a study on postal transportation in the later imperial postal system(s), see Giele 2015.
108 *Húnán shěng wénwù kǎogǔ yánjiūsuǒ* 2017: 16, 21.

A thorough analysis of the large cache of completely blank strips and tablets has yet to be conducted by the archeological team. Although these writing carriers lack textual content, their material characteristics should not be overlooked, as they could offer valuable information through radiocarbon testing, mass spectrometry and other methods. Unfortunately, the exact original positions of each document fragment or cluster within the layers of well J1 are unknown. This renders the reliable reconstruction and sequencing of these materials in specific cases extremely difficult, if not impossible.[109]

2.3 Handling of Documents

Invaluable to the reconstruction efforts is the abundance of documents containing the exact date, rank and names of individuals involved in the administrative process. At the end of certain entries, the signatory clause '*shǒu* 手', preceded by a slash, introduces the signature of the individual in charge of handling or writing the document (or parts of it). The person assigned the task of '*shǒu* 手' may have been responsible for composing or copying a text, or transcribing a speech act. Alternatively, or additionally, they may have been involved in organizing, binding, carving and archiving the documents.[110]

Intellectual scepticism dominates debates about the meaning of '*shǒu* 手', especially since the discovery of Hàn dynasty manuscripts at Jūyán 居延 and Zǒumǎlóu 走馬樓 dating back to the Wú dynasty. The signatures in these manuscripts suggest that administrative officials were most likely responsible for the textual content rather than the physical production of the texts.[111] Yet, the use of '*shǒu* 手' as a scribal signature by clerical personnel appears to have been a distinctive feature of Qín bureaucracy in particular regions of the empire, a practice that was discontinued in later dynasties.[112] Tab. 2.1 provides a comprehensive list of officials entrusted with this responsibility in the *Lǐyē* texts recovered from layers 5, 6 and 8 of well J1.

---

109 For a list of fragments that seem to match one another, see the appendix in Chén Wěi 2012: 480–486.

110 Lewis 1999; It is difficult to define the exact rate of literacy in early imperial dynasties, as it heavily depends on context (Giele 2009: 138).

111 See Giele 2005; Hú Píngshēng 2004; Xíng Yìtián 2012b; Zhāng Chūnlóng and Lóng Jīngshā 2003.

112 For a discussion of the signatory clause 'by the hand of' (*shǒu* 手) in the "Qín clay documents" (*Qín wǎshū* 秦瓦書), refer to Guō Zǐzhí 1986. For an analysis of '*shū* 書', '*xiě* 寫' and '*shǔ* 署', three terms used in early imperial administrative and legal texts to refer to acts of writing, see Staack 2019.

Tab. 2.1: List of officials in *Lǐyē* texts from layers 5, 6 and 8 assigned the task of '*shǒu* 手' (literally, 'by the hand of').[113]

No.	Personal Name	Year(s) of entry	Occurrence: Manuscript number(s)[114]
1	Bān 般	35th year of King Zhèng [212 BC]	**6:** 8-827, 8-1002+8-1091, 8-1055+8-1579, 8-1211, 8-1706, 8-2053
2	Bǎo 葆	–	**1:** 8-657r
3	Bèi 貝	28th year of King Zhèng [219 BC]	**2:** 8-767r, 8-1562r
4	Bīng 兵	27th year of King Zhèng [220 BC]	**1:** 8-63r
5	Bīng 冰	–	**1:** 8-60r+8-656r+8-665r+8-748r
6	Bǐsǐ 彼死	–	**1:** 8-647r
7	Cāo 操	26th year of King Zhèng [221 BC]	**1:** 8-1452r
8	Chǎn 產	–	**1:** 8-1020
9	Cháo 朝	28th year of King Zhèng [219 BC]	**1:** 8-1463r
10	Chōu 瘳	35th year of King Zhèng [212 BC]	**11:** 8-783, 8-785, 8-790, 8-811+8-1572, 8-888+8-936+8-2202, 8-984, 8-1361, 8-1678, 8-1933, 8-2186, 8-2200
11	Chū 初	32nd year of King Zhèng [215 BC]	**1:** 8-1443r
12	Chǔ 處	31st–32nd year of King Zhèng [216–215 BC]	**2:** 8-152r, 8-173r
13	Chuò 綽	–	**1:** 8-740r
14	Cuì 萃	30th year of King Zhèng [217 BC]	**1:** 8-141r+8-668r
15	Cuó 痤	32nd year of King Zhèng [215 BC]	**2:** 8-145r, 8-478
16	Dài 帶	–	**2:** 8-1259r, 8-1281
17	Dé 得	31st year of King Zhèng [216 BC]	**4:** 8-125, 8-212+8-426+8-1632, 8-216+8-351, 8-1894
18	Dìng 定	–	**1:** 8-66+8-208

113 Tab. 2.1 is based on the texts transcribed and revised in Chén Wěi 2012. To avoid inaccuracies, instances where 'by the hand of' (*shǒu* 手) is directly preceded by a missing text fragment have not been taken into account.
114 'R' stands for the '*recto*' of the manuscript.

Tab. 2.1 (continued)

No.	Personal Name	Year(s) of entry	Occurrence: Manuscript number(s)
19	Èr 貳	26th year of King Zhèng [221 BC]	**1:** 8-163r
20	Fù 富	31st year of King Zhèng [216 BC]	**5:** 8-262, 8-1266, 8-1300, 8-1545, 8-1621
21	Găn 感	29th, 31st year of King Zhèng [218 BC, 216 BC]	**29:** 8-4, 8-48, 8-184, 8-217, 8-261, 8-270, 8-326, 8-521, 8-763, 8-766, 8-1066, 8-1084, 8-1128, 8-1153+8-1342, 8-1177, 8-1192, 8-1239+8-1334, 8-1247, 8-1286, 8-1375, 8-1511r, 8-1540, 8-1580, 8-1584, 8-1642, 8-1652, 8-1938, 8-2245, 8-2249
22	Găn 敢	32nd year of King Zhèng [215 BC]	**1:** 8-2247
23	Gàn 贛	1st year of Second Emperor [209 BC]	**2:** 8-653r, 8-1050
24	Gǒu 狗	–	**1:** 8-1094
25	Guò 過	33rd year of King Zhèng [214 BC]	**3:** 8-761, 8-2487, 8-2548
26	Hé 和	–	**1:** 8-61+8-293+8-2012
27	Héng 横	32nd year of King Zhèng [215 BC]	**3:** 8-1069r+8-1434r+8-1520r, 8-1226, 8-2481
28	Hú 狐	35th year of King Zhèng [212 BC]	**1:** 8-769r
29	Huăn 緩	–	**1:** 8-611
30	Huà 華	35th year of King Zhèng [212 BC]	**3:** 8-820, 8-1008+8-1461+8-1532, 8-1454+8-1629
31	Hùn 圂	33rd year of King Zhèng [214 BC]	**2:** 8-154r, 8-904+8-1343
32	Jī 畸	–	**2:** 8-118, 8-864
33	Jiā 嘉	1st year of Second Emperor [209 BC]	**1:** 5-1
34	Jiān 兼	26th year of King Zhèng [221 BC]	**1:** 8-63
35	Jiāo 驕	–	**1:** 8-657r
36	Jìng 敬	35th year of King Zhèng [212 BC]	**2:** 8-770r, 8-880
37	Jiù 就	–	**2:** 8-137r, 8-2466
38	Jū 居	31st year of King Zhèng [216 BC]	**1:** 8-1559r
39	Jù 巨	–	**1:** 8-2035r

Tab. 2.1 (continued)

No.	Personal Name	Year(s) of entry	Occurrence: Manuscript number(s)
40	Kān 堪	30th year of King Zhèng [217 BC]	**3:** 8-76, 8-754r+8-1007r, 8-2030r
41	Kè 客	–	**1:** 6-6
42	Lián 連	30th year of King Zhèng [217 BC]	**1:** 8-141r+8-668r
43	Lǚ 履	33rd–34th year of King Zhèng [214–213 BC]	**3:** 8-143r, 8-768r, 8-1561r
44	Mǎo 卯	–	**1:** 8-1523r
45	Mù 目	–	**2:** 8-112, 8-1998r
46	Píng 平	34th year of King Zhèng [213 BC]	**2:** 8-1449r+8-1484r, 8-1527r
47	Qì 氣	28th, 32nd year of King Zhèng [219 BC, 215 BC]	**3:** 8-75r+8-166r+8-485r, 8-140, 8-157
48	Qiáng 強	–	**1:** 8-1824
49	Qióng 邛	29th, 30th year of King Zhèng [218–217 BC]	**3:** 8-645r, 8-1515r, 8-1628r
50	Qū 祛	–	**1:** 8-677r
51	Qū 詘	–	**5:** 8-1122, 8-1148, 8-1466r, 8-1483r, 8-1492r
52	Què 卻	–	**3:** 8-843, 8-1238, 8-1947
53	Quē 缺	–	**1:** 8-1014
54	Ráo 饒	–	**2:** 8-739r, 8-1436r
55	Rén 壬[115]	26th, 31st–32nd, 34th [221 BC, 216–215 BC, 213 BC]	**9:** 8-157r, 8-183r+8-290r+8-530r, 8-756, 8-764, 8-1324, 8-1421, 8-1516r, 8-1574+8-1787, 8-2246
56	Rúyì 如意	35th year of King Zhèng [212 BC]	**2:** 8-1565r, 8-2190r
57	Shàng 尚	28th, 30th, 32nd year of King Zhèng [219 BC, 217 BC, 215 BC]	**6:** 8-62r, 8-75r+8-166r+8-485r, 8-130r +8-190r+8-193r, 8-136r+8-144r, 8-722, 8-1490r+8-1518r

115 Judging from the calligraphy, the official named Rén 壬, who was active in 221 BC (manuscript 8-1516), was not the same Rén 壬 who wrote from 216 BC onward.

Tab. 2.1 (continued)

No.	Personal Name	Year(s) of entry	Occurrence: Manuscript number(s)
58	Shěn 沈	35th year of King Zhèng [212 BC]	**5:** 8-886, 8-1346, 8-1399, 8-1554r, 8-2234
59	Tián 恬	34th year of King Zhèng [213 BC]	**1:** 8-1525r
60	Tián 田	–	**1:** 8-1991
61	Tuó 佗	–	**1:** 8-1697
62	Wǔ 俉	–	**1:** 8-140r
63	Wǔ 午	–	**1:** 8-175r
64	Wú 吾	–	**1:** 8-1980
65	Xiē 歇	34th year of King Zhèng [213 BC]	**3:** 8-755r, 8-759, 8-1523
66	Xiè 謝	–	**1:** 8-1073r
67	Xīn 欣	30th, 32nd year of King Zhèng [217 BC, 215 BC]	**5:** 8-155, 8-158r, 8-164r+8-1475r, 8-178r, 8-890+8-1583
68	Xū 胥	–	**1:** 8-140
69	Yán 言	31st year of King Zhèng [216 BC]	**2:** 8-921, 8-1560r
70	Yí 瓯	34th year of King Zhèng [213 BC]	**1:** 8-1549
71	Yì 義	–	**2:** 8-1447r, 8-2036r
72	Yǐ 齮	28th year of King Zhèng, 1st year of Second Emperor [209 BC, 219 BC][116]	**2:** 5-1, 8-1563r
73	Yuán 元	–	**1:** 8-60r+8-656r+8-665r+8-748r
74	Zǎng 駔	–	**1:** 8-76
75	Zé 擇	35th year of King Zhèng [212 BC]	**1:** 8-169r+8-233r+8-407r+8-416r+8-1185r
76	Zhāng 章	31st year of King Zhèng [216 BC]	**1:** 8-648r
77	Zhōng 忠	–	**1:** 8-72
78	Zhú 逐	28th–30th year of King Zhèng [219–217 BC]	**4:** 8-686r+8-973r, 8-701r, 8-1563, 8-1566r

116 The term 'first year' (*yuán nián* 元年) in the text refers to the first year of the Second Emperor in 209 BC (Chén Wěi 2012: 1).

Tab. 2.1 (continued)

No.	Personal Name	Year(s) of entry	Occurrence: Manuscript number(s)
79	X (pronunciation unknown) 犯	28th year of King Zhèng [219 BC]	**1:** 8-75r+8-166r+8-485r
80	X (pronunciation unknown) 廬	26th–27th year of King Zhèng [221–220 BC]	**2:** 8-63, 8-135
81	X (pronunciation unknown) 脊	35th year of King Zhèng [212 BC]	**10:** 8-348, 8-405, 8-865, 8-898+8-972, 8-902, 8-1517r, 8-1771, 8-1809, 8-1973, 8-2269
82	X (pronunciation unknown) 雒	34th year of King Zhèng [213 BC]	**1:** 8-487+8-2004
83	X (pronunciation unknown) 龡	35th year of King Zhèng [212 BC]	**1:** 8-673+8-2002
84	X (pronunciation unknown) 郤	29th year of King Zhèng [218 BC]	**2:** 8-781+8-8-1102, 8-1524r
85	X (pronunciation unknown) 盂	–	**1:** 8-1839
Total		**26th–35th year of King Zhèng, 1st year of Second Emperor [221–212, 209 BC]**	**208**

In total, the *Lǐyē* documents from layers 5, 6 and 8 identify at least 85 individuals across 208 documents who were responsible for the production and handling of official texts in and around Dòngtíng Commandery.[117] The manuscripts span from 221 to 212 BC, as well as the first year of the Second Emperor's reign in 209 BC.

117 For an analysis of 680 different surnames discovered in excavated Qín documents, see Lǐ Shìchí 2017.

Whereas most of those identified are categorized as either 'Secretaries' (*shǐ* 史) or 'Assistants' (*zuǒ* 佐), their skills and duties were not confined to these titles.[118]

Strictly speaking, the term 'scribe' is too narrow to describe clerical personnel, as it fails to capture the full range of responsibilities these individuals undertook. Moreover, the term '*shǐ* 史', which at times means 'scribe', particularly in received literature from the Eastern Zhōu and later periods, is more accurately translated as 'Secretary' in the context of the *Lǐyē* documents.[119] "Scribal literacy" or "administrative literacy," categories proposed by William V. Harris and Ma Tsang Wing, respectively, or Michael Nylan's notion of "rude literacy" – denoting a group of specially trained workers capable of reading and writing characters in short, formulaic sentences, and handling the materiality of manuscripts – may more aptly describe the responsibilities of *Lǐyē* manuscript officials, without limiting them to a single skill set or rank.[120]

Given the sheer number of individuals responsible for the correspondence between Lǐyē and other imperial institutions of Qiānlíng County and Dòngtíng Commandery, the documents provide us with important data on early script types, styles and hands.[121] Unfortunately, there is little scholarly consensus regarding the specific calligraphic features.[122] Liú Gāng, for example, argues that the characters in these documents are emblematic of: 1) Chǔ script in rare cases, 2) 'ancient clerical script'

118 See Xíng Yìtián 2012a. Ma Tsang Wing identified 56 instances of official titles preceding their names elsewhere in the same document. He concluded that 10.5% of these were 'Incumbents' (*shǒu* 守), 29.8% 'Secretaries' (*shǐ* 史) (including 'Prefectural Secretaries' (*lìngshǐ* 令史), 'Judiciary Secretaries' (*yùshǐ* 獄史), and one 'Secretary to the (County) Military Commandant' (*wèishǐ* 尉史)), 57.9% 'Assistants' (*zuǒ* 佐), and the remaining 1.8% fell into other categories (Ma Tsang Wing 2017: 325–327). Qín material culture often featured production marks used to monitor quality and quantity, and as an administrative tool for resource distribution and mass mobilization. For a study on marking practices in the production of Qín terracotta warriors, see, for example, Lǐ Xiùzhēn, et al. 2016.

119 On the etymology of '*shǐ* 史', see Behr 2004: 16. On the broader history of '*shǐ* 史' during the Eastern Zhōu periods, see Vogelsang 2007.

120 Harris 1991: 7–8; Ma Tsang Wing 2017: 331; Nylan 2011. These views are reconciled by Muchou Poo, arguing that many individuals across various social strata possessed basic reading and writing skills in the Qín and Hàn dynasties. However, high cultural literacy, or the number of fully literate individuals, was, in relative numbers, very low among the lower classes (Poo 1998: 181–182). In addition to reading, writing and accounting, Ma extends the concept of "administrative literacy" to encompass skills related to the physical alteration of documents, such as the carving of 'notches' (*kèchǐ* 刻齒) into the *Lǐyē* 'tallies' (*quàn* 券) (Ma Tsang Wing 2017: 327–331). Other forms of handling material documents could have included organizing, binding, carving and archiving.

121 Richter 2006 differentiates between "types of script" (morphological qualities), "styles of script" (structural and non-structural or calligraphic features) and "hand" (different individuals or a single individual with different writing styles).

122 Liú Gāng 2012.

(*gǔ lì* 古隸), and 3) predominantly 'Qín clerical script' (*Qín lì* 秦隸).[123] The latter is believed to have evolved from the 'Qín seal script' (*Qín zhuàn* 秦篆) as a result of the Qín writing reforms and closely resembles records excavated at Shuìhǔdì 睡虎地.[124] Liú Gāng further subdivides the 'ancient clerical script' into three distinct variants:

2a) Similar to the script found on Chǔ 'silk manuscripts' (*bó shū* 帛書). In general, this variant is characterized by a high level of graphic complexity, more strokes per character and balanced brushwork with round or circular elements. Its overall asymmetry and flattened, rectangular composition impart a sense of spontaneity and natural casualness to the writing.
2b) Similar to the 'small seal script' (*xiǎo zhuàn* 小篆), this variant exhibits a less regular appearance, which may be attributed to the unique processing, function and materiality of the documents. It shares strong calligraphic similarities with the 'ancient elegance script' (*gǔ yǎ* 古雅) of the imperial Qín stone inscriptions found at Mount Tài 泰, Yì 嶧 and Lángyá 琅邪.[125]
2c) An ancient variant of the 'clerical script' (*lì shū* 隸書).[126] This variant is more consistent than 2a) and 2b) in terms of composition, size, horizontal orientation and symmetry, yet it remains less regular than the clerical script. For this reason, some scholars consider it an early stage of the 'Qín clerical script' (*Qín gǔ lì* 秦古隸).[127]

The study of scribal culture in ancient Chinese manuscripts has proven invaluable for exploring various facets of manuscript production, reconstruction, authentication, archiving and scribal training in Ancient China. It has also enabled

123 In Katsuya Onishi's opinion, clerical script and 'scribal writing' (*shǐ shū* 史書) should be differentiated. Clerical script is a "more vernacular script type which developed out of the clericization of seal script," while scribal writing "refers to a standardized clerical script" in terms of character forms and character selection that "had to conform to the Qín and Hàn writing systems." The scribal writing of the Qín and Early Hàn periods was hence a script influenced by regional Qín traditions. It was only from the mid-Western Hàn period onwards that scribal writing became "synonymous with Hàn clerical script" (Onishi 2018: 397–398).

124 Liú Gāng 2012.

125 See also Kern 2000.

126 Legend has it that 'clerical script' (*lì shū* 隸書) was invented by Chéng Miǎo 程邈, an educated prisoner of the First Emperor (Ch'en Chih-Mai 1966: 35). The character '*lì* 隸' means 'dependent', 'subordinate', 'servant' or 'slave' (*Hànyǔ dà cídiǎn* 漢語大詞典, Vol. 12, 2001: 175; Unger 1989: 61). Aesthetic properties of clerical script include more precise character alignment, greater uniformity in character composition and increased variation in character size and stroke thickness (see Hé Línyí 1998).

127 Liú Gāng 2012.

scholars to group excavated manuscript fragments into individual, local and regional writing systems.[128] It should be noted, however, that the classification and dating of scripts pose significant challenges.[129] Transitions between scripts are often fluid, complicating diachronic comparisons, particularly when characters are inscribed on different materials for various purposes across geographically distant regions. At best, visual juxtapositions of script criteria can help identify and recognize individual, local or regional tendencies at specific points in time.

The transition between different scripts in Ancient China was not abrupt; it resulted from long-term historical developments and transculturation processes within certain spheres of influence.[130] A unified writing system, inspired by the idea of strong political and administrative consolidation, reflects the envisioned microcosmic modality of an orderly worldview in which every element, graph and aesthetic feature had its designated place.[131] But to what extent did the Qín writing reform truly succeed in standardizing the form and meaning of characters? And how much variation from earlier scripts was incorporated into the new universal standard or, at least, tolerated by imperial institutions?[132]

When comparing the *Lǐyē* manuscripts, we observe that handwriting, character size, composition and alignment vary from text to text, and sometimes even within the same manuscript. A recent study by Chén Zhèngxián on the *Lǐyē* corpus and other excavated material from the Qín dynasty looks at graphic variants of '*qīng* 卿' (*khraŋ) and '*xiāng* 鄉' (*haŋ), '*lì* 吏' (*rəh) and '*shì* 事' (*s-rəʔ), '*shǔ* 鼠' (*nhaʔ) and '*yǔ* 予' (*laʔ) as well as '*dà* 大' (*dâs), '*tài* 太' (*thâs) and '*tài* 泰' (*thâs).[133] He argues that characters like '*tài* 太' and '*tài* 泰' or '*qīng* 卿' and '*xiāng* 鄉' serve as key benchmarks for distinguishing between the pre-imperial and imperial periods, as well as for assessing the extent of the writing reform completed under Lǐ Sī 李斯 (ca. 280–208 BC; served 246–208 BC). According to Chén Zhèngxián, the unification of the empire occurred in month 7 of the 25th year of King

---

128 See Smith 2011b. For more on detecting fake manuscripts through the analysis of textual errors, document material, provenance and calligraphy, see Hú Píngshēng 2010: 98–101; Xíng Wén 2016. For a study on scribal culture, workshops and scriptural calligraphy in early Buddhist manuscripts, see Tsui Chung-hui 2020. For a discussion of script unification based on early excavated and transmitted sources, see Galambos 2004, 2006.

129 Hé Línyí 1998; Lǐ Shǒukuí 2003; Wú Jiànwěi 2006.

130 Boltz 1994: 156–177; Ch'en Chih-Mai 1966: 35–36; Galambos 2006: 43.

131 See Harper 1999.

132 For criteria on discerning variations in handwriting on the *Guōdiàn* bamboo manuscripts (*Guōdiàn Chǔ jiǎn* 郭店楚簡) from the Warring States period, see Richter 2006: 132–147.

133 Chén Zhèngxián 2020. Phonological reconstructions are taken from Schuessler 2009: 55, 56, 78, 103, 233.

Zhèng (222 BC), while the writing reform occurred slightly later, between months 3 and 5 of the 26th year of King Zhèng (221 BC).[134]

Unsurprisingly, these two events did not unfold simultaneously. In fact, as Imre Galambos has convincingly demonstrated, the completion of the writing reform may have extended well into the Western Hàn dynasty.[135] Imre Galambos's argument can be supported by conducting a preliminary examination of the *Lǐyē* text entries under the name Rén 壬. As shown in Tab. 2.1, the name Rén 壬 occurs nine times as a handler of documents. Due to the differing handwriting of the Rén 壬 appearing in the 26th years of King Zhèng (221 BC) (Fig. 2.25), these entries could refer to two distinct individuals. The latter Rén 壬 (Fig. 2.26), active between the 31st and 34th years of King Zhèng (216–213 BC), was likely the "Assistant to Jìng, Incumbent of the Office of Agricultural Fields" (*tiánguān shǒu Jìng zuǒ* 田官守敬佐).[136] His handwriting features curved lines with round elements and an exceptionally consistent brushstroke, showing almost no variation in pressure or execution.[137] Given that this individual was producing texts in the later stages of the empire, his calligraphic expression and use of small seal script elements challenge Chén Zhèngxián's argument that the writing reform had been universally implemented by this point. In fact, other scribes active in the later years of the empire, such as Shěn 沈 (212–211 BC), Píng 平 (213 BC) and X (pronunciation unknown) 䏶 (212 BC), also exhibit a diversity of script type, script style and hand. As epigraphical evidence from the *Lǐyē* cache amply testifies, individual characters were written in a variety of ways (Fig. 2.27–2.29 and 2.30–2.32).

Future studies on handwriting in the *Lǐyē* texts will hopefully shed further light on the writing practices of the Dòngtíng area, as well as the origin of officials, their names, imperial affiliations, educational background, social status and potential for mobility within the Qín hierarchical order. Formal writing criteria could also serve as a powerful tool for identifying and matching unsigned document frag-

134 Alternatively, Katsuya Onishi explains the gradual replacement of '*tài* 太' with '*tài* 泰' from a regional rather than a diachronic perspective. '*Tài* 泰' is a variant of '*tài* 太' that emerged alongside the Qín notion of political unification (Onishi 2013: 141–142).

135 Galambos 2004, 2006.

136 This is observed on manuscripts 8-764, 8-1574+8-1787 and 8-2246.

137 From *Lǐyē* manuscripts 8-663, 8-1549 and 8-2501, we learn that white bird feathers could be used as writing utensils. On manuscript 8-1549, which is slightly damaged, officials from the 'Granary' (*cāng* 倉) sell 'white feathers for writing' (*bái hàn yǔ* 白翰羽) to, what appears to be, 'Section Commanders' (*zhǎng* 長) and 'minor girls' (*xiǎo nǚzǐ* 小女子). For a transcription and translation of manuscript 8-663, see chapter 4.3. For a transcription of manuscripts 8-1549 and 8-2501, see Chén Wěi 2012: 355, 475.

Fig. 2.25, 2.26: Manuscript 8-1516 (l), attributed to Rén 壬, active in the 26th year of King Zhèng (221 BC), and a detail of the *recto* of manuscript 8-756 (r), attributed to Rén 壬, active between the 31st and 34th year of King Zhèng (216–213 BC). Ink on wood. Dimensions unknown.[138]

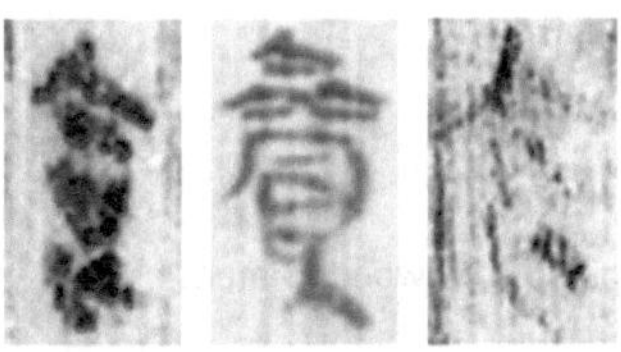

Fig. 2.27–2.29: Variants of the graph for 'food' (*shí* 食): manuscript 5.1 (l), attributed to Yǐ 齮, active in the 1st year of the Second Emperor (209 BC); manuscript 8-1517 (m), attributed to X (pronunciation unknown) 冑, active in the 35th year of King Zhèng (212 BC); manuscript 8-770 (r), attributed to Jìng 敬, active in the 35th year of King Zhèng (212 BC). Ink on wood. Dimensions unknown.[139]

ments that have yet to be collated, providing clues about scribal training, manuscript production, document circulation and archiving.

'*Fā* 發' is another term in the *Lǐyē* corpus that sometimes requires the addition of a personal name, the name of a government unit or an official title. It is occasionally separated from the preceding text by a slash mark. In most instances, '*fā* 發'

138 *Húnán shěng wénwù kǎogǔ yánjiūsuǒ* 2012a: 109, 192.
139 *Húnán shěng wénwù kǎogǔ yánjiūsuǒ* 2012a: 3, 113, 192.

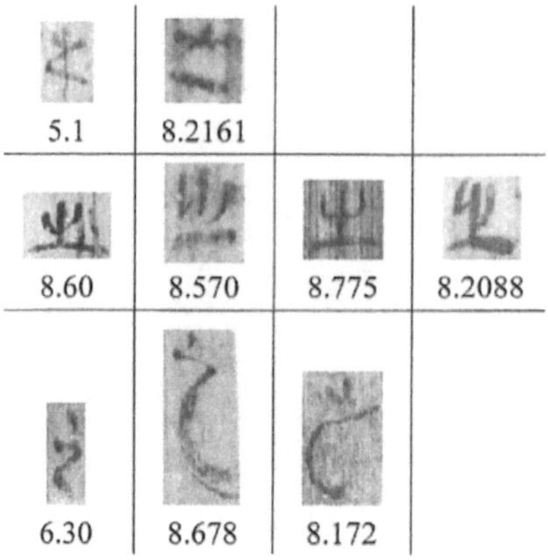

Fig. 2.30: Variants of the graph '*zhī* 之' on various *Lǐyē* manuscripts. Ink on wood. Dimensions unknown.[140]

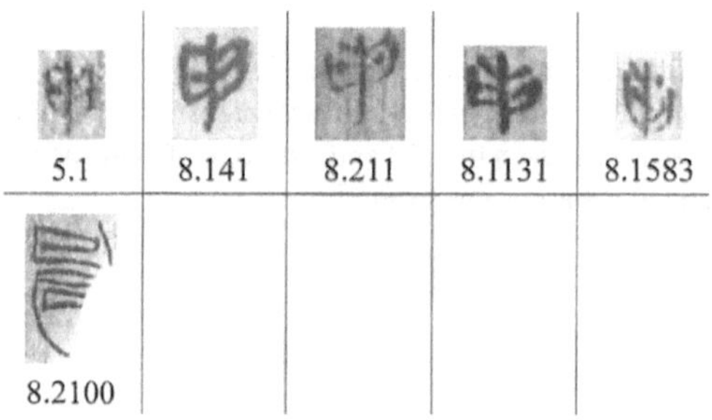

Fig. 2.31: Variants of the graph '*shēn* 申' on various *Lǐyē* manuscripts. Ink on wood. Dimensions unknown.[141]

arguably means 'to send', 'to be sent by', 'to dispatch', 'to be dispatched by', instead of 'to open' or 'to be opened by'.[142] As shown in the translations below, this reading is supported by the contextual and syntactic attributes found in many *Lǐyē* manuscripts. Furthermore, if '*fā* 發' exclusively meant 'to open' or 'to be opened by', one would expect it to appear in most *Lǐyē* documents received and opened by clerical personnel. However, in relative terms, only a minuscule number of documents throughout the entire archive contain a clause featuring a '*fā* 發' passive. Most of the

140 Yè Shūshān 2021: 221.

141 Yè Shūshān 2021: 237.

142 The term '*fā* 發' is not preceded by a slash mark in instances where the documents or resources were sent by a government unit, or when only an official title, without a personal name, is given. '*Fā* 發' is generally understood to mean 'to open' in other early manuscripts. See Lau and Staack 2016: 280, footnote 1301; Lóng Jīngshā 2001. For more information on the possible readings of '*fā* 發', see Chén Jiàn 2011a, 2011b.

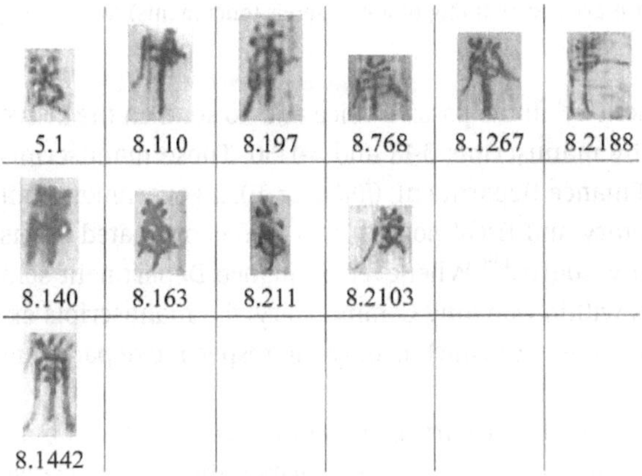

Fig. 2.32: Variants of the graph '*gēng* 庚' on various *Lǐyē* manuscripts. Ink on wood. Dimensions unknown.[143]

usages likely referring to postal dispatch occur on the 'postal document strips', as exemplified on manuscript 8-1130:

8-1130: 酉陽金布發.[144]

Sent by the Finance Department of Yǒuyáng (County).

In addition to the verb's passive usage, manuscripts 6-18, 8-303 and 8-922 demonstrate its transitive usage in other documents exchanged between institutions within Dòngtíng Commandery.

6-18 I–II: 遷陵金布┙ 發洞庭.[145]

The Finance Department of Qiānlíng (County)┙ sends (documents) to Dòngtíng (Commandery).

8-303: 遷陵主薄(簿)發洞庭.[146]

The Official in Charge of the Registers of Convicts in Qiānlíng (County) sends (documents) to Dòngtíng (Commandery).

Comment: The document is stained with black ink at the top.

8-922: 遷陵主倉發洞庭.[147]

---

143 Yè Shūshān 2021: 269.
144 Chén Wěi 2012: 281.
145 Chén Wěi 2012: 23.
146 Chén Wěi 2012: 131.
147 Chén Wěi 2012: 249.

The Official in Charge of the Granaries in Qiānlíng (County) sends (documents) to Dòngtíng (Commandery).

This reading of '*fā* 發' as 'to be sent by' in the passive voice and 'to send' in the active voice is further substantiated by manuscripts 6-18 and 8-1130. These manuscripts pertain to exchanges with the 'Finance Department' (*jīnbù* 金布), a government unit primarily responsible for monetary and fiscal policies that also coordinated assets and investments with the County Courts.[148] Whenever a Finance Department sent documents to higher authorities within Dòngtíng Commandery, the manuscripts explicitly mention – where decipherable – to which county the respective department belonged.

Hence, Finance Departments were not part of the Courts, contrary to the assumptions of most modern scholars who erroneously translate the passive clause "*tíng jīnbù fā* 廷金布發" in the *Lǐyē* corpus as "(the) Finance Department of the Court opens (the document)".[149] Instead, phrases like these should be considered inherent to the imperial Qín postal system, leading to the correct translation: "Sent from the (County) Court to the Finance Department." Finance Departments operated independently from the County Courts and government seats of a commandery, which is why on manuscript 6-18, for example, the Finance Department of Qiānlíng is separated by '*fā* 發' and distinguished from the government seat in Dòngtíng. On manuscript 8-1130, the addressee is missing, with only the Finance Department of Yǒuyáng identified as the sender, turning '*fā* 發' into a passive verb.

The '*fā* 發' in "*tíng jīnbù fā* 廷金布發" should also be read in the passive, and '*jīnbù* 布發' as an individual government unit; thus, the translation: "Sent from the (County) Court to the Finance Department."[150] The syntactic structure is even more evident on manuscript 8-978, where the "Commandery Governor sends (documents) to the Household Bureau" (*shǒufǔ hùcáo fā* 守府戶曹發). Translating this phrase as "the Household Bureau of the Commandery Governor opens (the document)" would make little sense, as Household Bureaus were units led by Incumbents at a lower administrative level.[151]

148 For more information on 'Finance Departments' (*jīnbù* 金布) and other 'Bureaus' (*cáo* 曹) or 'Offices' (*guān* 官) in the *Lǐyē* texts, see chapter five.

149 See Sūn Wénbó 2018: 86.

150 The phrase "*tíng jīnbù fā* 廷金布發," meaning "sent from the (County) Court to the Finance Department," appears seven times: namely on manuscripts 8-506, 8-799, 8-935, 8-969, 8-1166, 8-1297 and 8-1313. On two other manuscripts, 8-545 and 8-1183, the character following "*tíng jīnbù* 廷金布" is either damaged or missing. Given the parallel structure of many other texts, the missing character is very likely '*fā* 發'.

151 For more information on the various government units at the commandery and county levels, see chapter four.

As cited in 17 different manuscripts, only 11 individuals in and around Dòngtíng Commandery can be personally identified for dispatching documents.[152] The individuals Hùn 圂, Jiā 嘉, Lǚ 履, Qì 氣, Què 卻, Rén 壬, Xiè 謝 and Xīn 欣 were, in addition to transcribing conversations or copying written records, simultaneously assigned the responsibility of '*fā* 發'. Although their combined activity, in both concurrent or consecutive years, accounts for over 10% of all *Lǐyē* text entries, their official titles are provided in only 32 instances.[153] When personal names are used, they predominantly identify those individuals sending documents to the government seat in Dòngtíng Commandery.[154]

Five distinct government units appear in a total of 24 postal documents, ostensibly belonging to the collection of 'postal document strips'.[155] Documents, and

152 This includes "Cì, the Official in Charge of Èrchūn District" (*Èrchūn Xiāng zhǔ Cì* 貳春鄉主) (8-1548), Duān 端 (8-173), Hùn 圂 (8-221, 8-1069+8-1434+8-1520), Jiā 嘉 (5-1), Lǚ 履 (8-651, 8-1477, 8-1797), Qì 氣 (8-1559), Què 卻 (8-1317), Rèn 壬 (8-1525), Xiè 謝 (8-66+8-208), Xīn 欣 (8-152, 8-157, 8-196+8-1521, 8-1443+8-1455) and Zhù 翥 (8-1523). This and the following information is based on texts transcribed and revised in Chén Wěi 2012. To avoid inaccuracies, instances where '*fā* 發' is preceded by a missing fragment have not been taken into account.

153 The only exception is the official Xiè 謝, whose service dates are missing in both cases. Some personal names that appear in the *Lǐyē* manuscripts, like Pigsty (Hùn 圂) or Pneuma (Qì 氣), are quite peculiar. It may well be that some of these appellations were not given at birth but assigned to convey a person's physical or moral characteristics, status or life accomplishments. As such, they should be regarded as constructions of "meaningful epithets" (Goldin 2000). It also cannot be ruled out that some of these names may not carry any obvious meaning or may reflect local or regional customs that have yet to be adequately discussed.

154 These occurrences include the official titles 'Commandery Governor' (*shǒufù* 守府) (8-978), "Official in Charge of the Evaluations at the (County) Court" (*tíng zhǔ kè* 廷主課) (8-2198), "Official in Charge of the Food" (*zhǔ shí* 主食) (8-830+8-1010), "Official in Charge of the Granaries at the (County) Court" (*tíng zhǔ cāng* 廷主倉) (8-1228, 8-1294, 8-1366, 8-1628), "Official in Charge of the Granaries in Qiānlíng" (*Qiānlíng zhǔ cāng* 遷陵主倉) (8-579, 8-922), "Official in Charge of the Households at the (County) Court" (*tíng zhǔ hù* 廷主戶) (8-156, 8-266, 8-1142, 8-1249, 8-1395, 8-1607, 8-1650, 8-1752, 8-1955, 8-2547), "Official in Charge of the Commissioners of Public Works in Línyuán (County)" (*Línyuán zhǔ sīkōng* 臨沅主司空) (8-695), "Official in Charge of the Ordinances" (*zhǔ lìng* 主令) (8-601) and "Official in Charge of the Registers of Convicts in Qiānlíng" (*Qiānlíng zhǔ bù* 主簿) (8-303). On manuscript 8-978, the "Official in Charge of the Granaries in Qiānlíng," the "Official in Charge of the Commissioners of Public Works in Línyuán (County)" and the "Official in Charge of the Registers of Convicts in Qiānlíng" are mentioned exchanging correspondences with the government seat in Dòngtíng Commandery. Meanwhile, the Governor is mentioned exchanging correspondences with the 'Household Bureau' (*hùcáo* 戶曹).

155 This includes the '(County) Court' (*tíng* 廷) (8-1, 8-65, 8-263, 8-283, 8-506, 8-778, 8-878, 8-935, 8-952, 8-969, 8-1166, 8-1292, 8-1297, 8-1313, 8-1451, 8-1741, 8-1834, 8-1859, 8-2507), Qiānlíng (County) (*Qiānlíng* 遷陵) (8-264), the "Finance Department of Qiānlíng" (*Qiānlíng jīnbù* 遷陵金布) (6-18, 8-304), the 'Reinvestigation Bureau' (*fùcáo* 覆曹) (8-2550) and the "Finance Department of Yǒuyáng (County)" (*Yǒuyáng jīnbù* 酉陽金布) (8-1130). According to these manuscripts, the '(County)

possibly also messages or goods, arriving from institutions outside of Lǐyē were signed not with a personal name, but more generically with an official title or the name of the institution consigning the delivery.[156]

Further appearances of '*fā* 發' occur in the official title 'Crossbow Archer' (*fānǔ* 發弩) and in the term '*zìfā* 自發'. The latter arguably means 'to send/open something on one's own' and is often preceded by a personal name.[157] Notably, all three officials sending or opening private correspondences on their own are identified personally, with their names combined with the character '*jì* 季', as seen in Máojì 毛季, Mángjì 芒or Lǐjì 李季.[158]

> 8-272 I–II: 私進遷陵主吏┘ 毛季自發.[159]
>
> The privately submitted document [private letter] is sent/opened by the Official in Charge of the Low-Ranking Officials in Qiānlíng himself, Máojì.
>
> 8-1817: 私進令史芒季自發.[160]
>
> The privately submitted document is sent/opened by the Prefectural Secretary Mángjì himself.

Another term with usage comparable to '*fā* 發' (*pat) is '*bàn* 半' (*pans), which, in early manuscripts, usually means 'to halve' or 'to split in halves'. In the *Lǐyē* corpus, for reasons still unknown, '*bàn* 半' appears as a signatory expression on manuscripts dated before month 9 of the 30th year of King Zhèng (217 BC) and might refer to the act of breaking the seal of a document or document box.[161] At

---

Court' exchanged information with the, 'Household (Bureau)' (*hù(cáo)* 戶(曹)), the 'Finance Department' (*jīnbù* 金布), the 'Bureau of Ordinances' (*lìngcáo* 令曹), the 'Judiciary (Bureau) of the East' (*yùdōngcáo* 獄東(曹)) and the 'Bureau for Officials' (*lìcáo* 吏曹). The government seat in Qiānlíng, the "Finance Department of Qiānlíng," the 'Reinvestigation Bureau' (*fùcáo* 覆曹) and the "Finance Department of Yǒuyáng (County)" also appear to have exchanged documents with the government seat in Dòngtíng Commandery.

156 This applies, for example, to the '(County) Court' (*tíng* 廷), the 'Reinvestigation Bureau' (*fùcáo* 覆曹), the "Official in Charge of the Households at the (County) Court" (*tíng zhǔ hù* 廷主戶), the "Official in Charge of the Commissioners of Public Works in Línyuán" (*Línyuán zhǔ sīkōng* 臨沅主司空), and many more.

157 The title 'Crossbow Archer' (*fānǔ* 發弩) appears on manuscripts 8-141+8-668, 8-159, 8-761, 8-985, 8-1101, 8-1234, 8-1783+8-1852 and 8-1945. 'To send/open something on one's own' (*zìfā* 自發) appears on manuscripts 8-206, 8-272, 8-1065 and 8-1817.

158 'Máojì 毛季' appears on manuscripts 8-272, 8-1529 and 8-1694; 'Mángjì 芒季' on manuscripts 8-1065 and 8-1817; and 'Lǐjì 李季' on manuscript 8-206.

159 Chén Wěi 2012: 126.

160 Chén Wěi 2012: 395.

161 Chén Wěi 2012; Lóng Jīngshā 2001. Phonological reconstructions are taken from Schuessler 2009: 237, 261.

the same time, the verbal use of *'bàn* 半', in conjunction with personal names, might also represent an exoative version of its ascribed root *'bān* 搬' (*Cə-pˤan-s), meaning 'to move', which would make it similar to my proposed interpretation of *'fā* 發' as 'to send'.[162] In the overwhelming majority of cases, however, *'bàn* 半' is used in association with units of weight or volume, as seen in *"èr dǒu bàn dǒu* 二斗半斗" on tally 8-2249, meaning "2.5 bushels."

Shared responsibilities between officials of equal or differing status are also replicated at the semantic level. The term *'xíng* 行' has a variety of meanings apart from being a personal name. Depending on the context, it can mean 'to deliver', 'to follow up on', 'to inspect', 'to go' and 'to distribute'.[163] The reasons why both *'fā* 發' and *'xíng* 行' were used to designate certain kinds of postal dispatch or transportation are unclear. Perhaps, sending documents (or goods) in the sense of *'xíng* 行' meant utilizing the early stages of a complex imperial delivery network of 'postal units' (*yóu* 郵) that connected Qín counties and commanderies, whereas *'fā* 發' designated a more point-to-point courier service between individuals and government units. On manuscript 8-2550, *'xíng* 行' and *'fā* 發' even appear side-by-side:

> 8-2550 I–II: 遷陵以郵行┛ 覆曹發 • 洞庭.[164]
>
> Qiānlíng uses the imperial postal service to deliver (the documents).┛ The Reinvestigation Office [then] sends (the documents) to • Dòngtíng.

This reading of *'xíng* 行' as 'to send', 'to deliver' or 'to dispatch' is also supported on numerous manuscripts.[165] Some of these entries are translated below:

> 5-34 I–II: 酉陽以郵行┛ 洞庭.[166]
>
> Yǒuyáng uses the imperial postal service to deliver (the documents) to┛ Dòngtíng.
>
> 5-35 I–II: 遷陵洞庭, ┛ 以郵行.[167]
>
> From Qiānlíng to Dòngtíng,┛ use the imperial postal service to deliver (the documents).

---

162 My thanks to Prof. Dr. Wolfgang Behr for his comment on the possible link between the verbal usage of *'bàn* 半' and *'bān* 搬'.
163 Xíng 行 is used as a personal name on manuscripts 8-133 and 8-2210.
164 Chén Wěi 2012: 479.
165 This includes manuscripts 5-34, 5-35, 6-2, 8-12, 8-32, 8-63, 8-90, 8-413, 8-432, 8-704+8-706, 8-1147, 8-1464, 8-1553, 8-1685, 8-1840, 8-1876 and 8-2550. To avoid inaccuracies, instances where *'xíng* 行' is preceded by a missing fragment have not been taken into account.
166 Chén Wěi 2012: 15.
167 Chén Wěi 2012: 16. The graph *'tíng* 庭' on manuscript 8-413 is damaged and cannot be deciphered with certainty.

Comments: This text parallels the structure seen on manuscript 8-978, for example, where the verb is positioned at the end of the sentence, and both the sender and the recipient of the delivery are mentioned at the beginning. Manuscripts 8-32, 8-413, 8-1464, 8-1553 and 8-1685 are identical in this respect.

8-12: 遷陵以郵行 • 洞庭.[168]

Qiānlíng uses the imperial postal service to deliver (the documents) to • Dòngtíng.

8-63 V: [. . .] 即走申行司空.[169]

[. . .] Then Runner Shēn delivers (the documents) to the Commissioner of Public Works.

'Runners' (*zǒu* 走) who delivered documents (perhaps also messages or goods) are frequently mentioned in the *Lǐyē* corpus and other excavated materials.[170] Their function may have been similar to the 'Postal Runner' (*yóulìzú* 郵利足) cited on *Lǐyē* manuscripts 8-90 and 8-527, or the 'Light Foot' (*qīng zú* 輕足) in the "Statutes Concerning Arable Land" (*Tián lǜ* 田律) in the *Shuìhǔdì*.[171]

8-90: ☒遷陵以郵利足行洞庭, 急.[172]

☒ Qiānlíng uses the imperial postal service, rather than a Runner, to deliver (the documents) to Dòngtíng. Urgent.

A closer look at the "Statutes Concerning the Delivery of Documents" (*Xíng shū lǜ* 行書律) in the *Shuìhǔdì* reveals highly regulated procedures for the transportation of documents within Qín territories. When a document was deemed high priority, the character '*jí* 急' for 'urgent' was added to the entry. An inability to meet the delivery deadline was punished with a fine. Examples in the *Lǐyē* corpus further corroborate the use of this convention, where '*jí* 急' appears once or multiple times in succession, as seen on manuscript 8-1915. It states:

8-1915 I–II: 遷陵令【若】☒┘ 行急急急急☒[173]

Qiānlíng orders【Ruò】☒┘ deliver. Urgent, urgent, urgent, urgent ☒

We also learn from manuscript fragment 8-53 that the reception of documents may have been prioritized:[174]

168 Chén Wěi 2012: 31.
169 Chén Wěi 2012: 48.
170 See Lau and Staack 2016: 179. The term appears on manuscripts 8-133, 8-135, 8-373, 8-453, 8-652, 8-657, 8-742, 8-756, etc.
171 Chén Wěi 2012: 50–51.
172 Chén Wěi 2012: 60.
173 Chén Wěi 2012: 406.
174 Chén Wěi suggests that the text on manuscript fragment 8-53 may have been the result of a writing exercise (Chén Wěi 2012: 42). In any case, the limited evidence available does not allow for definitive conclusions about writing exercises, textual intensity markers or similar possibilities.

8-53: 受受受☐[175]

Receive, receive, receive ☐

8-704+8-706 rIV: ☐時都郵人羽行.☐[176]

☐ time, Postman Yǔ of Dū District delivers (the documents). ☐

8-1147 I–II: 貳春鄉┘ 以郵行.[177]

Èrchūn District┘ uses the imperial postal service to deliver (the documents).

8-1840 I–II: 遷陵以郵行┘ 洞庭.[178]

Qiānlíng uses the imperial postal service to deliver (the documents) to┘ Dòngtíng.

8-1876: ☐瘳行書遷陵, 因□[179]

☐ Chōu delivers the document to Qiānlíng, and □

The "Èrnián lǜlìng" is divided into 28 rubrics, each grouping statutes under specific headings. One of these rubrics is the "Statutes Concerning the Delivery of Documents" (*Xíng shū lǜ* 行書律), which is also present in the *Shuìhǔdì*.[180]

Individuals delivering documents to government units, or clerks serving these units, in the sense of '*xíng* 行', held various official titles. Those mentioned by personal name are, in most extant cases, ranked as '(Commandery) Governors' (*shǒufǔ* 守府), 'Assistants' (*zuǒ* 佐), 'Postmen' (*yóurén* 郵人), 'Runners' (*zǒu* 走) and '(female) Bondservants' (*lìqiè* 隸妾). In singular or more limited instances, one 'Military Commandant' (*wèi* 尉), one 'Commissioner of Public Works' (*sīkōng* 司空), '(male) Bondservants' (*lìchén* 隸臣), Members of a 'Unit of Five People' (*shìwǔ* (*wǔ*) 士五(伍)), as well as several names without official titles are recorded.[181] These

175 Chén Wěi 2012: 42.
176 Chén Wěi 2012: 207.
177 Chén Wěi 2012: 284.
178 Chén Wěi 2012: 398.
179 Chén Wěi 2012: 402
180 For more information on the delivery of documents in the Qín dynasty as reflected in the *Lǐyē* corpus, see Chén Zhìguó 2008.
181 'Governors' (*shǒufǔ* 守府) appear addressing the '(County) Court' (*tíng* 廷) (8-198+8-213+8-2013), 'Bureau of the (County) Military Commandant' (*wèicáo* 尉曹) (8-71), 'Lesser Treasury' (*shǎonèi* 少内) (8-60+8-656+8-665+8-748, 8-155) and 'adjacent counties' (*páng* 旁) (8-158). The Governors' names are Áng 卬 (8-1525), Chāng 昌 (8-198+8-213+8-2013), Jí 即 (8-768) and Kuài 快 (8-60+8-656+8-665+8-748, 8-71, 8-140, 8-155, 8-157, 8-158, 8-1560). One name is unknown due to a missing character (8-165). 'Assistants' (*zuǒ* 佐) appear addressing the '(County) Courts' (*tíng* 廷) (8-164+8-1475, 8-170, 8-890+8-1583) and 'adjacent counties' (*páng* 旁) (8-75+8-166+8-485). Their names are Jì 忌 (8-854), Píng 平 (8-1449+8-1484), Qì 氣 (8-75+8-166+8-485), Tián 恬 (6-21), Wǔ 午 (8-175), Xīn 欣 (8-164+8-1475, 8-890+8-1583), Xìn 信 (8-197) and Xuān 宣 (8-170). '(Female) Bondservants' (*lìqiè* 隸妾) appear in four instances. In one

texts typically form part of a longer 'report' (*gào* 告), often dated, that was exchanged between superior and inferior government units. The fact that Bondservants were delegated this task supports Hulsewé's theory concerning *Shuìhǔdì* strip 185, which posits that young and trustworthy Bondservants, both male and female, were assigned the responsibility of forwarding documents.[182]

Government units without personal names or titles appear as consignors, specifically the '(County) Court' (*tíng* 廷) of Qiānlíng, Èrchūn District and Yǒuyáng County.[183] With two exceptions, all recipients of these documents were government units within Dòngtíng Commandery or 'adjacent counties' (*páng* 旁). Unlike documents providing official titles and personal names, these manuscripts primarily contain brief one-line or two-line sentences. They are undated and may have functioned as reusable postal strips or text bundles frequently sent to a spe-

case, one "sends documents" (*xíng shū* 行書) to the Governor (8-904+8-1343). Their names are Guī 規 (8-69), Sūn 孫 (8-475+8-610, 8-1538) and Yì 益 (8-904+8-1343). 'Postmen' (*yóurén* 郵人) appear in four instances; their names are Chén 辰 (8-664+8-1053+8-2167), Dé 得 (8-154) and Yǔ 羽 (8-704+8-706). In one instance, the Postman is from Dū 都 District, but his name is unknown due to a missing character (8-62). 'Runners' (*zǒu* 走) appear in four instances, twice in conjunction with the 'Commissioner of Public Works' (*sīkōng* 司空) (8-63, 8-135); their names are Qū 屈 (8-1452), Shēn 申 (8-63) and Yìn 印 (8-657). One name is unknown due to a missing character (8-135). '(Male) Bondservants' (*lìchén* 隸臣) appear in three instances, once in conjunction with the '(County) Court' (*tíng* 廷) (8-686+8-973); their names are Fùjiě 負解 (8-686+8-973) and Jù 俱 (8-898+8-972). One name is unknown due to a missing character (8-1524). Two members of a 'Unit of Five People' (*shìwǔ (wǔ)*) 士五(伍)) appear twice, once in conjunction with 'adjacent counties' (*páng* 旁) (8-1516). The peculiar detail about these entries is that they include the provenance of the individuals, for example, "Shùn from (an unknown) township in Zǐguī 秭歸 (County)" (*Zǐguī* □ *lǐ shìwǔ (wǔ) Shùn* 秭歸□里士五(伍)順) (8-1516) and "X a member of a Unit of Five People, from the township of Gāo" (*Gāo lǐ shìwǔ (wǔ)* □ 高里士五(伍)□) (8-75 +8-166+8-485). Three individuals are mentioned by name only; Chōu 瘳 in conjunction with Qiānlíng County (8-1876), and Gǎn 感 (8-1511) or Zhāo 昭 (8-1510) with no addressee specified. A 'Military Commandant' (*wèi* 尉) is mentioned once, also without an addressee (8-1951). To avoid inaccuracies, instances where '*xíng* 行' is preceded by a missing fragment have not been taken into account.

182 Hulsewé 1985: 86. Strip 185 is part of the *Shuìhǔdì* text "Statutes Concerning the Delivery of Documents" (*Xíng shū lǜ* 行書律).

183 The '(County) Court' (*tíng* 廷) appears once in conjunction with the 'Household Bureau' (*hùcáo* 戶曹) (8-1318). Qiānlíng County appears a total of 25 times, 21 of which are in conjunction with Dòngtíng Commandery (5-35, 6-2, 6-12, 8-32, 8-90, 8-115+8-338, 8-134, 8-182, 8-320, 8-362+8-390, 8-371+8-622, 8-413, 8-432, 8-504+8-563, 8-527, 8-555, 8-1464, 8-1553, 8-1685, 8-1837, 8-1840). Èrchūn District (*Èrchūn xiāng* 貳春鄉) is referenced on manuscript 8-1147, and Yǒuyáng County appears to have exchanged documents with Dòngtíng Commandery (5-34). To avoid inaccuracies, instances where '*xíng* 行' is preceded by a missing fragment have not been taken into account.

cific region.[184] Of the extant 32 wooden strips, 21 were exchanged between Qiānlíng County and Dòngtíng Commandery. Several other documents may also have been exchanged between Qiānlíng and Dòngtíng, though they are damaged, with the sections detailing the sending and receiving units missing.

Instances of '*xíng* 行', meaning 'to follow up on', can be found on at least seven manuscripts.[185] A few examples are listed below:

8-159 rIV: [. . .] 以洞庭發弩印行事□□ [. . .][186]

[. . .] use the seal of the Crossbow Archers in Dòngtíng to follow up on the matter □□ [. . .]

8-657 IV: [. . .] 以蒼梧尉印行事. [. . .][187]

[. . .] use the seal of the Cāngwú Military Commandant to follow up on the matter. [. . .]

Comment: Cāngwú was a commandery adjacent to Dòngtíng Commandery, annexed from Chǔ territory.[188]

8-830+8-1010: [. . .] 以沅陽印行事.[189]

[. . .] use the seal of Yuányáng (County) to follow up on the matter.

When used in the sense of 'to inspect', '*xíng* 行' mainly appears in the verb-object collocation "to inspect the ancestral temple" (*xíng miào* 行廟), as evidenced on manuscripts 8-138+8-174+8-522+8-523 and 8-681:

8-138+8-174+8-522+8-523 II: 失期. 行廟者必謹視中□各自署廟所質日. [. . .][190]

Missed the (right) time. The one who inspects the temple must carefully watch the middle □ each proceeds from the event calendar in the registered temple. [. . .]

Comment: 'Event calendars' (*zhì rì* 質日) were effective tools for planning, coordinating and supervising the duties of officials during the Qín and Hàn eras.[191]

8-138+8-174+8-522+8-523 rAII: 十一月己巳[6], 令史慶行廟.[192]

In month 11 on a *jǐsì*, Prefectural Secretary Kàng inspects the temple.

184 On the imperial postal system and document delivery in the *Lǐyē* corpus and other excavated texts, see Chén Zhìguó 2008.
185 This applies to manuscripts 8-159, 8-462+8-685, 8-657, 8-759, 8-830+8-1010, 8-1516 and 8-1523.
186 Chén Wěi 2012: 96.
187 Cài Wànjìn 2008; Chén Wěi 2011a; 2012: 193.
188 Yóu Yìfēi 2015.
189 Chén Wěi 2012: 234.
190 Chén Wěi 2012: 78.
191 Chén Wěi 2018.
192 Chén Wěi 2012: 78.

8-138+8-174+8-522+8-523 rAIII: 十二月戊辰[5], 令史陽行廟.[193]

In month 12 on a *wùchén*, Prefectural Secretary Yáng inspects the temple.

8-138+8-174+8-522+8-523 rCIII: 二月壬寅[39], 令史釦行廟.[194]

In month 2 on a *rényín*, Prefectural Secretary Kòu inspects the temple.

8-138+8-174+8-522+8-523 rDII: 五月丙辰[53], 令史上行廟.[195]

In month 12 on a *bǐngchén*, Prefectural Secretary Shàng inspects the temple.

8-138+8-174+8-522+8-523 rDIV: 六月癸巳[30], 令史除行廟.[196]

In month 6 on a *guǐsì*, Prefectural Secretary Chú inspects the temple.

The alternate reading of '*xíng miào* 行廟' as "conducting a temple sacrifice" can be challenged on several grounds.[197] Firstly, a commentary by the Later Hàn scholar Zhèng Xuán 鄭玄 (127–200 CE) on a passage in the "Musical Records" (*Yuèjì* 樂記) of the *Book of Rites* (*Lǐjì* 禮記) glosses '*xíng* 行' as 'to observe' or 'to inspect' (*shì* 視). Secondly, according to the *Shuìhǔdì* section in "Day Books on the Heavenly Stem Jiǎ" (*Rì shū jiǎ zhǒng* 日書甲種), it was legally inadmissible to perform sacrifices on the '3rd day' (*yín* 寅), or to pray on the '6th day' (*sì* 巳) of each month. This would make the aforementioned reading inconsistent with entries rAII, rCIII and rDIV recorded on manuscript 8-138+8-174+8-522+8-523.

Regarding ritual continuity from the Late Zhōu to the Qín dynasty, manuscripts 8-138+8-174+8-522+8-523 and 8-681 highlight the significance of ancestral temples and the institutionalized obligations associated with them. Manuscript 8-138+8-174+8-522+8-523 lists 14 officials in charge of "inspecting the ancestral temple" (*xíng miào* 行廟) in the 26th year of King Zhèng (221 BC), 12 of whom are identified as 'Prefectural Secretaries' (*lìngshǐ* 令史), while the remaining two are 'Secretaries' (*shǐ* 史).[198] Tablet 8-461 records similarities between the sacrifices offered to the Emperor and those offered to the 'Celestial Thearch' (*Tiāndì* 天帝). Not only does this suggest the deification of the Emperor, but also indicates that

193 Chén Wěi 2012: 78.
194 Chén Wěi 2012: 78.
195 Chén Wěi 2012: 78.
196 Chén Wěi 2012: 78.
197 Chén Wěi 2012: 79; Fàn Yúnfēi 2016.
198 The names of the twelve 'Prefectural Secretaries' (*lìngshǐ* 令史) are Chú 除, Fàn 犯, Fū 夫, Gèng 更, Kòu 釦, Mòyá 莫邪, Qìng 慶, Shàng 上, Wéi 韋, Xíng 行, Yáng 陽 and X (pronunciation unknown) 𢉊. The names of the two 'Secretaries' (*shǐ* 史) are Kòu 釦 and Róngfū 戎夫.

the Qín dynasty may have been governed on the basis of religious authority or divine right.[199]

On manuscript 8-161+8-307, the verb-object collocation "*xíng tián* 行田宇" should be understood as "to distribute the farming territory":[200]

> 8-161+8-307 IV: 庚申[57], 潁陰相來行田宇.[201]
>
> On a *gēngshēn*, Xiāng of Yǐngyīn (County) comes to distribute the farming territory.

In some *Lǐyē* documents, the character '*xíng* 行' conveys the meaning 'to go'. Manuscript 8-439+8-519+8-537, an arrest warrant issued for an escaped "convict in a Unit of Five People" (*tú wǔ* 徒伍) named Liáokě 繚可, illustrates this point. Dated to the 25th year of King Zhèng (222 BC), the first line reads:

> 8-439+8-519+8-537 I: 廿五年九月己丑[26], 將奔命校長周爰書: 敦長買、什長嘉皆告曰: 徒士五(伍)右里繚可, 行到零陽廡谿橋亡, 不智(知) □□☒ ┛ [. . .][202]
>
> In the 25th year (of King Zhèng) [222 BC], month 9 on a *jǐchǒu*, Zhōu, Section Commander with an Emergency Deployment Troop, writes a protocol: Mǎi, Commanding Official of the Emergency Deployment Troop, and Jiā, Commander of a Unit of Ten People, unanimously report: Liáokě, a convict in a Unit of Five People from the township of Yòu, went to the bridge of Wǔxī in Língyáng and absconded. We do not know □□☒┛ [. . .]

The observations above are based on an analysis of the *Lǐyē* texts from layers 5, 6 and 8. Manuscripts found in other layers of well J1, as well as future excavation and research efforts on wells J2 and J3, may either confirm or refute these findings. When interpreting words, individual characters, sentences or passages, it is important to frame the discussion within the context of the *Lǐyē* corpus. It is also crucial to remember that these documents were local to a peripheral region of the empire and thus represent only a small segment of Qín society. Even though direct comparisons with other excavated or transmitted sources from around the same period may prove beneficial, these sources were possibly intended for en-

199 For a translation and more detailed discussion of tablet 8-461, see chapter 4.1. For discussions on religious authority in the Qín dynasty, see Kern 2000; Puett 2002; Yóu Yìfēi 2015.
200 Chén Wěi 2012: 98. Chén Wěi partly derives this reading of '*xíng tián* 行田' from strip 239 of the "Statutes On Agricultural Fields" (*Tián lǜ* 田律) in the "Èrnián lǜlìng", where, in a similar context, '*xíng* 行' is understood as 'to bestow upon' or 'to direct' (*zhǐshòu* 指授). The "Statutes On Households" (*Hù lǜ* 戶律) in the same corpus contain regulations on 'receiving' (*shòu*受) 'farming territory' (*tiányǔ* 田宇). The preceding manuscripts are heavily damaged and do not provide additional context.
201 Chén Wěi 2012: 97.
202 Chén Wěi 2012: 149.

tirely different purposes or limited in circulation to specific individuals, groups or communities. Such variations in function and accessibility could distort findings in a way that is detrimental to our understanding of these administrative texts. By consciously avoiding, or at least minimizing, such comparisons, we allow for a more exclusive appreciation of the collection as a self-contained entity with its own nuanced modes of interaction that might otherwise be underestimated or overlooked.

3 Calendar and Timekeeping Systems

Postal and logistic services in Qiānlíng County maintained lively contact with other regions via extensive land and water networks. Petty records detailing the creation, dispatch or reception times of documents indicate that effective communication was vital to the Qín dynasty, especially for land cultivation, bureaucratic operations and upholding cooperation across different counties and commanderies.[203] A sense of how accurate time was measured and recorded is illustrated in the last two lines on the *recto* of manuscript 8-60+8-656+8-665+8-748:

> 8-60+8-656+8-665+8-748 rIII–IV: 六月庚辰[17]水十一刻刻下六, 守府快行少内.☒⅃ 六月乙亥[12]水十一刻刻下二, 佐同以來. ノ元手.☒[204]
>
> In month 6 on a *gēngchén*, at water below mark 6 of 11 marks, Governor Kuài dispatches (the document) to the Lesser Treasury. ☒⅃ In month 6 on a *yǐhài*, at water below mark 2 of 11 marks, (the document) is brought by Assistant Tóng. / By the hand of Yuán. ☒
>
> Comment: Between '*gēngchén* 庚辰' and '*yǐhài* 乙亥' within the same month and year, there are 56 positions, which cannot be correct and could indicate either a clerical error or a missing interrelation between the two lines.

Calendars and established timekeeping methods provided the foundation for communication within the imperial government. Almost every year, events in the *Lǐyē* corpus are systematically recorded.[205] The explicit reference to exact dates and times allows us to confine the *Lǐyē* documents to the period between the 25th year of King Zhèng in 222 BC (one year before the Qín unification) and the 2nd year of the Second Emperor in 209 BC (one year before the inglorious collapse of the Qín dynasty). To clarify, the *Lǐyē* corpus does not introduce a new official calendar but rather continues the administrative tradition of the Warring States period. Entries identify the year in relation to the ruler's ascension to power and employ sequential numbering of lunar months, starting with the '1st month' (*zhēng yuè* 正月 or *duān yuè* 端月) and ending with 'month 12' (*shí'èr yuè* 十二月) of an annual cycle. Additionally, the day of the respective month is expressed as dictated by the 'sexagenary cycle' (*gānzhī* 干支).[206] The Warring States notion

203 For a discussion of the 'event calendars' (*zhì rì* 質日) compiled by local authorities in the Qín and Hàn dynasties, refer to Chén Wěi 2018.

204 Chén Wěi 2012: 43.

205 For a discussion of calendar and planetary calculations in ancient Greece (ca. 80–50 BC), see Bitsakis and Hartz 2015; Iversen 2017; Lobell 2007.

206 For more information on the origins of the 'sexagenary cycle' (*gānzhī* 干支) in early Chinese astronomy and hemerology, see Smith 2011a, 2020.

https://doi.org/10.1515/9783110791853-003

that "all phenomena and human activity were linked in microcosmic synchronicity" seems to have been adopted and appropriated by the Qín dynasty.[207]

Interestingly, the reference year to King Zhèng's ascension to power did not change after he declared himself the First Emperor and was maintained throughout his reign until his succession by the Second Emperor. After unification, the reference date for any record written during the First Emperor's reign remains, without exceptions, the year of his ascension to the throne as a 13-year-old King in 246 BC. An entry on manuscript 8-757 from King Zhèng's 34th regnal year (213 BC), regarding the cultivation of 'agricultural fields' (*tián* 田) between his 26th and 29th year of reign (221–218 BC), reads:[208]

> 8-757 II: [. . .] 廿九年田廿六年盡廿八年當田 [. . .][209]
>
> [. . .] In the 29th year (of King Zhèng) [218 BC], the agricultural fields [were cultivated], (but) from the 26th year to the 28th year (of King Zhèng) [221–219 BC], the agricultural fields should have been (cultivated also) [. . .]

Five entries in the *Lǐyē* corpus record the Emperor's '1st year' (*yuán nián* 元年) of reign.[210] Based on the context, including the officials' names and titles as well as the total time span covered by the *Lǐyē* texts, these records refer to the first year of the Second Emperor. There is no apparent correlation between the layers in which the documents were found and the temporal sequence of text entries. The earliest or latest documents were not located at the top or bottom of well J1 but were found in layers 5 to 17. Layer 5, which was the highest layer containing Qín documents, also held the largest cache of Chǔ bamboo fragments. Claims of missing documents from King Zhèng's 36th regnal year (211 BC) can be challenged on the basis of at least three texts.[211] For instance, manuscript 8-500, uncovered in layer 8, is dated to the First Emperor's year of passing, in other words, his 37th regnal year (210 BC). This entry must have been added before news of his death

207 Harper 1999: 831.

208 The 34th year of King Zhèng is recorded in the first line of manuscript 8-755, which forms part of a larger text spanning multiple writing carriers.

209 Chén Wěi 2012: 217.

210 Refers to manuscripts numbers 5-1, 6-3, 8-653, 8-860 and 8-2131.

211 See Yates 2012–2013: 303. Proof to the contrary can be found on manuscripts 8-1041+8-1043, 8-1437 and 8-2234.

had reached Qiānlíng or Dòngtíng, or before the Second Emperor's succession to the throne was announced.[212]

Nowhere are the Emperors named personally, even though there is no indication that public use of Qín Shǐhuáng's personal name 'Zhèng 政' was taboo during his reign as First Emperor. References to the '1st month' (*zhēng yuè* 正月) are common, as verified on manuscript 8-253, for example:

8-253: 尉曹卅四年正月已事☒ [213]

The Bureau of the Military Commandant, in the 34th year (of King Zhèng) [213 BC], 1st month, the finished matter ☒

Only after the First Emperor's death may a ban on his personal name 'Zhèng 政' have been enforced more rigorously in official documents. Manuscript 6-3 dates back to the last year of the First Emperor or the first year of the Second Emperor (209 BC) and replaces the term '*zhēng yuè* 正月' with '*duān yuè* 端月' to denote the '1st month' or 'beginning month' of the year:

6-3: [. . .] 元年端月癸卯[40]朔□□, 司空□□受倉□☒ [214]

[. . .] In the first year (of the Second Emperor), beginning month [month 1], new moon on a *guǐmǎo*, □□, the Commissioner of Public Works □□ receives the Granary □☒

One possible reason '*zhēng/zhèng* 正' (*teŋ/teŋh) was avoided and replaced with '*duān* 端' (*tôn) could be its homophony with 'Zhèng 政' (*teŋh) at the time.[215]

Lǐ Xuéqín (1933–2019) argued that examining tabooed characters is one of the most reliable methods for determining the age of early writings.[216] The replacement of '*zhèng* 正' with '*duān* 端' is frequent in the *Shuìhǔdì* texts, particularly in the "Book of Discourses" (*Yǔ shū* 語書), where phrases like 'to rectify oneself' (*zì zhèng* 自正), 'to rectify the public' (*gōng zhèng* 公正) and 'to rectify' (*jiǎo zhèng* 矯正) give way to the synonyms '*zì duān* 自端', '*gōng duān* 公端' and '*jiǎo duān* 矯端'. Similarly, the term and official title 'Township Principal' (*lǐzhèng* 里正) is repeatedly interchanged with the compound noun '*lǐdiǎn* 里典' in the "Qín lǜ shíbā zhǒng" and "Models for Sealing and Investigating" (*Fēng zhěn shì* 封診式) from the same corpus.

212 Apparently, two other *Lǐyē* texts from layers 7 and 9 are dated to the 37th year of King Zhèng (210 BC) (Yates 2012–2013: 302).

213 Chén Wěi 2012: 122.

214 Chén Wěi 2012: 18.

215 Phonological reconstructions are taken from Schuessler 2009: 138, 270.

216 Lǐ Xuéqín 1981: 337–338.

Some scholars contend that certain *Shuìhǔdì* documents were originally compiled during the Late Warring States period and transmitted through written sources to the Qín empire.[217] In this context, the use, avoidance or substitution of the term '*zhèng* 正' is often treated as a criterion for dating texts. Its presence is generally considered an indication that the text was compiled in the Late Warring States, as observed in works like "Statutes Concerning Crosschecking" (*Xiào lǜ* 效律) and "Miscellaneous Excerpts from Qín Statutes" (*Qín lǜ zá chāo* 秦律雜抄). Conversely, its absence supposedly dates the text to the very late Qín state or the imperial Qín period.

However, based on my own research of the *Lǐyē* corpus, relying solely on the examination of taboo characters to determine the chronological sequencing of written material is not entirely reliable. One could argue that the practice of tabooing particular terms spread gradually, much like the unification of the writing system, and that the factors of this process were more geographical than temporal. Also, the use of either '*zhēng/zhèng* 正' or '*duān* 端' in official documents seems to have been less rigid than previously assumed.[218] '*Zhēng yuè*', designating the '1st month' of the annual cycle, appears 32 times across 24 manuscripts written in the 29th year of King Zhèng (218 BC) and between his 31st and 35th years (216–212 BC) (Tab. 3.1). In contrast, '*duān yuè* 端月', a synonym of '*zhēng yuè*', appears only three times across three different documents excavated from layers 5, 6 and 8 – roughly ten times less frequent than '*zhēng yuè*'. One instance is dated to the 26th year of King Zhèng (221 BC), another to his final year as First Emperor (or the first year of the Second Emperor) (209 BC), and the third occurs on a heavily damaged manuscript with an unidentified date.[219]

I am not convinced that the appearance of '*duān yuè*' on fragment 8-213 is part of manuscript 8-198+8-213+8-2013.[220] Moreover, Chén Wěi's argument that the Second Emperor introduced the term only *after* his father's death seems unfounded.[221] The use of '*duān yuè*' in the 26th year of King Zhèng (221 BC) occurs on the earlier manuscript 8-138+8-174+8-522+8-523 and is more plausibly attributed to the First Emperor's ascension to the imperial throne, as well as the proclamation of his successor in 209 BC. Textual evidence in the *Lǐyē* corpus is too scarce to be conclusive, but dated

217 Huáng Shèngzhāng 1979. For a study on taboo characters in the Qín dynasty see, for example, Venture 2012.

218 See Mittag 2003: 550—551.

219 This refers specifically to manuscripts 6-3, 8-138+8-174+8-522+8-523 and 8-1555.

220 See Yates 2012–2013: 303.

221 See Chén Wěi 2014b.

Tab. 3.1: Appearances of terms for the '1st month' (*zhēng yuè* 正月 and *duān yuè* 端月) of the annual cycle in the *Lǐyē* texts from layers 5, 6 and 8 (sorted by term, occurrence(s) and manuscript number).

'zhēng yuè 正月'			'duān yuè 端月'		
Occurrence(s)	**Manuscript number**[222]	**Year [conversion]**	**Occurrence(s)**	**Manuscript number**	**Year [conversion]**
1	8-157	32nd year of King Zhèng [215 BC]	1	6-3	1st year of Second Emperor [209 BC]
3	8-157r	32nd year of King Zhèng [215 BC]	1	8-138r +8-174r +8-522r +8-523r	26th year of King Zhèng [221 BC]
1	8-175r	–	1	8-1555	–
1	8-197	34th year of King Zhèng [213 BC]			
1	8-197r	34th year of King Zhèng [213 BC]			
1	8-212+8-426+8-1632	31st year of King Zhèng [216 BC]			
1	8-214	33rd year of King Zhèng [214 BC]			
1	8-253	34th year of King Zhèng [213 BC]			
1	8-259	35th year of King Zhèng [212 BC]			
1	8-464	–			
1	8-474+8-2075	31st year of King Zhèng [216 BC]			
1	8-478	32nd year of King Zhèng [215 BC]			
2	8-651	33rd year of King Zhèng [214 BC]			

222 'R' stands for the '*recto*' of the manuscript.

Tab. 3.1 (continued)

'zhēng yuè 正月'			**'duān yuè 端月'**		
Occurrence(s)	**Manuscript number**	**Year [conversion]**	**Occurrence(s)**	**Manuscript number**	**Year [conversion]**
1	8-651r	33rd year of King Zhèng [214 BC]			
1	8-744	–			
1	8-764	31st year of King Zhèng [216 BC]			
2	8-765r	35th year of King Zhèng [212 BC]			
1	8-839+8-901+8-926	35th year of King Zhèng [212 BC]			
2	8-925+8-2195	31st year of King Zhèng [216 BC]			
1	8-1207+8-1255+8-1323	33rd year of King Zhèng [214 BC]			
1	8-1241	31st year of King Zhèng [216 BC]			
1	8-1246	29th year of King Zhèng [218 BC]			
1	8-1457+8-1458	35th year of King Zhèng [212 BC]			
1	8-1580	–			
1	8-1738	35th year of King Zhèng [212 BC]			
1	8-2203	–			
1	8-2453	34th year of King Zhèng [213 BC]			
	TOTAL			**TOTAL**	
32	**27**	**29th, 31–35th year of King Zhèng [218, 216–212 BC]**	**3**	**3**	**26th year of King Zhèng [221 BC], 1st year of Second Emperor [209 BC]**

manuscripts discovered in a Qín tomb at Zhōujiātái 周家台 and those acquired by the Yuèlù Academy may lend additional support to the hypothesis.[223]

Manuscript 8-141+8-668 from the 30th year of King Zhèng (217 BC) and manuscript 8-157 from the 32nd year of King Zhèng (215 BC) are particularly revealing regarding the widely accepted view that the official title 'Township Principal' (*lǐdiǎn* 里典) is a tabooed version of '*lǐzhèng* 里正'.[224]

8-141+8-668 I–III, rI: 卅年十一月庚申[57]朔丙子[13], 發弩守涓敢言之: 廷下御史書曰縣┛ □治獄及覆獄者, 或一人獨訊囚, 嗇夫長、丞、正、監非能與┛ □□殹, 不參不便. 書到尉言. • 今已到, 敢言之. └十一月丙子[13]旦食, 守府定以來. ノ連手. 萃手.

In the 30th year (of King Zhèng) [217 BC], month 11, new moon on a *gēngshēn*, *bǐngzǐ* [day 17], Juān, Head of the Crossbow Archers, dares to say: The (County) Court sent the document of the Prosecutor to the lower authorities. It states: Concerning the counties┛ □ the one who executes the lawsuit and the one who reviews the lawsuit, if only one person interrogates a prisoner, the Commanding Official of the Overseers, the Prefect, the Administrators and the Supervisors are not the ones able to┛ □□ if there are not three, it is not according to the procedure. The document has reached the Military Commandant and has been communicated orally. • Now that it has already arrived, he dares to communicate orally └In month 11 on a *bǐngzǐ* [day 17], at morning meal, (the document) is brought by Governor Dìng. / By the hand of Lián. By the hand of Cuì.

Comments: 'Administrators' ((*tíng wèi*) *zhèng* (廷尉)正) and 'Supervisors' ((*tíng wèi*) *jiān* (廷尉)監) also appear on strip 184 of the "Zòuyànshū" and were officials within the Imperial Court.[225] Since this manuscript deals with interrogation procedures at the county level, I assume that the 'Administrators' (*zhèng* 正) and 'Supervisors' (*jiàn* 監) mentioned are officials operating at commandery and/or county level. For a discussion of Supervisors, see chapter four.

'*Gǎn yán zhī* 敢言之' (dare to say, dare to report) is a respectful, formulaic expression or bureaucratic marker used in reports orally recited by a subordinate official or agency to a higher authority.

Lau and Staack translate '*sèfū* 嗇夫' as 'Overseer of a Prefecture'.[226] As confirmed by the *Lǐyē* corpus and Yínquèshān Hàn tomb bamboo strips (*Yínquèshān Hàn mù zhújiǎn* 銀雀山漢墓竹簡), 'Overseers' (*sèfū* 嗇夫) were employed across various units of the empire, including 'Granaries' (*cāng* 倉), 'Warehouses' (*kù* 庫), 'Stables' (*jiù* 廄), 'marketplaces' (*shì* 市) and 'kitchens' (*chú* 廚).[227]

223 See Chén Wěi 2014b. For more information on the *Zhōujiātái* 周家台 documents discovered in 1993 in Húběi Province, refer to *Húběi shěng Jīngzhōushì Zhōuliángyùqiáo yízhǐ bówùguǎn* 2001; Wáng Guìyuán 2007.

224 See also Beck 1987.

225 Lau and Staack 2016: 311.

226 Lau and Staack 2016: 114.

227 Lander 2015: 304.

8-157 I–III: 卅二年正月戊寅[15]朔甲午[31], 啓陵鄉夫敢言之: 成里典、啓陵┘ 郵人缺. 除士五(伍)成里匄、成, 成爲典, 匄爲郵人, 謁令┘ 尉以從事. 敢言之.[228]

In the 32nd year (of King Zhèng) [215 BC], 1st month, new moon on a *wùyín*, *jiǎwǔ* [day 17], the Overseer of Qǐlíng District dares to say: The (positions of the) Principal of the township of Chéng and the Postman┘ of Qǐlíng (District) are vacant. Appoint Gài from a Unit of Five People and Chéng from the township of Chéng. Chéng becomes Principal, and Gài becomes Postman. Request an ordinance┘ of the Military Commander to follow up on the matter. (He) dares to say.

The official title 'Overseer of a District' (*xiāng fū* 鄉夫) is an abbreviation of '*xiāng sīfū* 鄉嗇夫' in this context.[229] As outlined by Wáng Huànlín, '*fū* 夫' could also be interpreted as the name of an official in Qǐlíng District.[230] While his proposed reading of the character is syntactically plausible, I find the original interpretation more convincing for two reasons: 1) official titles and responsibilities are usually prioritized over personal names in these types of documents, and 2) '*xiāng fū* 鄉夫' is a common abbreviation for '*xiāng sīfū* 鄉嗇夫' in other Qín and Hàn dynasty manuscripts.[231]

From the readings of manuscripts 8-141+8-668 and 8-157, it can be concluded that '*zhèng* 正' was not avoided in imperial correspondence. In fact, the former manuscript shows that the term was an official title, possibly subordinate to the 'Commanding Official of the Overseers' (*sèfūzhǎng* 嗇夫長) at the '(County) Court' (*tíng* 廷) and the 'Prefect' (*chéng* 丞). On the latter manuscript, '*zhēng* 正' – used as a temporal designation – appears alongside the official title 'Township Principal' (*lǐdiǎn* 里典). This would not occur if '*lǐdiǎn* 里典' were a tabooed variant of '*lǐzhèng* 里正'. On strips 201 and 329 of the "Èrnián lǜlìng", the term is again listed as an official title with '*diǎn* 典', occupying a similar or even higher hierarchical position.

Zhāngjiāshān, "Èrnián lǜlìng", "Statutes On Cash" (*Qián lǜ* 錢律), strip 201: 盜鑄錢及佐者: 棄市. 同居不告: 贖耐. 正、典、田典、伍人不告: 罰金四兩.[232]

For the one who thievishly casts cash, as well as the one who assists: cast (the criminal) away in the marketplace. For the co-resident who does not make a denunciation: (order him or her) to redeem shaving. For the Township Principal, the Township Administrator and people in the (mutually responsible) Unit of Five People who do not make a denunciation: fine four ounces of gold.[233]

228 Chén Wěi 2012: 94.
229 *Húnán shěng wénwù kǎogǔ yánjiūsuǒ, Xiāngxī Tǔjiāzú Miáozú zìzhìzhōu wénwù chǔ* 2003.
230 Wáng Huànlín 2007: 52.
231 Chén Wěi 2012: 95.
232 Barbieri-Low and Yates 2015: 632.
233 Translation taken from Barbieri-Low and Yates 2015: 635, with minor modifications. Barbieri-Low and Yates translate '*zhèng* 正' as 'Village Chief', and '*diǎn* 典' as 'Chief of the Fields'.

Zhāngjiāshān, "Èrnián lǜlìng", "Statutes On Households" (*Hù lǜ* 戶律), strip 329: 數在所正、典弗告：與同罪.[234]

When the Chief and the Township Principal of the place (where people are) counted do not report, (they) are likewise liable to punishment.

A closer look at the calendar format has led to noteworthy inferences about the status of certain terms and expressions in the *Lǐyē* cache. In most cases, homophones of the Emperor's personal name were not strictly prohibited in administrative writing but continued as part of the official vocabulary.

The calendar system adopted in the *Lǐyē* texts was developed and established during the Warring States period. In its most complete form, it records the regnal year of King Zhèng (x), the month, starting from 'month 1' (*zhēng yuè* 正月 or *duān yuè* 端月) of 12 or 13 months (including the intercalary month 9 at irregular intervals every 2–3 years) (y), the 'new moon day' (*shuò(rì)* 朔(日)) (z1), and the day of the event or document entry (z2) (Tabs. 3.2, 3.3). Each annual cycle began in 'month 10' (*shí yuè* 十月). Building upon this format, so-called 'event cal-

Tab. 3.2: Calendar structure based on the *Lǐyē* manuscripts from layers 5, 6 and 8.

Year x [conversion]	**Month y**	**Day z1, z2**
25th–37th year of King *Zhèng* [222–210 BC]	'Month 1' (*zhèng yuè* 正月) to 'month 12' (*shíèr yuè* 十二月) of the lunar calendar with 'intercalary month 9' (*hòu jiǔ yuè* 後九月) every two to three years; the annual cycle began in 'month 10' (*shí yuè* 十月).	Definition of the 'new moon' (*shuò(rì)* 朔(日)) reference day and calculation of the event day in accordance with the sexagenary cycle.
1st year of the Second Emperor [209 BC]	'Month 1' (*zhèng yuè* 正月 or *duān yuè* 端月) to 'month 12' (*shíèr yuè* 十二月) of the lunar calendar with 'intercalary month 9' (*hòu jiǔ yuè* 後九月) in the first year of the Second Emperor's reign; the annual cycle began in 'month 10' (*shí yuè* 十月); in the excavated manuscripts, only reference to 'month 1' (*duān yuè* 端月), 'month 4' (*sì yuè* 四月), 'month 7' (*qī yuè* 七月), 'month 8' (*bā yuè* 八月) and 'intercalary month 9' (*hòu jiǔ yuè* 後九月) is made.	Definition of the 'new moon' (*shuò(rì)* 朔(日)) reference day and calculation of the event day in accordance with the sexagenary cycle.

234 Barbieri-Low and Yates 2015: 798.

endars' (*zhì rì* 質日), limited to a narrower range of tasks, were employed for planning, coordinating and supervising the specific duties of government officials in the Qín and Hàn dynasties.

For modern readers, the ordering of days based on the sexagenary cycle is not intuitive and requires careful calculation. This complexity is best illustrated by a few text excerpts:

8-73 I: 卅四年後九月壬辰[29]朔壬寅[39], 司空☐ [235]

In the 34th year (of King Zhèng) [213 BC], intercalary month 9, new moon on a *rénchén*, *rényín* [day 11], the Commissioner of Public Works ☐

8-1514 I: 廿九年四月甲子[1]朔辛巳[18], 庫守悍敢言之 [. . .][236]

In the 29th year (of King Zhèng) [218 BC], month 4, new moon on a *jiǎzǐ*, *xīnsì* [day 18], Warehouse Incumbent Hàn dares to say [. . .]

8-1515 I: 卅年十月辛卯[28]朔乙未[32], 貳春鄉守綽敢告司空主 [. . .][237]

In the 30th year (of King Zhèng) [217 BC], month 10, new moon on a *xīnmǎo*, *yǐwèi* [day 5], Chuò, Incumbent of Èrchūn District, dares to report the Commissioner of Public Works [. . .]

8-1560 I: 卅一年後九月庚辰[17]朔辛巳[18], 遷陵丞昌謂倉嗇夫 [. . .][238]

In the 31st year (of King Zhèng) [216 BC], intercalary month 9, new moon on a *gēngchén*, *xīnsì* [day 2], Chāng, Governor of Qiānlíng, addressed the Overseer of the Granary [. . .]

8-2194 I: [. . .] 卅二年三月丁丑[14]朔癸巳[30], 貳 ☐ [239]

[. . .] In the 32nd year (of King Zhèng) [215 BC], month 3, new moon on a *dīngchǒu*, *guǐsì* [day 17], Èr ☐

8-2441 I: 卅三年十月甲辰[41]朔庚申[57], 啓陵☐ [240]

In the 33rd year (of King Zhèng) [214 BC], month 10, new moon on a *jiǎchén*, *gēngshēn* [day 17], Qǐlíng ☐

235 Chén Wěi 2012: 55.
236 Chén Wěi 2012: 342.
237 Chén Wěi 2012: 343.
238 Chén Wěi 2012: 359.
239 Chén Wěi 2012: 443.
240 Chén Wěi 2012: 469.

To calculate the exact day, one must first identify the '1st day of the lunar month', or 'new moon day' (*shuò(rì)* 朔(日)), and then count the days between this 1st day (or new moon day) and the day on which the event was recorded.[241] For example, the entry on manuscript 8-1514 occurred on a '*xīnsì* 辛巳', which is 18 days after the new moon day '*jiǎzǐ* 甲子' in month 4 of the 29th year of King Zhèng (218 BC). Counting from *jiǎzǐ* in position 1 to *xīnsì* gives 18 days, making '*xīnsì*' equivalent to 'day 18' (see Tab. 3.3).

On manuscript 8-1515, the 1st day of month 10 in the 31st year of King Zhèng (216 BC) is a '*xīnmǎo* 辛卯' in position 28. The actual day of the entry is a '*yǐwèi* 乙未' in position 32. Therefore, *yǐwèi* is equal to day 5 of month 10. Manuscript 8-1560 is dated to the 2nd day of the intercalary month 9 in the 31st year of King Zhèng (216 BC). In this month, the new moon day was a '*gēngchén* 庚辰' in position 17. '*Xīnsì* 辛巳' follows immediately in position 18, and is thus equal to 'day 2'. For manuscripts 8-2194 and 8-2441, '*dīngchǒu* 丁丑' in position 14 and '*guǐsì* 癸巳' in position 30 (8-2194), as well as '*jiǎchén* 甲辰' in position 41 and '*gēngshēn* 庚申' in position 57 (8-2441), are each 17 days apart. The new moon day on manuscript 8-73 is a '*rénchén* 壬辰' in position 29, making '*rényín* 壬寅' equal to 'day 11'.

There are also examples where reference to the new moon day is omitted, as seen on manuscript 8-1041+8-1043. In such cases, either the official recording the entry left it out, or the event day itself was day 1. Cross-referencing with other manuscripts usually helps resolve these ambiguities. For example, with manuscript 8-1173+8-1420, we can determine that the 1st day of month 1 in the 36th year of King Zhèng (211 BC) was a '*bǐngxū* 丙戌'.

> 8-1173+8-1420: [. . .] 卅六年十一月丙戌[23]朔壬辰[29]☑ [242]
>
> [. . .] In the 36th year (of King Zhèng) [211 BC], month 11, new moon on a *bǐngxū*, *rénchén* [day 7] ☑

Every day can be identified using the method described above, although there are a few manuscripts where these calculations do not align. Manuscript 8-768, for instance, presents an unresolved issue. It cites the new moon day as a '*gēngzǐ* 庚子' in position 37, and the event day as a '*dīngwèi* 丁未' in position 44, which would make '*dīngwèi* 丁未' equal to 'day 8'. Strangely, the last entry containing the day and time of the delivery is identified as a '*yǐsì* 乙巳' in position 42. This implies that the document was dispatched before the entry was made. Chén Wěi's transcription of the pa-

241 For a discussion and reconstruction of the 'new moon days' (*shuò(rì)* 朔(日)) during the reigns of the First and Second Emperor's, see Xǔ Míngqiāng 2013b.
242 Chén Wěi 2012: 287.

leographic material seems to be accurate, yet this contradiction has been overlooked.[243] This discrepancy could be either a miscalculation by the official in charge or a mismatch between the *recto* and *verso* of the document.

Manuscript 8-1449+8-1484 cites the new moon day in intercalary month 9 of the 34th year of King Zhèng (213 BC) as a '*rénxū* 壬戌' in position 59. However, in line with manuscripts 8-73, 8-838 and the "Event Calendar of the 34th Year" (*Sānshísì nián zhì rì* 三十四年質日) in the *Yuèlù* strips, the new moon day for this particular month and year is a '*rénchén* 壬辰' in position 29.[244] Asserting the new moon day as a '*rénxū* 壬戌' in this case is problematic, as it would place the entry one day earlier, on a '*xīnyǒu* 辛酉' in position 58. Chén Wěi acknowledges the new moon day provided in the *Lǐyē* and *Yuèlù* manuscripts, but leaves the inconsistency on the *recto* unaddressed. Specifically, '*yǐmǎo* 乙卯' in position 52 is four days ahead of the new moon day '*wùwǔ* 戊午' in month 10, as cited on manuscript 8-183+8-290-8-530. Manuscript 8-1449+8-1484 reads:[245]

> 8-1449+8-1484 I–II, rI: 卅四年後九月壬戌〈辰〉[29]朔辛酉[58], 遷陵守丞茲敢┙ 言之: 遷陵道里毋蠻更者. 敢言之. └十月己卯[16]旦, 令佐平行. 平手.[246]
>
> In the 34th year (of King Zhèng) [213 BC], intercalary month 9, new moon on a *rénchén, xīnyǒu* [day 30], Zī, Deputy Prefect of Qiānlíng, dares┙ to say: The marches and townships of Qiānlíng do not have Mán people assuming military service. (He) dares to say. └In the morning of month 10 on a *jǐmǎo* [day 18], Prefectural Assistant Píng dispatches (the document). By the hand of Píng.
>
> Comments: According to Chén Wěi, the 'marches and townships' (*dào lǐ* 道里) of Qiānlíng can be understood as the territories overseen by the officials of Qiānlíng.[247] In transmitted literature, '*dào lǐ* 道里' is generally understood as 'villages/townships along the way' or 'lands'.[248]
>
> In Hàn culture, '*Mán gèng* 蠻更' signifies "Mán people assuming a turn of duty or statute labor."[249] Here, '*gèng* 更' is interpreted as 'turn of duty' and appears in *Shuìhǔdì* strips 13–14, 54, 109 and 134–135.[250]
>
> Consistent with a passage in the "Event Calendar of the 35th Year" (*Sānshíwǔ nián zhì rì* 三十五年質日) found in the *Yuèlù* documents, the new moon day in month 10 of the 34th year of King Zhèng (213 BC) was a '*rénxū* 壬戌', which places '*jǐmǎo* 己卯' on 'day 18'.[251]

---

243 Compare with Chén Wěi 2012: 222; *Húnán shěng wénwù kǎogǔ yánjiūsuǒ* 2012a: 114.
244 Compare with Chén Wěi 2012: 328.
245 Chén Wěi 2012: 106.
246 Chén Wěi 2012: 328.
247 Chén Wěi 2012: 328.
248 Lǔ Jiāliàng 2013: 104–106; Lycas 2019: 155–156.
249 Chén Wěi 2012: 328.
250 Hulsewé 1985.
251 Compare with Xǔ Míngqiāng 2013b.

Unless these contradictory day calculations and labels were intentionally produced for reasons unknown, these examples indicate that the *Lǐyē* corpus contains several clerical errors that either went unnoticed or were deliberately left uncorrected by the various government officials handling, examining and archiving the documents. These inconsistencies may highlight the practical challenges and human errors inherent in the daily operations of ancient bureaucracies, providing insights into the administrative complexities and potential leniencies within the Qín dynasty's record-keeping practices.

Tab. 3.3: The 60 days of the 'sexagenary cycle' (*gānzhī* 干支).

1	*jiǎzǐ* 甲子	2	*yǐchǒu* 乙丑	3	*bǐngyín* 丙寅	4	*dīngmǎo* 丁卯	5	*wùchén* 戊辰
6	*jǐsì* 己巳	7	*gēngwǔ* 庚午	8	*xīnwèi* 辛未	9	*rénshēn* 壬申	10	*guǐyǒu* 癸酉
11	*jiǎxū* 甲戌	12	*yǐhài* 乙亥	13	*bǐngzǐ* 丙子	14	*dīngchǒu* 丁丑	15	*wùyín* 戊寅
16	*jǐmǎo* 己卯	17	*gēngchén* 庚辰	18	*xīnsì* 辛巳	19	*rénwǔ* 壬午	20	*guǐwèi* 癸未
21	*jiǎshēn* 甲申	22	*yǐyǒu* 乙酉	23	*bǐngxū* 丙戌	24	*dīnghài* 丁亥	25	*wùzǐ* 戊子
26	*jǐchǒu* 己丑	27	*gēngyín* 庚寅	28	*xīnmǎo* 辛卯	29	*rénchén* 壬辰	30	*guǐsì* 癸巳
31	*jiǎwǔ* 甲午	32	*yǐwèi* 乙未	33	*bǐngshēn* 丙申	34	*dīngyǒu* 丁酉	35	*wùxū* 戊戌
36	*jǐhài* 己亥	37	*gēngzǐ* 庚子	38	*xīnchǒu* 辛丑	39	*rényín* 壬寅	40	*guǐmǎo* 癸卯
41	*jiǎchén* 甲辰	42	*yǐsì* 乙巳	43	*bǐngwǔ* 丙午	44	*dīngwèi* 丁未	45	*wùshēn* 戊申
46	*jǐyǒu* 己酉	47	*gēngxū* 庚戌	48	*xīnhài* 辛亥	49	*rénzǐ* 壬子	50	*guǐchǒu* 癸丑
51	*jiǎyín* 甲寅	52	*yǐmǎo* 乙卯	53	*bǐngchén* 丙辰	54	*dīngsì* 丁巳	55	*wùwǔ* 戊午
56	*jǐwèi* 己未	57	*gēngshēn* 庚申	58	*xīnyǒu* 辛酉	59	*rénxū* 壬戌	60	*guǐhài* 癸亥

In addition to the years, months and days of the imperial calendar, the hours of the day were also subject to meticulous timekeeping. Qín officials relied on a double-hour clock system marked by scaled water clocks, or clepsydrae, which had hour 'notches' or 'marks' (*kè* 刻) carved into their sides.[252] Unfortunately, no water clock has yet been discovered in or around Lǐyē, but a replica, based on Hàn dynasty findings, is on display at the Lǐyē Museum of Qín Slips (*Lǐyē Qín jiǎn bówùguǎn* 里耶秦簡博物館) (Figs. 3.1, 3.2).

A slow, constant flow of water into the clepsydra caused a rod floating inside to indicate the passing of time relative to the notches and water level. As the Hàn dynasty philosopher and politician Huán Tán 桓譚 (43 BC–28 CE) observed, the working accuracy of clepsydrae was influenced by weather conditions, atmospheric humidity

252 One of the earliest references to simpler forms of clepsydrae in Ancient China is arguably found in a passage from the *Book of Odes* (*Shījīng* 詩經) (see Needham and Wang 1959: 315; Needham 2000). A particularly notable mention of a clepsydra is in the *Rites of Zhōu* (*Zhōulǐ* 周禮) (see Needham and Wang 1959: 319). 'clepsydra' is of Greek origin (*κλεψύδρα*) and literally means 'water-stealer' (Needham and Wang 1959: 313).

Fig. 3.1: Replica of a water clock displayed at the Lǐyē Museum of Qín Slips.[253]

and air temperature. For these reasons, these devices had to be regularly calibrated using sundials during the day and star constellations at night to ensure precise functioning.[254]

> *New Discourses* (*Xīn lùn* 新論), “On Separate Matters” (*Lí shì* 離事): 余前爲郎. 典漏刻. 燥, 濕, 寒, 溫, 輒異度. 故有昏明晝夜. 晝日參以晷景. 夜分參以星宿. 則得其正.[255]
>
> Formerly, when I served as a Gentleman, I was in charge of the marks on the clepsydra. If (conditions varied between) dry and humid, cold and warm, then there were different (ways to) measure (the marks) on the sides. Therefore, (in order to) have (the timings) for dusk and dawn, daylight and nighttime, I checked against the solar shadow in the daytime, and in the night part (of the clepsydra circle), I checked against the stellar lodges. Thus, I attained the correct (measurement).[256]

---

253 My photograph, taken at the Lǐyē Museum of Qín Slips (*Lǐyē Qín jiǎn bówùguǎn* 里耶秦簡博物館) on May 14, 2019.

254 Needham 1995: 321–322.

255 Huán Tán 1977: 44.

256 Translation is taken from Cullen 2018: 184, with minor modifications. My *pīnyīn* transcriptions.

Fig. 3.2: Monochrome photograph of a Western Hàn dynasty bronze water clock, dated to 27 BC, excavated in Inner Mongolia in 1977. Dimensions: 479 mm x 187 mm. The side inscription reads: "Bronze clepsydra (from) Qiānzhāng. One weighs 32 'catties' (*jīn* 斤). Manufactured in the 2nd year of the Hépíng period in month 4" *(Qiānzhāng tónglòu yī zhòng shìèr jīn Hépíng èr nián sì yuè zào* 千章銅漏一重卅二斤河平二年四月造).[257]

Comment: Cullen notes that "in the context of observations of the 'stellar lodges' (*xīngxiù* 星宿), '*dù* 度' seems very likely to be referring to [. . .] a measure of movement of the sun and other heavenly bodies [. . .]."[258]

Joseph Needham (1900–1995) argues that, from the Late Zhōu to the Former Hàn dynasty, a day was divided into 12 equal 'double-hours' (*shí* 時) and 100 equal 'quarters' (*kè* 刻).[259] In reference to Robin Yates, double-hours were divided into 'quarters' (*kè* 刻), with each quarter subdivided into 10 'marks' (*kè* 刻).[260] Yates has also published a time chart for Qín officials, comprising 12 work and leisure hours, which is on display at the Lǐyē Museum of Qín Slips (Fig. 3.3).

According to this chart, 'midnight' (*yèbàn* 夜半), 'cock crow' (*jīmíng* 雞鳴) and 'dawn' (*píngdàn* 平旦) were consecutive double-hours between 11 pm and 5 am. At 5 am, Qín officials started their 12-hour workday, which lasted until 5 pm. The early evening spanned from 5 pm to 9 pm and included 'sunset' (*rìrù* 日入) and 'dusk' (*huánghūn* 黃昏), followed by the 'settling of people' (*réndìng* 人定) from 9 pm to 11 pm. Conversions into modern time only make sense if we assume that

257 Cullen 2018: 184. Photograph is taken from Cullen 2018: 185. For a discussion of this particular water clock, see Chén Měidōng 1989.

258 Cullen 2018: 186.

259 Needham and Wang 1959: 322.

260 Yates 2012–2013: 302. Robin Yates translates '*kè* 刻' as both 'quarters' and 'marks'.

Chrononym	Name	Qin Time for Work	Modern Time	Activity
zi 子	*yeban* 夜半		23:00–01:00	
chou 丑	*jiming* 雞鳴		01:00–03:00	
yin 寅	*pingdan* 平旦		03:00–05:00	
mao 卯	*richu* 日出	*Shuixia yi zhi erke* 水下一至二刻[a]	05:00–07:00	Office open
chen 辰	*shishi* 食時	*Shuixia san zhi sike* 水下三至四刻	07:00–09:00	Morning meal
si 巳	*yuzhong* 隅中	*Shuixia wu zhi liuke* 水下五至六刻	09:00–11:00	
wu 午	*rizhong* 日中	*Shuixia qi zhi bake* 水下七至八刻	11:00–13:00	
wei 未	*riyi* 日昳	*Shuixia jiu zhi shike* 水下九至十刻	13:00–15:00	
shen 申	*bushi* 晡時	*Shuixia shi zhi shiyi ke* 水下十至十一刻	15:00–17:00	Office closed; Evening meal
you 酉	*riru* 日入		17:00–19:00	
xu 戌	*huanghun* 黃昏		19:00–21:00	
hai 亥	*rending* 人定		21:00–23:00	Bedtime

Fig. 3.3: Time schedule for Qín dynasty officials according to the table by Robin Yates, on display at the Lǐyē Museum of Qín Slips.[261]

the imperial Qín workday consistently began at 5 am and did not depend, as is arguably much more likely, on the actual time of 'sunrise' (*rìchū* 日出). For simplicity's sake, I will continue to use these time conversions throughout the remainder of the book.

Judging from the manuscripts transcribed by Chén Wěi, the *Lǐyē* corpus does not entirely align with the time formats described above. While double-hours are utilized, they are neither assigned specific chrononyms nor divided into 100 equal 'quarters'. The further division of 'quarters' into 10 'marks' is also absent from the texts. Instead, water clocks seem to have displayed 11 one-hour 'marks' (*kè* 刻), dividing the time between 5 am and 5 pm, or the 12 hours starting from sunrise.[262] Some *Lǐyē* texts mention the complete daytime scale, consisting of '11 marks' (*shíyī kè* 十一刻), followed by the actual time, specified as "mark below mark x" (*kè xià* x 刻下X). Others omit the scale and state the time directly, using phrases like "water below mark x" (*shuǐ xià* x *kè* 水下X刻), "water below x" (*shuǐ*

261 Yates 2012–2013: 302.

262 This hypothesis can be compared with similar theories on timekeeping in the early empires, as presented in Lǐ Xuéqín 2003.

xià x 水下X) or "water below below x" (*shuǐ xià xià* x 水下下X). The following examples illustrate these timekeeping variants in the *Lǐyē* documents.[263]

Instances of "water below mark x of 11 marks" (*shuǐ shíyī kè kè xià* x 水十一刻刻下X):

8-60+8-656+8-665+8-748 rIII–IV: 六月庚辰[17]水十一刻刻下六, 守府快行少內.☐┘ 六月乙亥[12]水十一刻刻下二, 佐同以來. ノ元手. ☐[264]

In month 6 on a *gēngchén*, at water below mark 6 of 11 marks, Governor Kuài dispatches (the document) to the Lesser Treasury. ☐┘ In month 6 on a *yǐhài*, at water below mark 2 of 11 marks, (the document) is brought by Assistant Tóng. / By the hand of Yuán. ☐

Comments: Since neither the year nor the 1st day of the lunar month is specified, it is not possible to reconstruct the numerical values for the days '*gēngchén*' and '*yǐhài*'. The 'Lesser Treasuries' (*shǎonèi* 少內) were responsible for the finances of the counties.[265] Based on my reconstructions of timekeeping within the *Lǐyē* corpus, as detailed above, the phrase "mark below mark 6 of 11 marks" converts to the period between 11:00 am and 11:59 am, whereas "water below mark 2 of 11" corresponds to the period between 7:00 am and 7:59 am.

8-71 I–II, rI: 丗一年二月癸未[20]朔丙戌[23], 遷陵丞昌敢言之: 遷☐┘ 佐日備者, 士五(伍)梓潼長親欣補, 謁令☐ └二月丙戌[23]水十一刻刻下八, 守府快行尉曹. ☐ [266]

In the 31st year (of King Zhèng) [216 BC], month 2, new moon on a *guǐwèi*, *bǐngxū* [day 4], Chāng, Prefect of Qiānlíng, dares to say: Qiān ☐┘ Assistant comes to the end of his term. Xīnbǔ from a Unit of Five People of ChángX [pronunciation unknown] (township) in Zǐtóng, requests an ordinance ☐ └In month 2 on a *bǐngxū* [day 4], at water below mark 8 of 11 marks, Governor Kuài dispatches (the document) to the Bureau of the Military Commandant. ☐

Zǐtóng 梓潼 was the name of a county in Shǔ Commandery (*Shǔ jùn* 蜀郡), with its government seat in what is now modern-day Sìchuān Province.[267] It is also mentioned on manuscript 8-1445, in conjunction with the township of Wǔchāng 武昌.[268] The township of ChángX [pronunciation unknown] is not known from any other sources, so this interpretation should be approached with caution. The phrase "water below mark 8 of 11 marks" converts to the period between 1:00 pm and 1:59 pm.

---

263 These and the following conversions into modern time are derived from Fig. 3.3 in Yates 2012–2013: 302.
264 Chén Wěi 2012: 43.
265 For a discussion of the 'Lesser Treasury' (*shǎonèi* 少內) based on transmitted textual passages, see Hulsewé 1985: 195–200.
266 Chén Wěi 2012: 54.
267 Chén Wěi 2012: 54. For a discussion of the counties and commanderies included in the *Lǐyē* corpus, see chapter four.
268 Chén Wěi 2012: 327.

8-154 I–II, rI: 卅三年二月壬寅[39]朔朔日, 遷陵守丞都敢言之: 令曰恒以┛ 朔日上所買徒隸數. • 問之, 毋當令者, 敢言之. └二月壬寅[39]水十一刻刻下二, 郵人得行. 圂手.[269]

In the 33rd year (of King Zhèng) [214 BC], month 2, new moon on a *rényín*, new moon day [day 1], Dū, Deputy Governor of Qiānlíng, dares to say: The ordinance states to regularly submit (a document)┛ to the higher authorities on the 1st day (of the lunar month) with the number of convict servants that have been bought. • Clarify whether there is no corresponding ordinance. (He) dares to say. └In month 2 on a *rényín* [day 1], at water below mark 2 of 11 marks, Postman Dé dispatches (the document). By the hand of Hùn.

"Water below mark 2 of 11 marks" converts to the period between 7:00 am and 7:59 am (see manuscript 8-60+8-656+8-665+8-748 above for comparison).

8-155 I–II: 四月丙午[43]朔癸丑[50], 遷陵守丞色下: 少內謹案致之. 書到言, 署金布發, 它如┛ 律令. ノ欣手.ノ四月癸丑[50]水十一刻刻下五, 守府快行少內.[270]

In month 4, new moon on a *bǐngwǔ*, *guǐchǒu* [day 8], Sè, Deputy Prefect of Qiānlíng, (sends a document to) the lower authorities: The Lesser Treasury should investigate and establish firmly. The document needs to be communicated orally upon delivery, and then forwarded by all Finance Departments. Everything else (should be handled) in accordance with the┛ statutes and ordinances. / By the hand of Xīn. / In month 4 on a *guǐchǒu* [day 8], at water below mark 5 of 11 marks, Governor Kuài dispatches (the document) to the Lesser Treasury.

This excerpt reveals that the 'Finance Department' (*jīnbù* 金布) was a county-level unit subordinate to the 'Lesser Treasury' (*shǎonèi* 少內).[271] The phrase "water below mark 5 of 11 marks" converts to the period between 10:00 am and 10:59 am.

8-164+8-1475 I–III, rI: □□年後九月辛酉[58]朔丁亥[24], 少內武敢言之: 上計☒┛ □□而後論者獄校廿一牒, 謁告遷陵將計丞☒┛ 上校. 敢言之. ☒ └☒九月丁亥[24]水十一刻刻下三, 佐欣行廷. 欣手.☒ [272]

□□ year, intercalary month 9, new moon on a *xīnyǒu*, *dīnghài* [day 27], Wǔ of the Lesser Treasury dares to say: Submit the statistics to the higher authorities ☒┛ □□ and judgments in the future, when (settling) criminal cases, examine the 21 official documents, respectfully report the statistics of Qiānlíng to the Governor ☒┛ submit the evaluation results to the higher authorities. (He) dares to say. ☒ └ ☒ in month 9 on a *dīnghài* [day 27], at water below mark 3 of 11 marks, Assistant Xīn dispatches (the document) to the Court. By the hand of Xīn. ☒

269 Chén Wěi 2012: 93.
270 Chén Wěi 2012: 94.
271 For more information on this manuscript text, refer to *Húnán shěng wénwù kǎogǔ yánjiūsuǒ* 2012b: 19.
272 Chén Wěi 2012: 100.

Chén Wěi believes that the missing character before 'year' (*nián* 年) is 'first' (*yuán* 元), which would date the manuscript to the first year of the Second Emperor (209 BC).[273] According to Zhāng Péiyú, the first year of the Second Emperor did not have an intercalary month, unlike the 29th year of King Zhèng (218 BC), as recorded on manuscript 8-1450: "In the 29th year (of King Zhèng) [218 BC], intercalary month 9 on a *xīnwèi*" (*niànjiǔ nián hòu jiǔ yuè xīnwèi* 廿九年後九月辛未).[274] Xǔ Míngqiāng further notes that the 1st day of intercalary month 9 in the 29th year of King Zhèng was a '*xīnyǒu* 辛酉'. We can further determine that an official named Xīn was active in the 32nd year of King Zhèng (215 BC) (see Tab. 2.1).[275] As a result, this entry was most likely recorded in the 29th year of King Zhèng, and Xīn was active from at least the 29th to the 32nd regnal year (218–215 BC).

During the Warring States period, as well as the Qín and Hàn periods, imperial officials compiled 'statistics' (*jì* 計) at the end of the year. These statistics contained comprehensive data on households and population within their jurisdiction, as well as information on taxes, criminal records, lawsuits, and other relevant matters, which were submitted to the Courts. This process was known as "submission of statistics to the higher authorities" (*shàng jì* 上計). For more on these statistics, see chapters four and six.

The phrase "water below mark 3 of 11 marks" converts to the period between 8:00 am and 8:59 am.

Instances of "water below mark x" (*shuǐ xià* x *kè* 水下X刻):

5-22: 獄東曹書一封, 丞印, 詣無陽. • 九月己亥[36]水下三刻, □□以來.[276]

There is one document from the Judiciary Bureau of the East with the seal of the Prefect. It is addressed to Wúyáng. • In month 9 on a *jǐhài*, at water below mark 3, (the document) is brought by □□.

In a collection of documents titled "Models for Sealing and Investigating" (*Fēng zhěn shì* 封診式) from the *Shuìhǔdì* texts, strip 48 already features '*fēng* 封' as a measure word for 'documents' (*shū* 書).[277]

Wúyáng was a county in the Qín and Hàn dynasties.[278] It also appears on manuscript 8-1555, in conjunction with Zhòngyáng 衆陽, one of its districts.[279] Wúyáng County belonged to Wǔlíng 武陵 Commandery, with its government seat located in what is now the northeast of Zhǐjiāng Dòng Minority Autonomous County (*Zhǐjiāng Dòngzú Zìzhìxiàn* 芷江侗族自治縣) in Húnán Province.

The phrase "water below mark 3" converts to the period between 8:00 am and 8:59 am (see manuscript 8-164+8-1475 above for comparison).

273 Chén Wěi 2012: 100.
274 Zhāng Péiyú 1987.
275 Xǔ Míngqiāng 2013b.
276 Chén Wěi 2012: 13.
277 Lǐ Jiànpíng and Zhāng Xiǎnchéng 2009: 73.
278 Chén Wěi 2012: 13.
279 Chén Wěi 2012: 357.

8-453 AI–IV, B: A 尉曹書三封, 令印.ᒧ 其一詣銷,ᒧ 一丹陽,ᒧ 一□陵.ᒧ B 廿八年九月庚子[37]水下二刻, 走祿以來.[280]

There are three documents from the Bureau of the Military Commandant with the seal of the ordinance.ᒧ Of these (documents), one (is addressed) to Xiāo,ᒧ one to Dānyáng (County)ᒧ and one to □líng.ᒧ B In the 29th year (of King Zhèng) [218 BC], month 9 on a *gēngzǐ*, at water below mark 2, (the document) is brought by Runner Lù.

Xiāo was a county in Nán 南 Commandery during the Qín dynasty and the Early Hàn dynasty.[281] This is its only mention in the *Lǐyē* corpus. The county's government seat was located near modern-day Jīngmén 荊門, a prefecture-level city in Húběi Province.[282] Dānyáng was another county of the imperial Qín dynasty, with its government seat situated in the northeast of present-day Dāngtú 當塗 County, Ānhuī Province.[283] It also referenced on manuscripts 8-430 and 8-1807, the latter linking the township of Xiàlǐ 下里 to Dānyáng.[284]

The phrase "water below mark 2" converts to the period between 7:00 am and 7:59 am (see manuscripts 8-60+8-656+8-665+8-748 and 8-154 above for comparison).

8-1155: 獄東曹書一封, 丞印, 詣泰守府. 廿八年九月己亥[36]水下四刻, 隸臣申以來.[285]

There is one document from the Judiciary Bureau of the East with the Prefect's seal. It is addressed to the Governor. In the 28th year, month 9 on a *jǐhài*, at water below mark 4, (the document) is brought by Bondservant Shēn.

The phrase "water below mark 4" converts to the period between 9:00 am and 9:59 am.

Instances of "water below x" (*shuǐ xià* x 水下X):

8-78 rII: ⧄□酉水下盡, 隸臣匕隹以⧄ [286]

⧄□ *yǒu*, water at exhaustion, (male) Bondservant X [pronunciation unknown] ⧄

The term '*yǒu* 酉' possibly forms part of a day name. When the water reached the bottom of the scale at 5 pm, this was called 'exhaustion' (*jìn* 盡). This expression also appears on manuscript 9-1867.[287]

The structural parallelism in these sentences evokes the Bondservant X (pronunciation unknown) "bringing (the document)" (*yǐ lái* 以來) at the end of the text.

280 Chén Wěi 2012: 152.
281 Chén Wěi 2012: 152.
282 Chén Wěi 2011b.
283 Chén Wěi 2012: 147.
284 Chén Wěi 2012: 394.
285 Chén Wěi 2012: 285.
286 Chén Wěi 2012: 57.
287 Yates 2012–2013: 302.

8-1671: ☒□□□水下一, 隸妾【强】☒[288]

☒□□□ at water below one, (female) Bondservant【Qiáng】☒

Instance of "water below below x" (*shuǐ xià xià* x 水下下X):

8-1510 rIII: 三月己酉[46]水下下九, 佐赾以來. ノ釦手.[289]

In month 3 on a *jǐyǒu*, at water below (mark) 9, (the document) is brought by Assistant X [pronunciation unknown]. / By the hand of Kòu.

The commencement of each water clock cycle was labeled as 'beginning' (*qǐ* 起), whereas its end was referred to as 'exhaustion' (*jìn* 盡).[290] Consequently, the marks '0' and '12' do not appear in the *Lǐyē* corpus, as they would be redundant on the clepsydra scale. Dividing a 12-hour cycle into 11 marks is logical only when the top and bottom of the clepsydra are incorporated into the timekeeping system (Fig. 3.4).

The time is always given at the end of an entry, typically written by an unidentified individual on the left side of the *recto* (Fig. 3.5). The main text begins in the top right corner of the front side of the writing carrier and usually includes the 'year' (*nián* 年), 'month' (*yuè* 月), 'new moon day' (*shuò(rì)* 朔(日)) and the actual day of the entry, without the time. The officials responsible for producing the texts are identified by their signature at the bottom of the *verso*. The line recording the month, day and exact time at the end of the text, without reiterating the year or the new moon day, suggests that logistical decisions regarding document movement or personnel reassignment had already been acknowledged by the authorities and were in progress. This method, in addition to concluding the case at hand, likely facilitated efficient document retrieval.

All surviving entries featuring the designations "water below mark x" (*shuǐ xià* x *kè* 水下X刻), "water below x" (*shuǐ xià* x 水下X) or "water below below x" (*shuǐ xià xià* x 水下下X) date from the 26th to the 30th year of King Zhèng (221–217 BC). In comparison, ten out of the eleven entries (or 91%) containing "water below mark x of 11 marks" (*shuǐ shíyī kè kè xià* x 水十一刻刻下X) were written between the 30th and the 33rd year of King Zhèng (217–214 BC). Only one entry (8-1452) dates back to the 26th year of King Zhèng (221 BC), confirmed by the presence of the official Cāo 操, who was active during that year (see Tab. 2.1). It is reasonable to assume that the more detailed designation, "water below mark x of 11 marks"

288 Chén Wěi 2012: 376.
289 Chén Wěi 2012: 341.
290 See *Lǐyē* manuscripts 8-792, 9-1867 and 12-1799. See also Mǎ Yí 2005; Yates 2012–2013: 302.

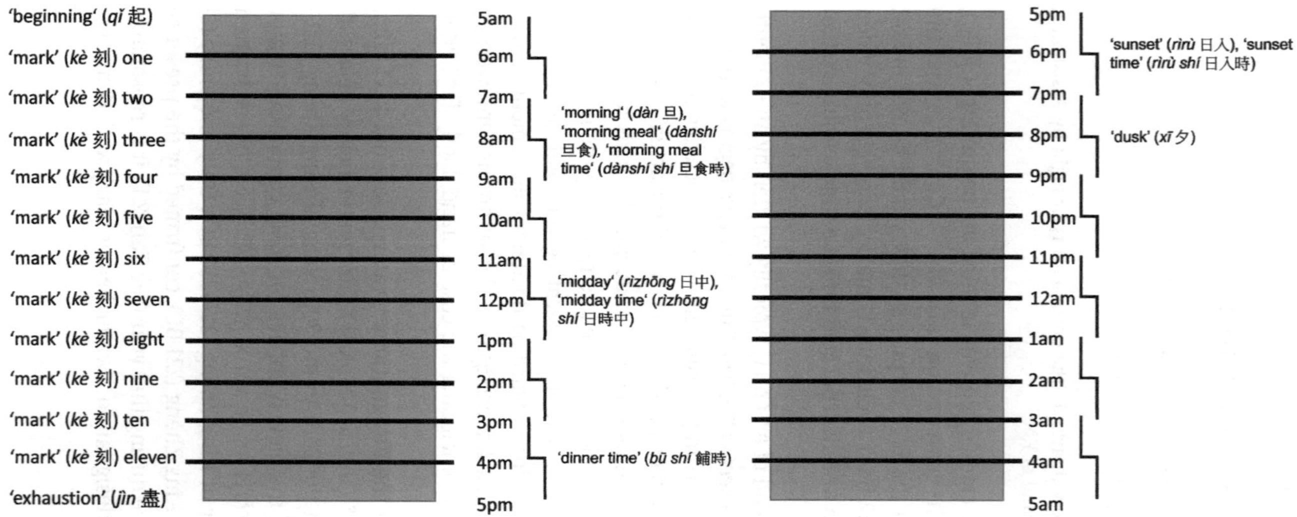

Fig. 3.4: Chart of a 24-hour cycle (5 am to 5 am, or sunrise to sunrise) with the addition of double-hours on Qín dynasty clepsydrae, as reflected in the *Lǐyē* manuscripts from layers 5, 6, 8, 9 and 12. The top and bottom sections of the chart mark the 'beginning' (*qǐ* 起) and 'exhaustion' (*jìn* 盡) of the 12-hour working day.

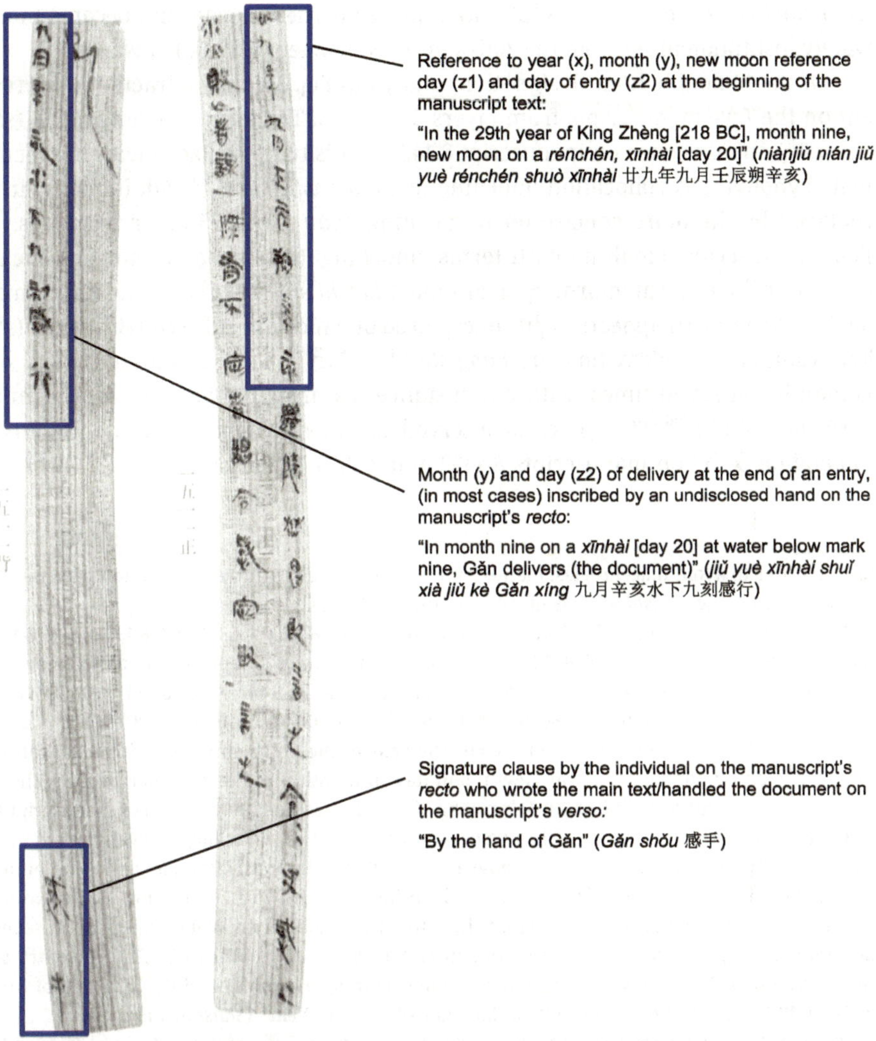

Fig. 3.5: Monochrome photograph of manuscript 8-1511, *verso* (r) and *recto* (l). The *verso* was written by Gǎn 感, who recorded his name on the *recto's* bottom left. The exact time is written on the *recto's* top left by an undisclosed person.[291]

291 For a transcription of manuscript 8-1511, see Chén Wěi 2012: 341–342.

(*shuǐ shíyī kè kè xià* x 水十一刻刻下X), replaced earlier formats and became more widely institutionalized from the 30th year of King Zhèng (217 BC) onwards.

Of the 12 double-hour designations shown in Fig. 3.3, only a fraction are present on the *Lǐyē* manuscripts from layers 5, 6 and 8. The term 'morning meal' (*shí shí* 食時) appears once on manuscript 8-1432, but is twice replaced with the seemingly synonymous collocation 'morning meal' (*dànshí* 旦食).[292] Additionally, it is replaced by the more concise term 'morning' (*dàn* 旦) in 25 other instances.[293] Four manuscripts combine both terms, 'morning' (*dàn* 旦) and 'morning meal' (*shí shí* 食時), to form 'morning meal time' (*dànshí shí* 旦食時).[294] The term 'midday' (*rìzhōng* 日中) appears eight times, three of which are suffixed with 'time' (*shí* 時), resulting in 'midday time' (*rìzhōng shí* 日中時).[295] The term 'sunset' (*rìrù* 日入) is mentioned seven times, with one instance featuring the variant 'sunset time' (*rìrù shí* 日入時).[296] '*Xī* 夕' may have served as a monosyllabic substitute for 'dusk' (*huánghūn* 黄昏) on manuscripts 8-1432 and 8-1823. Manuscript 8-728+8-1474 con-

---

292 In order to avoid inaccuracies, manuscript 8-716 is not taken into consideration because the character following '*dànshí* 旦食' is illegible (*dànshí* □旦食□).

293 'Morning meal' (*dànshí* 旦食) appears on manuscripts 8-141+8-668, 8-157, 8-130+8-190+8-193, 8-657, 8-664+8-1053+8-2167, 8-716, 8-770 and 8-1886. 'Morning' (*dàn* 旦) appears on manuscripts 5-1, possibly 6-21 (the preceding characters are illegible), 8-63, 8-140, 8-142, 8-143, 8-158, 8-170, 8-196+8-1521, 8-197, 8-645, 8-651, 8-672, 8-686+8-973, 8-736, 8-768, 8-1432 (twice), 8-1449+8-1484, 8-1482, 8-1515, 8-1523, 8-1525 (twice), 8-1559 and 8-1560. The year of the entry on manuscript 8-140 can be reconstructed by knowing that the new moon day in month 9 of the 31st year of King Zhèng (216 BC) was a '*gēngxū* 庚戌' (see Xǔ Míngqiāng 2013b). In line one on the *verso* of manuscript 8-738, the 'new moon day' (*shuò* 朔) is a '*gēngzǐ* 庚子' in position 37, while the event day is on a '*dīngwèi* 丁未' in position 44, making '*dīngwèi* 丁未' 'day 8'. Strangely, the last entry containing the day and time of Governor Jí's departure is identified as a '*yǐsì* 乙巳' in position 42. This would imply that the document was sent before the entry was made. Chén Wěi's transcription seems accurate, although he does not address this discrepancy (see Chén Wěi 2012: 222). The entry on manuscript 8-1449+8-1484 was made in the 1st lunar month, or month 10, of the 35th year of King Zhèng (212 BC). According to the "Event Calendar of the 35th Year" (*Sānshíwǔ nián zhì rì* 三十五年質日) in the *Yuèlù* corpus, the new moon day was a '*rénxū* 壬戌', making '*jǐmǎo* 己卯' 'day 18' (compare with Xǔ Míngqiāng 2013b). The text on manuscript 8-1523 does not explicitly mention the year, although manuscript 8-759 cites the new moon day in month 7 of the 34th year of King Zhèng (213 BC) as a '*jiǎzǐ* 甲子' in position 1.

294 This refers to manuscripts 8-130+8-190+8-193, 8-157, 8-664+8-1053+8-2167 and 8-1886. Regarding the dating on manuscript 8-1886, manuscript 8-525 provides a clue: it indicates that the new moon day in month 9 of the 30th year of King Zhèng (217 BC) was a '*bǐngchén* 丙辰', making '*bǐngzǐ* 丙子' correspond to 'day 21' (see Chén Wěi 2012: 173).

295 'Midday' (*rìzhōng* 日中) appears on manuscripts 8-51, 8-152, 8-157, 8-173 and 8-252, and 'midday time' (*rìzhōng shí* 日中時) on manuscripts 8-890+8-1583, 8-1069+8-1434+8-1520 and 8-1439.

296 'Sunset' (*rìrù* 日入) appears on manuscripts 8-69, 8-1459, 8-1468, 8-1538, 8-1554 and 8-1971, and 'sunset time' (*rìrù shí* 日入時) on manuscript 8-274+8-2138.

tains the term 'dinner time' (*bū shí* 餔時). However, due to its poor physical condition, cross-comparisons cannot be made, nor can definitive conclusions be drawn about its use and significance in an administrative context. In all cases discussed, the more accurate clepsydra timescale is absent and replaced by corresponding designations.

8-728+8-1474 I, rI: ☒獄南曹書二封, 遷陵印: 一洞庭泰守府, 一洞庭尉府. • 九月☒ └☒己亥[36] 餔時, 牢人誤以來.☒ [297]

☒ There are two documents by the Judiciary Bureau of the South with the seal of Qiānlíng: One (is addressed to) the Governor of Dòngtíng, and one (is addressed to) the administrative unit of the Military Commandant of Dòngtíng. • Month 9 ☒ └☒ on a *jǐhài*, at dinner time, (the document) is brought by Prisoner Wù. ☒

Comments: As indicated by this manuscript, the 'Judiciary Bureau of the South' (*yùnáncǎo* 獄南曹) was a county level institution that maintained correspondence with the Governor and Military Commandant of Dòngtíng. This Bureau is also mentioned on manuscripts 8-1760 and 8-1886. While manuscript 8-1760 provides limited information containing only the characters '*yùnáncǎo* 獄南曹', manuscript 8-1886 reveals that letters sent on behalf of the Judiciary Bureau of the South necessitated the "seal of the (County) Prefect" (*chéng yìn* 丞印). The 'Judiciary Bureau of the East' (*yùdōngcǎo* 獄東曹) fulfilled similar obligations in Qiānlíng County.[298]

Outside the official working day, and excluding the aforementioned 'dinner time' (*bū shí* 餔時), 'sunset' (*rìrù* 日入) and 'dusk' (*xī* 夕), the method for measuring hours and double-hours remains unclear. Entries recording the latter hours of the day using the 11-mark scale are found on manuscripts 8-1510, 8-1511 and 8–78. It is possible that refilling the clepsydra at 5 pm (or 12 hours after sunrise) provided a consistent reference point throughout the night until the start of the following working day at 5 am (or sunrise).

In summary, the calendar and timekeeping systems utilized by the Qín dynasty were paramount to the large-scale standardization and synchronization of administrative processes. These systems underpinned the processing of information and the creation, dispatch and reception of documents. Moreover, timekeeping introduced a level of consistency to government operations and individual daily routines. This consistency is evident in the construction and maintenance of public sites, the cultivation of agricultural land and the clear demarcation between work and leisure time.

297 Chén Wěi 2012: 211.
298 See manuscripts 5-22, 8-273+8-2138, 8-959+8-1291, 8-996 and 8-1155.

Future analysis of the *Lǐyē* corpus regarding calendar and timekeeping systems may offer additional insights into the duration of various processes within the imperial administration. It may also clarify the geographical distances between administrative units, the strategic priorities of the imperial government, the dating of damaged or incomplete documents and issues related to the unification of the early empire, including political, writing, land and other reforms.

4 Administrative Geography

4.1 Central Government

Building and maintaining the empire required not only the standardization of names, titles and terminology but also the organization of vast populations under a complex hierarchy of institutions and processes. A consistent technical vocabulary, combined with a coordinated administrative structure where individuals were employed based on merit, laid the foundation for the effective, semi-independent execution of common activities and objectives. From this perspective, manuscript 8-461 offers intriguing insights into administrative language and political representation, advocating for imperial unity through the portrayal of authoritative figures (Fig. 4.1).

In terms of physical dimensions, the tablet measures 27.4 cm in length, 12.5 cm in width, and 0.6 cm in height, making it slightly longer and substantially wider than other *Lǐyē* wooden tablets, which typically average 23 cm in length and 1.4 to 5 cm in width.[299] The manuscript is fragmented into six main pieces, with additional damage noted at the upper right corner, lower right corner, lower left corner and bottom side – all of which are missing. The layout and formatting of the surviving section are rarely seen in other Qín dynasty texts. It consists of two horizontal columns on the *verso*, with the upper half containing 24 lines and 101 legible characters, and the bottom half 32 lines and 174 legible characters. The longest lines, AXXII, BVI and BXXXII, each consist of eight characters. In sum, 56 lines with 275 regularly arranged characters, written in a form resembling 'small seal script' (*xiǎo zhuàn* 小篆), are decipherable on the *verso*.

On the *recto*, remaining legible characters are written vertically along the direction of the wood grain and are split into two lines with three characters each. This side of the tablet, distinguished by its layout, brushwork and use of 'clerical script' (*lìshū* 隸書), presents a less formal appearance compared to the *verso*. The recurring formulaic phrase "dares to say" (*gǎn yán zhī* 敢言之) in the first line, frequently employed to report matters to higher authorities, seems out of place here and could have been written by someone else for a different purpose. Overall, the tablet contains 58 lines and 281 characters across both sides.

After the First Emperor proclaimed the unification of annexed territories in 221 BC, he is believed to have undertaken major social, political and structural reforms, some of which may be echoed on the *verso* of the tablet. Unfortunately, the missing first and last sections omit crucial details such as the year, month, time

299 *Húnán shěng wénwù kǎogǔ yánjiūsuǒ* 2007: 179.

https://doi.org/10.1515/9783110791853-004

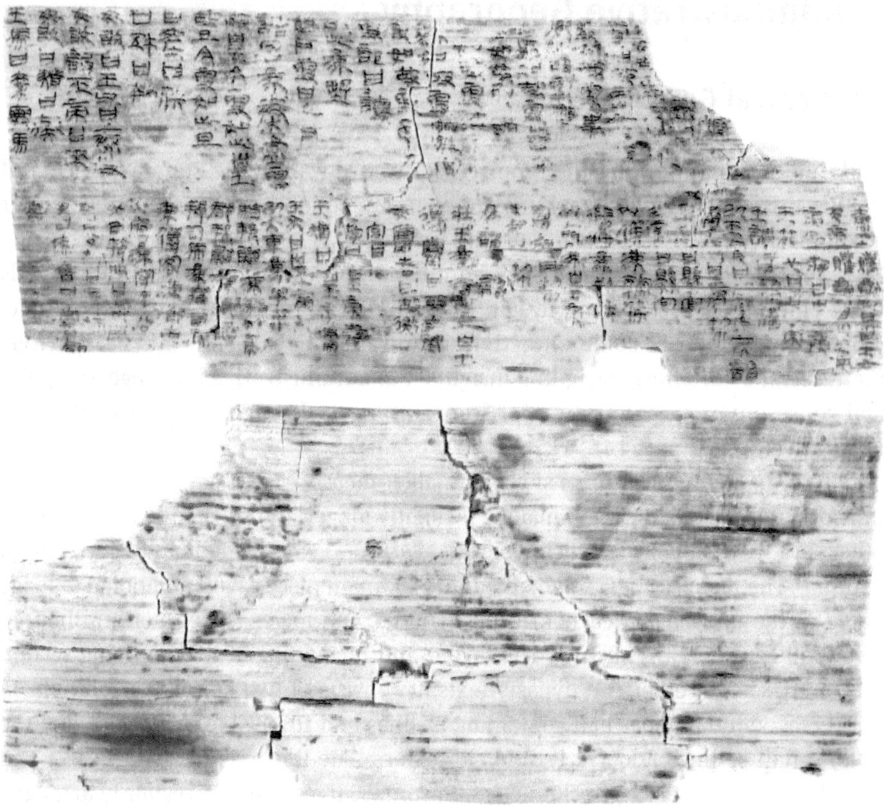

Fig. 4.1: Monochrome photograph of manuscript 8-461, *verso* and *recto*. Ink on wood. Dimensions: 12.5 x 27.4 x 0.6 cm.[300]

and possibly the identity of the official responsible for the document. This absence makes it impossible to determine the tablet's precise age or chronological relation to other manuscripts. The content primarily focuses on the rectification of official titles and terms in daily use, which should be understood within the context of the transition from a Qín state, among others states, to a unified empire. Accordingly, the text was likely composed between 221 and 209 BC.

> 8-461 AI–BXXXII, rI–II: A ☒□.┘ ☒□.┘ ☒假□.┘ ☒□錢□.┘ □【如故】更□□.┘ □如故□□□.┘ □如故更事.┘ □如故更□.┘ □□如故更□□.┘ □如故【更】□.┘ □如故更□.┘ □如故更廢官.┘ □如故更予□.┘ 更詑曰謾.┘ 以此爲野.┘ 歸戶更曰乙戶.┘ 諸官爲秦盡更.┘ 故皇今更如此皇.┘ 故旦今更如此旦.┘ 曰產曰族.┘ 曰⿰卒斤曰荆.┘ 毋敢曰王父曰泰父.┘ 毋敢謂巫帝曰巫.┘ 毋敢曰

300 *Húnán shěng wénwù kǎogǔ yánjiūsuǒ* 2012a: 68–69.

豬曰彘.┙ 王馬曰乘輿馬.┙ B 泰【王】觀獻曰皇帝.┙ 天帝觀獻曰皇帝.┙ 帝子游曰皇帝.┙ 王節弋曰皇帝.┙ 王譴曰制譴.┙ 以王令曰【以】皇帝詔. ┙ 承【命】曰承制.┙ 王室曰縣官.┙ 公室曰縣官.┙ 內侯爲輪(倫)侯.┙ 徹侯爲【死〈列〉】侯.┙ 以命爲皇帝.┙ 受(授)命曰制.┙ □命曰制.┙ 爲謂□詔.┙ 莊王爲泰上皇.┙ 邊塞曰故塞.┙ 毋塞者爲故徼.┙ 王宮曰□□□.┙ 王游曰皇帝游.┙ 王獵曰皇帝獵.┙ 王犬曰皇帝犬.┙ 以大車爲牛車.┙ 騎邦尉爲騎□尉.┙ 郡邦尉爲郡尉.┙ 邦司馬爲郡司馬.┙ 乘傳客爲都吏.┙ 大府爲守□公.┙ 毋曰邦門曰都門.┙ 毋曰公坾曰□坾.┙ 毋曰客舍曰賓【飤】.┙ 舍. └敢言之.┙ • 九十八.[301]

A ⊠□.┙ ⊠□.┙ ⊠ borrow □.┙ ⊠□ coins □.┙ □【as before】changes to □□.┙ □ as before □□□.┙ □ as before changes to affair.┙ □ as before changes to □.┙ □□ as before changes to □□.┙ □ as before 【changes to】□.┙ □ as before changes to □.┙ □ as before changes to 'Abolished Offices'.┙ □ as before changes to 'awarding □'.┙ Change (the term) 'to cheat' and say 'to deceive'.┙ Make this (the basis of) the outskirts.┙ (The term) 'returning households' is changed and referred to as 'second households'.┙ All offices are without exception changed by the Qín.┙ (The term) 'Past August' now changes to 'This August'.┙ (The term) 'past dawn' now changes to 'this dawn'.┙ Say 'birth', and say 'relatives'.┙ Say '*wú*' [a self-referential term for a married man] and say '*jīng*' [used by a married man to courteously refers to his wife]┙ Do not dare to say 'Patron of the King', (but) say 'Patron of the Highest'.┙ Do not dare to name (someone) a 'Diviner Thearch', (but) say 'Diviner'.┙ Do no dare to say 'pig', (but) say 'swine'.┙ (The term) 'King's horse' is called 'chariot horse'.┙ B 'Sacrificial offerings to the great 【Kings】' are called 'sacrificial offerings to the Emperor'.┙ 'Sacrificial offerings to the Celestial Thearch' are called 'sacrificial offerings to the Emperor'.┙ The 'processions of the Celestial Thearch' are called the 'processions of the Emperor'.┙ The 'Kings' bird huntings' are called the 'Emperor's bird huntings'.┙ 'Accusations by the Kings' are called 'accusations by (a system of) decrees'.┙ 'By means of the Kings' ordinances' is called '【by means】 of the Emperor's edicts'.┙ 'Accepting the 【mandate】' is called 'accepting (a system of) decrees'.┙ The 'Royal Courts' are called the 'County Offices'.┙ The 'Homes of Dukes' are called the 'County Offices'.┙ The title 'Marquis of the Interior' becomes 'Marquis of Human Relations'.┙ The title 'Marquis of the Twentieth Rank' becomes '【Holder of the Twentieth Rank】'.┙ 'By means of the mandate' becomes '【by means of】 the Emperor's edicts'.┙ 'To receive the mandate' is called '(to receive a system of) the decrees'.┙ □ 'mandate' is called 'decrees'.┙ 'To act' is called '□ edict'.┙ 'King Zhuāng' becomes the 'Highest August'.┙ The 'border fortifications' are called the 'fortifications of the past'.┙ The '(places with) no fortifications' are called the 'frontiers of the past'.┙ The 'palace of the King' is called '□□□'.┙ The 'processions of the Kings' are called the 'processions of the Emperor'.┙ The 'Kings' hunting' is called the 'Emperor's hunting'.┙ The 'Kings' hunting dogs' are called the 'Emperor's hunting dogs'.┙ 'By means of the big carriage' means '(by means of the) cow's carriage'.┙ The title 'Cavalry Commandant of a State' becomes 'Cavalry Commandant of a □'.┙ The title 'Military Commandant of a Commandery and State' becomes 'Military Commandant of a Commandery'.┙ The title 'State Minister of War' becomes 'Commandery Commissioner of War'.┙ The 'traveler on the carriage' acts like an 'officer from the capital'.┙ The title 'Grand Commandery Governor' becomes 'Head □ Public'.┙ Do not say the 'gates of the states', (but) say the 'gates of the capital'.┙ Do not say the 'public [character unknown]', (but) say '□ [character unknown]'.┙ Do not say 'guests' house', (but) say '【visitors'】 house'. └(He) dares to say.┙ • Ninety-eight.

301 Chén Wěi 2012: 155–157.

Comments: The character '*jīng* 荆', seemingly used by married men to courteously refer to their wives, also appears in two other *Lǐyē* texts; once in 'Gùjīng 故荆' on manuscript 8-135, designating the former state of Chǔ 楚, and once in 'Jīngshān March' (*Jīngshān dào* 荆山道) on manuscript 8-1516.[302] Chén Wěi reads '*jīng* 荆' as referring to the former state of Chǔ, and '*wú* 䢋' (*C.ŋˤaʔ) as the former state of Wú 吳 (*ŋʷˤa), with the latter two being homoeophones.[303] Both interpretations, in which the terms function as state names or pronouns, are justified.

A messenger from a 'Marquis of a Dependent State' (*bāng hóu* 邦侯), who 'entered' (*rù* 入) the Qín state would be considered 'guest' (*kè* 客) in the region.[304] The graphs '*tài* 泰' (*thâs), '*tài* 太' (*thâs) and '*dà* 大' (*dâs) were interchangeable in early texts.[305]

An entry in the Hàn dynasty dictionary *Local Expressions* (*Fāngyán* 方言) differentiates between '*zhū* 豬' and '*zhì* 彘' as follows:

> *Local Expressions* (*Fāngyán* 方言), "Eighth Chapter" (*Dì bā* 第 八): 豬, 北燕朝鮮之間謂之豭, 關東西或謂之彘, 或謂之豕. 南楚謂之豨. 其子或謂之豚, 或謂之貕, 吳揚之間謂之豬子.[306]
>
> The 'pig' (*tra) is called '*jiā*' (*krâ) in the north between the state of Yān and the kingdom of Zhāoxiān. In the east and west of the mountain passes, some call it 'swine' (*drats < r-lats), and some call it '*shǐ*' (*lheʔ). South of the state of Chǔ, it is called '*xī*' (*həi, həiʔ). Some call its offspring '*tún*' (*lûn), and some call it '*xī*' (*[g]ˤe).[307] Between the state of Wú and the Yáng region, it is called the offspring of a 'pig' (*tra).[308]

It appears that the term 'swine' (*zhì* 彘) was a regional expression used in both the eastern and western regions of the Hàn and possibly Qín dynasties.

In the following passage of the *Shǐjì*, William H. Nienhauser translates the terms '*mìng* 命' as 'order', '*zhì* 制' as 'decree', '*lìng* 令' as 'ordinance', and '*zhào* 詔' as 'edict'.

---

302 Chén Wěi 2012: 73, 344.

303 Chén Wěi 2012: 157. The phonological reconstruction of '*wú* 䢋' was suggested by Prof. Dr. Wolfgang Behr. '*Wú* 䢋' is a discontinued character where '*wǔ* 五' (*C.ŋˤaʔ) replaced '*wǔ* 午' (*[m].qhˤaʔ) as the phonetic element from the Warring States period onward, as initial uvulars (*m.qh-), in this case '*wǔ* 午' (*[m].qhˤaʔ), gradually began to disappear.

304 Lau and Staack 2016: 176–177. The term 'guest' (*kè* 客) might also have referred to merchants who traveled to Qín territories to sell their goods (Lau and Staack 2016: 194). For a further discussion on the status of foreigners in Ancient China, see Pú Mùzhōu 2005.

305 Wáng Lì 2003: 178. Phonological reconstructions are taken from Schuessler 2009: 233.

306 Huá Xuéchéng 2011: 545.

307 The phonological reconstruction of '*xī* 貕' in Old Chinese is *[g]ˤe, according to the Baxter and Sagart reconstruction system. In Middle Chinese, '*xī* 貕' is pronounced *hej in Middle Chinese and is a complete homophone of '*xī* 奚'. These conclusions are based on a consultation with Prof. Dr. Wolfgang Behr.

308 Phonological reconstructions, where available, are taken from Schuessler 2009: 48, 53, 125, 234, 287, 338.

Shǐjì, "Annals of Qín Shǐhuáng" (*Qín Shǐhuáng běnjì* 秦始皇本 纪): 王爲泰皇. 命爲制、令爲詔、天子自稱曰朕.[309]

> The King was called 'His Primeval Majesty'. His orders were called 'decrees', and his ordinances 'edicts'. The Son of Heaven used 'The Secluded One' when referring to himself.[310]

This excerpt further illustrates that by the early imperial period, the term '*mìng* 命' had acquired a strong normative component and was devoid of the religious connotations it carried during the Shāng and Zhōu dynasties.[311] As the terminology provided on manuscript 8-461 is broad and raises concerns about the 'Son of the Thearch' (*Dìzǐ* 帝子), who was appointed by the 'heavenly mandate' (*tiānmìng* 天命) in the Late Zhōu period, I suggest that '*mìng* 命' and '*zhì* 制' could refer to a synoptic comparison between the pre-Qín 'mandate' and the complex, post-Qín system of 'decrees'.[312] The manuscript attempts to deliberately avoid former official Zhōu titles and any references to heaven.

The modification of terms and titles appears to be normative in nature. The text does not follow a specific order but begins with outdated vocabulary that is adjusted to fit the customary exigencies of the emerging empire and its new administrative system.[313]

Firstly, the text states that "all Offices" (*zhū guān* 諸官) are to be changed under imperial rule, including the 'Abolished Offices' (*fèiguān* 廢官) and the 'County Offices' (*xiànguān* 縣官). It proceeds by replacing the verb 'to cheat' (*yí* 詑) with 'to deceive' (*mán* 謾) and specifies that these changes should also be implemented in remote regions of the empire. It then goes on to define the title 'Emperor' and outlines the terms and institutions associated with the Imperial Court.

Terms and titles to be avoided in administrative writing included: 'King' (*wáng* 王), 'King Zhuāng' (*Zhuāng Wáng* 莊王), 'Thearch' (*Dì* 帝) (for any title

309 See Qiān Sīmǎ and Yáo Zǔ'ēn 1988: 1–4; Zhāng Dàkě 1990: 112.

310 Translation is taken from Nienhauser 1994: 136, with minor modifications.

311 On the so-called 'heavenly mandate' (*tiānmìng* 天命) in the Shāng and Zhōu dynasties, see Goldin 2017: 125–126; Kominami 1992.

312 On the limits of human agency in the notion of '*mìng* 命' in Early China, as reflected in case studies of received and excavated texts, see Puett 2005; Valmisa 2019.

313 Regarding this observation, Prof. Dr. Wolfgang Behr has commented that there appears to be a rough ordering by assonance of the Qín target word: '*yě* 野' (*lAʔ), '*hù* 戶' (*m-*qʕaʔ), '*fù* 父' (*[N-p](r)aʔ), '*wū* 巫' (*C.m(r)[o, a]), '*zhì* 彘' (*[l]r[a][t]-s) and '*mǎ* 馬' (*mʕraʔ). If ordering by assonance of the Qín target word applies, it would be intriguing, as it would mean that many other tablets of this kind exist.

other than the 'August Thearch' (*Huángdì* 皇帝),[314] 'Celestial Thearch' (*Tiāndì* 天帝), 'Son of the Thearch' (*Dìzǐ* 帝子), 'state' (*bāng* 邦), 'pig' (*zhū* 豬), the 'Royal Courts' (*wáng shì* 王室) and 'Homes of Dukes' (*gōng shì* 公室). Additionally, the adjective 'big' (*dà* 大) was to be avoided when used in conjunction with 'carriages' (*chē* 車) and 'horses' (*mǎ* 馬) directly associated with the August Thearch. The 'Emperor's hunting dogs' (*Huángdì quǎn* 皇帝犬), however, remained an acceptable expression.

Whenever the Emperor is mentioned in the *Lǐyē* corpus, he is referred to as '*Huángdì* 皇帝', literally meaning 'August Thearch'.[315] In the Qín dynasty, '*Huángdì* 皇帝' came to denote an earthly ruler and was the standard designation for 'Emperor'.[316] The term appears only once in the texts, specifically on manuscript 8-461, which makes it extremely rare and valuable.

Imperial authority is also reflected in manuscripts that date text entries in relation to the Emperor's regnal year. This practice granted legitimacy and chronological traceability to documents, while also imposing accountability on the officials and institutions involved in administrative processes. Manuscript 8-461 illustrates the Emperor's ambition to govern the conquered territories of the Warring States, a task he nominally undertook from 221 BC onwards. As the text suggests, several new, standardized practices emerged as a result of adopting, appropriating and eventually replacing norms from earlier periods.[317]

Not all pre-imperial conventions and traditions were abolished during the Qín dynasty. In light of the *Lǐyē* manuscripts, the discontinuous nature of Qín polities, as asserted by Shāng Yāng 商鞅 (390–338 BC), the reformer and architect of the Qín state, seems oversimplified. According to Yuri Pines, the authors of the *Book of Lord Shāng* (*Shāng jūn shū* 商君書) criticized entrenched officials for their refusal to break with the past.[318] Believing that the new empire needed to abolish tradition in order to thrive, they advocated for the establishment of a "fourth dynasty" (*sì dài* 四代):

> *Book of Lord Shāng* (*Shāng jūn shū* 商君書), "Opening that which is Blocked" (*Kāi sāi* 開塞): 今世彊國事兼并, 弱國務力守; 上不及虞、夏之時, 而下不修湯、武. 湯、武之道塞, 故萬乘莫不戰, 千乘莫不守. 此道之塞久矣, 而世主莫之能廢也, 故三代不四.[319]

314 On the nature and function of '*Dì* 帝' in Zhōu dynasty bronze inscriptions and in early canonical texts, see Itō and Takashima 1996.

315 Pines 2017a: 86.

316 Allan 1984.

317 On the borrowing and modifying of certain aspects of Zhōu cultural traditions, see Kern 2000.

318 Pines 2008: 82–85.

319 Liáng Wànrú and Ráo Zōngyí 2017: 151.

> Nowadays, strong states engage in conquests and annexations, while the weak are committed to forceful defense. Above, they do not reach the times of Yǔ (Shùn) and Xià; below, they do not embrace (the ways of the Kings) Tāng and Wǔ. (The ways of) Tāng and Wǔ are blocked; hence, every state of ten thousand chariots is engaged in (offensive) war, and every state of one thousand chariots is engaged in defense. These ways have been blocked for a long time, and contemporary rulers are unable to act otherwise; hence, the Three Dynasties lack the fourth.[320]

Pines views Shāng Yāng's alleged insistence on the necessity of change as decontextualized. I agree and contend that, within the context of political transformation, the Qín dynasty maintained a strong sense of historical continuity, as evidenced by its aspiration for a model of unification based on the idealized legacy of earlier rulers. Why else would Shāng Yāng retain the numbering system and speak of a fourth dynasty if not for comparative purposes? Although he does not call for the restoration of former paragons, he reveals himself as a proponent for both novelty and tradition. The concentrated authority of the imperial sovereign, imposed on all states and ethnic groups, aimed to subdue both territorial and cultural divisions. Additionally, its purpose was to fight endemic violence, counter legal transgressions and appropriate or reshape the political power of the annexed states.

> *Shǐjì*, "Annals of Qín Shǐhuáng" (*Qín Shǐhuáng běnjì* 秦始皇本紀): 天下共苦戰鬪不休, 以有侯王. 賴宗廟, 天下初定, 又復立國. 是樹兵也, 而求其寧息, 豈不難哉! 廷尉議是.[321]

> All-under-Heaven suffered endlessly from bitter strife and warfare because of Kings and overlords. Now, with the help of my ancestral temple, All-under-Heaven was pacified at last. To re-establish states meant to sow weapons: And was it, thereby, not difficult to demand peace? The view of the Commandant of Justice [Lǐ Sī] was correct.[322]

The unified realm envisioned by the First Emperor preserved features of the previous Three Dynasties while incorporating changes introduced by the Qín sovereigns. This contradicts Pines' assertion that the Zhōu model of government was "of no value for the Qín."[323] 'Commanderies' (*jùn* 郡) and 'Counties' (*xiàn* 縣), administered by 'Governors' (*shǒufǔ* 守府) and 'Prefects' (*chéng* 丞), replaced the system of 'fiefs' (*fēngjiàn* 封建).[324] 'Lords' (*jūn* 君) were substituted by 'officials' (*lì*

---

320 Pines 2008: 82. My *pīnyīn* transcriptions.

321 Zhāng Dàkě 1990: 114.

322 Translation is taken from Pines 2008: 84–85. My *pīnyīn* transcriptions.

323 Pines 2008: 84. My *pīnyīn* transcription.

324 For more information on the concept of 'fiefdom' (*fēngjiàn* 封建) as a political ideology and decentralized system of government during the latter part of the Zhōu dynasty, see Murthy 2006.

吏), and legal arbitrariness was superseded by the rule of law. All 'Offices' (*guān* 官) were changed to Qín offices, with old titles, including that of the Emperor, reappropriated to align with the new imperial government. This dualistic coexistence of novelty and continuity is illustrated in the following excerpts from manuscript 8-461:

> 8-461 AXVIII: 故皇今更如此皇.[325]
>
> (The term) 'Past August' now changes to 'This August'.
>
> 8-461 AXIX: 故旦今更如此旦.[326]
>
> (The term) 'past dawn' now changes to 'this dawn'.

Past terminologies were not simply discarded but instead underwent a process of transformation, adaptation and renewal. Conventional titles and expressions from earlier periods became redefined and associated with a new era of government:

> 8-461 BI-XIII: 泰【王】觀獻曰皇帝.┘ 天帝觀獻曰皇帝.┘ 帝子游 曰皇帝.┘ 王節弋曰皇帝.┘ 王譴曰制譴.┘ 以王令曰【以】皇帝詔. ┘ 承【命】曰承制.┘ 王室曰縣官.┘ 公室曰縣官.┘ 內侯爲輪(倫)侯. ┘ 徹侯爲【死〈列〉】侯.┘ 以命爲皇帝.┘ 受(授)命曰制.[327]
>
> 'Sacrificial offerings to the great【Kings】' are called 'sacrificial offerings to the Emperor'.┘ 'Sacrificial offerings to the Celestial Thearch' are called 'sacrificial offerings to the Emperor'.┘ The 'processions of the Celestial Thearch' are called the 'processions of the Emperor'.┘ The 'Kings' bird huntings' are called the 'Emperor's bird huntings'.┘ 'Accusations by the Kings' are called 'accusations by (a system of) decrees'.┘ 'By means of the Kings, ordinances' is called '【by means】 of the Emperor's edicts'.┘ 'Accepting the【mandate】' is called 'accepting (a system of) decrees'.┘ The 'Royal Courts' are called the 'County Offices'.┘ The 'Homes of Dukes' are called the 'County Offices'.┘ The title 'Marquis of the Interior' becomes 'Marquis of Human Relations'.┘ The title 'Marquis of the Twentieth Rank' becomes '【Holder of the Twentieth Rank】'.┘ 'By means of mandate' becomes '(by means of) the Emperor'.┘ 'To receive the mandate' is called '(to receive) the decrees'.

Ritual demonstrations of devotion and reverence were no longer directed toward the 'King' (*wáng* 王), the 'Celestial Thearch' (*Tiāndì* 天帝) or the 'Son of the Thearch' (*Dìzǐ* 帝子), but to the 'August Thearch' (*Huángdì* 皇帝). This shift turned the Emperor into a deified ruler and bearer of supreme authority who occupied the liminal space between earth and sky. Places of spiritual worship dedicated to the Emperor and his father arguably came to embody the ethos of the first empire, a

325 Chén Wěi 2012: 156.
326 Chén Wěi 2012: 156.
327 Chén Wěi 2012: 156.

temporal nexus, where past, present and future rulers converged. These spaces channeled the continuum of history into the present, blurring the boundaries between the material and spiritual realms.[328]

The primary efforts of the Qín in constructing the new empire centered on a thorough reworking of symbolic state legitimacy, necessitating a rethinking of the interrelationships between the empire, its social hierarchy and individual identity. In the context of rituals, the Emperor functioned as both the sovereign ruler and the intermediary for ancestral worship. In this sense, sacrificial offerings and the veneration of ancestral lineage carried both spiritual and political significance and established a historical foundation for imperial authority.[329] Other rituals, such as the Emperor's 'processions' (*yóu* 游), incorporated various performances designed to reflect the standards and cultural practices of earlier rulers. The continuous display of control over material and immaterial resources not only heightened the visibility of imperial power but also reinforced the social hierarchy through repetitive acts of commemoration.[330]

> 8-461 BXVI: 莊王爲泰上皇.[331]
>
> 'King Zhuāng' becomes the 'Highest August'.

This line formally exalts the First Emperor's deceased father with the title 'Highest August' (*Tàishàng Huáng* 太上皇), a narrative reinforced by a passage in the *Shǐjì*.[332] Starting from the 26th year of King Zhèng (221 BC), the formal designation 'Highest August' became an archetype of filial piety in later imperial periods, awarded to the father of reigning, retired and deceased Emperors.

I would not go as far as Wu Hung, who claims that the First Emperor "broke radically with the Western Zhōu concept of religious and political authority" by positioning "himself in the social and religious hierarchy unequaled by any other being, even his ancestors."[333] Nevertheless, by praising and immortalizing his father's accomplishments, he rhetorically legitimized his own place in the dynastic lineage and asserted himself as the rightful heir to the imperial throne. In this way,

328 For a discussion of the Medieval English doctrine of the King's two bodies, which was perceived to consist of a worldly, transient "body natural" and an immortal, transcendent "body politic" linked with the King's successors and predecessors, see Kantorowicz 1957.
329 Kern 2009: 153.
330 On ancestral sacrifice in the Western Zhōu dynasty, see Kern 2009. For a study on the cognitive aspects of early Chinese ancestral ritual, see Brashier 2011.
331 Chén Wěi 2012: 156–157.
332 Liú Qìngzhù and Barbieri-Low 2007.
333 Wu 1988: 92. My *pīnyīn* transcription.

the Qín dynasty skillfully combined novelty *with* historical preservation to bring about the annexation of the former Warring States and justify the establishment of a new dynasty.

Given that the titles 'King', 'Celestial Thearch' and 'Son of the Thearch' were no longer accepted, the designation 'August Thearch' seems to have emerged to retain political power and compensate for the partial loss of historical practices caused by the transition away from the pre-imperial multi-state order.[334] In a narrower sense, manuscript 8-461 complements the famous inscriptions on the stone stelae at Mount Yì 嶧, attributed to the First Emperor, where he extends his accumulated merits to his ancestors after ascending the throne.[335] The opening lines of the inscription read:

> *Shǐjì*, "Annals of Qín Shǐhuáng" (*Qín Shǐhuáng běnjì* 秦始皇本紀): 皇帝立國, 維初在昔, 嗣世稱王, 討伐亂逆, 威動四極, 武義真方, 戎臣奉詔, 經時不久, 滅六暴強, 廿有六年, 上薦高號, 孝道顯明, 既獻泰成, 乃降專惠, 親巡遠方. [. . .][336]
>
> The Emperor established the state. Originally, in times past, he inherited the throne and was designated King. He launched punitive attacks against the rebellious and recalcitrant. His might shook the four extremities: martial order and righteous behavior [now] stood upright. The military officers received the imperial orders. Through a passage of time not long, they exterminated the six cruel and violent ones. In his 26th year, he proffered a noble proclamation to those above and his filial way was brilliantly shining. As soon as He had presented this grand accomplishment, He sent out radiating grace, and personally toured the distant regions. [. . .][337]

The Second Emperor is the first sovereign in the *Shǐjì* to revisit the stone stelae and add his own inscriptions to praise his father's accomplishments.[338] In keeping with the same text, he also erected seven 'ancestral temples' (*zǔ miào* 祖廟) dedicated to his father in his first regnal year (209 BC):

> *Shǐjì*, "Annals of Qín Shǐhuáng" (*Qín Shǐhuáng běnjì* 秦始皇本紀): 古者天子七廟, 諸侯五, 大夫三, 雖萬世世不軼毀. 今始皇爲極廟, 四海之内皆獻貢職, 增犧牲, 禮咸備, 毋以加. 先王廟或在西雍, 或在咸陽. 天子儀當獨奉酌祠始皇廟.[339]
>
> In ancient times, the Son of Heaven (had) seven ancestral temples, the Dukes (had) five, five individual temples, and Grand Officials (had) three. Even though (these ancestral temples)

---

334 For a study on the choice of 'august' (*huáng* 皇) as an epithet, refer to Gorodetskaya 2008 and Yú 1962.

335 For a discussion of the stone inscriptions attributed to the First Emperor, see Kern 2000.

336 Kern 2000: 10–13.

337 Translation is taken from Kern 2000: 10–13, with minor modifications.

338 Kern 2000: 146.

339 Zhāng Dàkě 1990: 131.

persisted for ten thousand generations, they were neither attacked nor destroyed. Then, the First Emperor built the Temple of the Absolute, and (people) from within the four seas all gave their tributes and offerings. They increased the sacrifices and libations. The rituals were perfectly prepared. Nothing could be added. The ancestral temples of the previous Kings were either located in Xīyōng or Xiányáng. The ceremonies of the Son of Heaven exclusively offered alcoholic beverages and sacrifices in the Temple of the First Emperor.

The passage is ambiguous regarding the exact structure and purpose of the seven temples, although it alludes to the possibility – derived from the *Lǐyē* and *Yuèlù* documents – that the building of ancestral temples may have been commissioned by the First Emperor eleven years earlier, in 221 BC. One of the earliest known references to temples in Qín dynasty texts mentions the 'Memorial Temples of the Highest August' (*Tàishàng Huáng cí miào* 泰上皇祠廟) on *Yuèlù* strip 325:[340]

Yuèlù, strip 325: • 泰上皇祠廟在縣道者☑[341]

• The Memorial Temples of the Highest August in the counties and marches ☑

Unfortunately, the strip is undated, and its ending is missing. Additional information from *Lǐyē* manuscript 8-461, indicating that the last King of Qín held the title 'Highest August' after 221 BC, suggests that the establishment of memorial temples dedicated to the 'Highest August' had already been authorized by the First Emperor in commemoration of his father. According to the *Yuèlù* manuscripts, officials began to "inspect memorial temples" (*xíng miào* 行廟) in the second year of the Second Emperor (208 BC).[342] Interestingly, manuscript 8-138+8-174+8-522+8-523 already specifies the existence of this institutional practice as early as 221 BC:[343]

8-138+8-174+8-522+8-523 I–II: 廿六年六月壬子[49], 遷陵□【丞】敦狐爲令史更行廟詔: 令史行☑┘ 失期. 行廟者必謹視中□各自署廟所質日. [. . .][344]

In the 26th year (of King Zhèng) [221 BC], month 6 on a *rénzǐ*, Dūnhú, 【Prefect】 of Qiānlíng □, orders Prefectural Secretary Gèng to inspect the temple and instructs: (I) order the Prefectural Secretary to inspect ☑ ┘ missed the (right) time. The one who inspects the temple must carefully watch the middle □ each proceeds from the event calendar in the registered temple. [. . .]

Not only do these findings shed new light on the Qín "legalist" tradition, but also highlight customs that had formerly varied across different states and were preserved in succeeding periods, starting with the First Emperor. During the Western

340 Fàn Yúnfēi 2016.
341 Chén Sōngcháng 2015: 226.
342 Chén Sōngcháng 2015: 226.
343 For a full transcription and translation of manuscript 8-138+8-174+8-522+8-523, see chapter six.
344 Chén Wěi 2012: 78.

Hàn dynasty, for instance, rulers such as Emperor Gāozǔ 高祖 (256–195 BC), Emperor Wén 文 (202–157 BC) and Emperor Wǔ 武 (156–87 BC), for whom memorial temples were posthumously constructed, were also bestowed the title 'Highest August' (*Tàishàng Huáng* 太上皇).[345]

On manuscript 8-461, the physical borders of the expanding empire also undergo a process of erosion:

8-461 BXVII–XVIII: 邊塞曰故塞.┘ 毋塞者爲故徼.[346]

The 'border fortifications' are called the 'fortifications of the past'.┘ The '(places with) no fortifications' are called the 'frontiers of the past'.

To a large degree, Qín society was heterogeneous, characterized by a combination of various cultural units.[347] The Emperor's program of territorial expansion thus aimed at unprecedented political and cultural integration. During the formation of the empire, old frontiers and fortifications at the borders of former states, where dissimilarities between ethnic groups and inhabited spaces persisted, were deemphasized. Power relations were redefined as discursive constructs of social transformation, expressed through both voluntary or involuntary subordination.

In the effort to consolidate power within a unified system of government, symmetries and impartialities in relation to collective goods were enacted firmly, if not violently. Strategic and tactical access to other territories, meteorological and geological conditions, the availability of natural resources and other public assets all determined a region's positioning within the larger structure. The central administration may have initially experienced a revenue decline due to hierarchical reconfigurations in more peripheral areas. Therefore, a comprehensive system of regulations was required to redistribute resources, compensate for losses and uphold security at both the center and the outskirts of the empire. This project of dismantling local state borders in the name of imperial unification is arguably reflected in line BXXIX of manuscript 8-461:

8-461 BXXIX: 毋曰邦門曰都門.[348]

Do not say the 'gates of the states', (but) say the 'gates of the capital'.

---

345 Liú Qìngzhù and Barbieri-Low 2007.

346 Chén Wěi 2012: 157.

347 For a study on the genetic diversity of the workers assigned to build the mausoleum of the First Emperor, see Xu Zhi, et al. 2008. For discussions on the notion of early contact between China and the Hellenistic kingdoms of Asia, see Nickel 2013.

348 Chén Wěi 2012: 157.

Ongoing conflicts and anti-Qín sentiment among native populations of the former Warring States likely stemmed from the imperial occupation of vast, previously hostile territories. Discontent was probably intensified by the demand for social cohesion imposed on these populations, as well as the unifying system of government that disrupted and undermined local boundaries, former markers of distinct cultural identities.[349] New borders and strategic hubs were created throughout the empire to prevent uneven development and structural inequality. The partitioning of territories into 'commanderies' (*jùn* 郡), 'counties' (*xiàn* 縣), 'marches' (*dào* 道) and 'districts' (*xiāng* 鄉), further subdivided into 'settlements' (*yì* 邑), 'townships' (*lǐ* 里), 'sections' (*xiào* 校), 'subsections' (*máo* 髳), 'posts' (*tíng* 亭) and 'postal units' (*yóu* 郵), provided a degree of stability amid subsequent identitarian conflicts.[350] This universal compartmentalization of society facilitated the dissemination of information and other forms of coercion, laying the groundwork for systems of taxation and bureaucratization. Ultimately, it enhanced the economic, military and exploitative capabilities of the Qín administration.

> 8-461 BVIII–XIII, BXXIV–XXVI: 王室曰縣官.┛ 公室曰縣官.┛ 内侯爲輪(倫)侯.┛ 徹侯爲【死〈列〉】侯.┛ 以命爲皇帝.┛ 受(授)命曰制.┛ [. . .] 騎邦尉爲騎□尉.┛ 郡邦尉爲郡尉.┛ 邦司馬爲郡司馬.┛ [351]
>
> The 'Royal Courts' are called the 'County Offices'.┛ The 'Homes of Dukes' are called the 'County Offices'.┛ The title 'Marquis of the Interior' becomes 'Marquis of Human Relations'.┛ The title 'Marquis of the Twentieth Rank' becomes '【Holder of the Twentieth Rank】'.┛ A mandate is a mandate (issued by) the Emperor.┛ 'To receive the mandate' is called '(to receive) the decrees'.┛ [. . .] The title 'Cavalry Commandant of a State' becomes 'Cavalry Commandant of a □'.┛ The title 'Military Commandant of a Commandery and State' becomes 'Military Commandant of a Commandery'.┛ The title 'State Minister of War' becomes 'Commandery Commissioner of War'.

The imposition of pre-defined norms and categorizations was central to constituting institutional order across territories and cultures. The 'mandate' (*mìng* 命) – a concept possibly traceable to the Shāng dynasty – was abolished and replaced by a system of 'decrees' (*zhì* 制).[352] Traditional titles such as 'Kings' (*wáng* 王) and 'Dukes'

---

349 For a study on the impact of military engagements on identity formation, see Smith 1981.

350 For a discussion of 'townships' (*lǐ* 里) and 'posts' (*tíng* 亭) in the Qín dynasty, see Lau and Staack 2016; Zhāng Jīnguāng 2004: 587–602. 'Marches' (*dào* 道) were regions similar to 'counties' (*xiàn* 縣) in size but inhabited by native tribes (Yates 2012–2013: 306). The term 'settlement' (*yì* 邑) appears, for example, on manuscript 8-1236+8-1791; the term 'subsection' (*máo* 髳) on manuscript 8-149+8-489, and the term 'postal unit' (*yóu* 郵) on manuscript 8-1432.

351 Chén Wěi 2012: 156, 157.

352 On the notion of the 'mandate' (*mìng* 命) in the Shāng and Zhōu dynasties see, for example, Goldin 2017.

(*gōng* 公) were equated with the relatively low-ranking 'County Offices' (*xiànguān* 縣官) and made subordinate to administrative units at the 'commandery' (*jùn* 郡) level. Other hereditary royal titles, like 'Marquis of the Interior' (*nèi hóu* 內侯) and 'Marquis of the Twentieth Rank' (*chè hóu* 徹侯), also underwent significant transformation. Even former 'Military Commandants of Commanderies and States' (*jùnbāng wèi* 郡邦尉) and 'State Ministers of War' (*bāng sīmǎ* 邦司馬) were assigned more senior positions than former Kings and Dukes, which could be indicative of the increased prominence of military institutions within the imperial regime.

Due to their broad applicability, legal practices constitute an immaterial framework that bridges the divide between foreign and indigenous, self and other. They provide a groundwork for social inclusion and thereby exemplify a key mechanism through which cultural development is achieved. In this respect, the expansion of the administrative apparatus and the promulgation of modified edicts by the Emperor were inevitable developments that can be understood as the reconfiguration of governance designed to accommodate a new social structure. Legal matters were no longer adjudicated by Kings but were managed through an intricate system of 'decrees' (*zhì* 制). The Emperor ceased to intervene in legalities at lower administrative levels, instead delegating these responsibilities to highly trained officials within judiciary institutions:

8-461 BV–VII: 王讉曰制讉.┘ 以王令曰【以】皇帝詔.┘ 承【命】曰承制.[353]

'Accusations by the Kings' are called 'accusations by (a system of) decrees'.┘ 'By means of the 'Kings' ordinances' is called '【by means of】 the Emperor's edicts.'┘ 'Accepting the 【mandate】' is called 'accepting (a system of) decrees'.

Methods of efficient bureaucracy were gradually perfected by the Qín to manage vast, often sparsely populated regions, some of which were remote and isolated. The exchange of official documents over extended periods and distances, together with the authority granted to officials stationed far from the central administration to carry out 'decrees' (*zhì* 制) and 'edicts' (*zhào* 詔), allowed for the effective handling of matters that would have otherwise been excessively time-consuming or resource-intensive.

Both transmitted and recovered sources, such as the *Lǐyē* texts, provide strong evidence for these developments.[354] The term '*zhì shū* 制書' is commonly

353 Chén Wěi 2012: 156.

354 For a discussion on the replacement of the 'mandate' (*mìng* 命) with 'decrees' (*zhì* 制) during the Qín dynasty, as evidenced by both transmitted and excavated sources, see Bǔ Xiànqún 1997.

found in historical records, as well as on bamboo manuscripts and stone stelae from the Hàn dynasty. Japanese scholar Osamu Ōba identified three methods by which official rulings were issued during the Hàn dynasty: 1) unilaterally by the Emperor, 2) through recommendations from officials authorized by the Emperor, or 3) by officials independently implementing the Emperor's policy directives.[355] The 'Document of Decrees' (*zhì shū* 制書) was one of many Qín innovations that were adopted and refined by the Hàn rulers.[356]

Similarly, pre-Qín 'orders' or 'ordinances' (*lìng* 令) evolved into edicts during the Hàn dynasty. The term '*zhào* 詔', from the Qín dynasty onwards, was long believed to have been used exclusively for imperial edicts.[357] The distinction between edicts and decrees is not always clear in available Qín and Hàn dynasty sources, and definitions can sometimes vary or even or contradict each other. Edicts issued during the Hàn period often included the formulaic phrase "proclaim throughout the world" (*bùgào tiānxià* 布告天下), emphasizing their role as public declarations. Generally, decrees are understood as documents issued by authorized officials, while edicts served as public notifications.

Decrees appear alongside edicts on *Lǐyē* manuscript 8-461. Concerning decrees, it instructs that the 'mandate' (*mìng* 命) should be referred to as 'decrees' (□ *mìng yuē zhì* □命曰制), that "accepting the【mandate】" should be called "accepting (a system of) decrees" (*chéng*【*mìng*】*yuē chéng zhì* 承【命】曰承制), that "to receive the mandate" should be referred to as "(to receive) the decrees" (*shòu (shòu) mìng yuē zhì* 受(授)命曰制), and that "accusations by the Kings" should be termed "accusations by (a system of) decrees" (*wáng qiǎn yuē zhì qiǎn* 王譴曰制譴).[358] Regarding edicts, the text specifies that "by means of the Kings' ordinances" should be referred to as "【by means】 of the Emperor's edicts" (*yǐ wáng lìng yuē*【*yǐ*】*Huángdì zhào* 以王令曰【以】皇帝詔), and – though a character is missing in the sentence – that "to act" should be referred to as "□ edict" (*wèi wèi* □ *zhào* 爲謂□詔).

On manuscript 8-138+8-174+8-522+8-523, '*zhào* 詔' takes on an alternate, broader meaning. In the excerpt below, a Prefect addresses his Prefectural Secretaries using '*zhào* 詔' in the sense of 'to instruct' or 'to order' (also meaning 'instruc-

355 Ōba 1991.

356 These and following observations are taken from Bǔ Xiànqún 1997.

357 Lau and Staack 2016: 286, footnote 1335.

358 Due to the parallel structure of the sentences on manuscript 8-461, I argue that the missing character in "□ *mìng yuē zhì*" is most likely '*yuē* 曰'. For a translation and discussion of manuscript 8-461, see chapter 4.1.

tion' or 'edict'). This suggests a more generalized usage of the term with wider applicability.[359]

> 8-138+8-174+8-522+8-523 I–II: 廿六年六月壬子[49], 遷陵□【丞】敦狐爲令史更行廟詔: 令史行☒┘ 失期.行廟者必謹視中□各自署廟所質日. [. . .][360]
>
> In the 26th year (of King Zhèng) [221 BC], month 6 on a *rénzǐ*, Dūnhú, 【Prefect】 of Qiānlíng, orders Prefectural Secretary Gèng to inspect the temple and instructs: (I) order the Prefectural Secretary to inspect ☒ ┘ missed the (right) time. The one who inspects the temple must carefully watch the middle □ each proceeds from the event calendar in the registered temple. [. . .]

Given the nature of decrees, specifically their role in conferring delegated authority and independence from the Emperor, and considering the juxtaposition of the former mandate of Kings with decrees on manuscript 8-461, I suggest that '*mìng* 命' and '*zhì* 制' could refer to a synoptic comparison between the pre-Qín mandate and the complex system of decrees developed during the imperial era.

The Emperor's 'processions' (*yóu* 游) further improved communication between the Emperor, officials and the empire's inhabitants. This ceremonial event existed long before its enactment by the Qín Kings and Emperors and can be traced back to the 'peripatetic Kings' (*wǎng wáng* 往王) of the Shāng dynasty.[361] Inspecting newly conquered territories, disseminating information and pacifying populations negatively impacted by reforms were just some of the reasons the First and Second Emperors risked traveling great distances, sometimes through contested regions.[362] Regardless of the means chosen to convey messages, the imperial processions ultimately fulfilled the canonical-political function of appropriating the wealth of former Kings and integrating various territories under Qín rule.

Evidence that these elaborate processions actually took place is verified by both transmitted and excavated sources. Manuscript 29A from the *Shuìhǔdì* 睡虎地 "Annals" (*Biānnián jì* 編年記), for example, recounts a first-hand experience of a local official witnessing King Zhāoxiāng 昭襄 of Qín (325–251 BC) passing

359 Lau and Staack 2016: 286.

360 Chén Wěi 2012: 78.

361 For a study on state power, peripatetic kingship and elite formation in early Chinese and Roman Civilization, see Scheidel 2015.

362 For an overview of processions in Early China, see Chén Shùguó 1991: 158–163, 234–243; Zhāng Róngmíng 2001.

through Ānlù 安陸, located at Dòngtíng Lake (*Dòngtíng hú* 洞庭湖) in Nán 南 Commandery, present-day Húběi Province, during his 29th regnal year (278 BC).[363] Dòngtíng Lake was bordered by Nán Commandery to the north, Cāngwú 蒼梧 Commandery to the north and east and belonged to Dòngtíng Commandery, which included Qiānlíng County and the township of Lǐyē.[364]

By drawing upon the customs of former Kings and Sages, the First Emperor continued many of the traditions of the Three Dynasties, thereby affirming his status as the supreme ruler. His processions across the annexed territories of former enemy states were intended to serve a similar legitimizing function. This was particularly significant after five centuries of political instability and nominal rule by the Kings of the Eastern Zhōu dynasty, who governed within a much more limited geographical scope.

To assert political authority, one must have the means to acquire wealth, property and natural resources, as well as the ability to disseminate information repeatedly, often through various methods and ceremonial displays of power that promote an idealized socio-political order. Inscriptions on stone stelae – permanent, commemorative markers of processions – were certainly one way of rendering these otherwise transitory events more enduring. Offering sacrifices and erecting altars at historically significant locations had a comparable impact. In particular cases, clearing or obstructing large areas to make way for the processional journey was also deemed necessary.[365] Animals symbolizing dominion over spirits or places were hunted and killed, and during his last procession, the First Emperor even ordered the killing of a whale to dispel the negative influence of a local demon. Manuscript 8-461 references the term 'hunting' twice: as 1) 'bird hunting' (*jié yì* 節弋), and 2) in a more general sense as 'hunting' (*liè* 獵), for which the 'Emperor's hunting dogs' (*Huángdì quǎn* 皇帝犬) may have been advantageous.[366]

> 8-461 BIV, BXX–XXII: 王節弋曰皇帝.┛ [. . .]王游曰皇帝游.┛ 王獵曰皇帝獵.┛ 王犬曰皇帝犬.[367]

> The 'Kings' bird huntings' are called the 'Emperor's bird huntings'.┛ [. . .] The 'processions of the Kings' are called the 'processions of the Emperor'.┛ The 'Kings' hunting' is called the 'Emperor's hunting'.┛ The 'Kings' hunting dogs' are called the 'Emperor's hunting dogs'.

363 *Shuìhǔdì Qín mù zhú jiǎn zhènglǐ xiǎo zǔbiān* 1990: 7; See also Huáng Àiméi 1997: 49; Mǎ Fēibǎi 1985: 399, 429; Mittag 2003: 562.

364 Guō Tāo 2017; Zhāng Xiūguī 1981: 102–103; Zhōu Zhènhè 2005: 63–67.

365 For more details on the symbolic punishment of Mount Xiāng 湘 by the First Emperor, see Sanft 2014a: 83, 90, 92–94.

366 Sanft 2014a: 78.

367 Chén Wěi 2012: 156, 157.

The incomplete state of manuscript 8-461 impedes efforts to fully reconstruct its intended audience, reach and purpose. Was it an authoritative text asserting a political order through imposed terminologies reflective of advancing, empire-wide transformations? Or did it cohere with a semantic imperative aimed at didactic instructions to be implemented by local clerical staff? The existing data on this partially preserved tablet is not yet sufficient to answer these questions.

According to the *Shuìhǔdì* documents, the highest institutional level in the Qín dynasty was the 'Imperial Court' (*tíng* 廷).[368] In other paleographic materials, '*tíng* 廷' can represent authorities at different administrative levels.[369] In the *Lǐyē* corpus, however, '*tíng* 廷' seems to refer almost exclusively to the government seat of a 'county' (*xiàn* 縣).[370] There are also a few examples where the term may denote interactions between Dòngtíng Commandery or counties therein and the Imperial Court, either directly or indirectly. Among these, 37 wooden strips discovered in layers 5, 6 and 8, otherwise devoid of additional information, contain the term by itself on the top half (Fig. 4.2).[371] These documents, characterized by their pointed tip at the bottom, are known as 'postal document strips' (*yóu shū jiǎn* 郵書簡).[372] Since additional textual or material references are missing, it cannot be entirely ruled out that these documents had a direct connection to the Imperial Court in the Qín capital city of Xiányáng 咸陽.

One document in the *Lǐyē* corpus explicitly mentions the city of Xiányáng 咸陽 in an interchange of documents between a 'Household Bureau' (*hùcáo* 戶曹), the Qín capital and three other counties.

> 8-1533 I–II: 戶曹書四封, 遷陵印, 一咸陽, 一高陵, 一陰密, 一競陵.┘ 廿七年五月戊辰[5]水下五刻, 走荼以來.[373]

---

368 Lau and Staack 2016: 311.

369 Lau and Staack 2016: 249.

370 See Chén Wěi 2012: 28, 41–42, 51.

371 This refers specifically to manuscripts 8-30, 8-774, 8-812, 8-832, 8-840, 8-862, 8-905, 8-928, 8-990, 8-1026, 8-1085, 8-1096, 8-1100, 8-1106, 8-1227, 8-1283, 8-1326, 8-1331, 8-1348, 8-1368, 8-1384, 8-1402, 8-1503, 8-1543, 8-1571, 8-1582, 8-1596, 8-1658, 8-1698, 8-1746, 8-1767, 8-1778, 8-1780, 8-1789, 8-1803, 8-1906 and 8-1963.

372 For more information on the so-called 'postal document strips' (*yóu shū jiǎn* 郵書簡), see chapter two.

373 Chén Wěi 2012: 352.

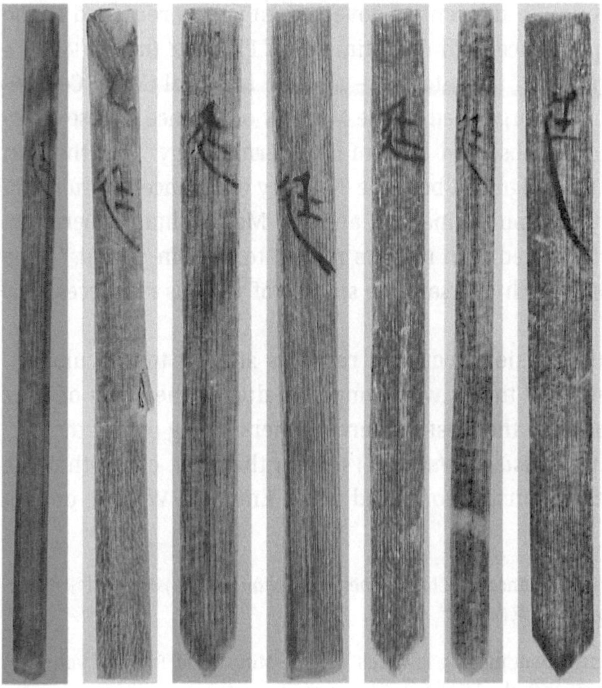

Fig. 4.2: Monochrome photographs of manuscripts 8-1026, 8-1100, 8-1331, 8-1368, 8-1503, 8-1658 and 8-1698 (from left to right). Ink on wood. Dimensions of manuscript 8-1026: 22.5 x 1.1 cm. Dimensions of manuscript 8-1100: 15.2 x 1.2 cm. Dimensions of manuscript 8-1331: 13.3 x 1.1 cm. Dimensions of manuscript 8-1368: 12.0 x 1.4 cm. Dimensions of manuscript 8-1503: 14.3 x 1.2 cm. Dimensions of manuscript 8-1658: 22.1 x 1.3 cm. Dimensions of manuscript 8-1698: 15.5 x 1.5 cm.[374]

There are four documents from the Household Bureau with the seal of Qiānlíng. One (is addressed to) Xiányáng, one to Gāolíng, one to Yīnmì and one to Jìnglíng.┘ In the 27th year (of King Zhèng) [220 BC], month 5 on a *wùchén*, at water below mark 5, (the document) is brought by Runner Tú.

Comments: Gāolíng 高陵, Yīnmì 陰密 and Jìnglíng 競陵 were counties in the Qín and Hàn dynasties.[375] The expression "water below mark 5" (*shuǐ xià wǔ kè* 水下五刻) converts to the period between 10:00 am and 10:59 am. For a discussion of the calendar and timekeeping systems used in the *Lǐyē* corpus, see chapter three.

374 *Húnán shěng wénwù kǎogǔ yánjiūsuǒ* 2012a: 139, 145, 166, 169, 189, 216, 219.
375 Chén Wěi 2012: 352. The dimensions are taken from Yáo Lěi 2015.

The exchange of information with the central government is also recorded in the *Lǐyē* corpus such as correspondence with the 'Ministry of Finance' (*nèishǐ* 内史) – literally translated as 'Secretary of the Interior' – and the Imperial Court. Centralized financial administration was organized by the Ministry of Finance, whose structure and hierarchic position are obscured by confusing terminology.[376] In the Hàn Dynasty, the term '*nèishǐ* 内史' designated both the 'Ministry of Finance' (*zhìsù nèishǐ* 治粟内史), responsible for the empire's finances, and the 'Metropolitan Superintendent' (*nèishǐ* 内史), an official tasked with matters related to the Qín capital.[377] The 'Main Treasury' (*dànèi* 大内), which oversaw the storage of various resources, was another central institution.[378]

Determining when this division occurred remains subject to speculation. Some scholars advocate the view that it was completed during the reign of King Zhèng or shortly after he became the First Emperor. Others, citing one of the earliest occurrences of the title '*zhìsù nèishǐ* 治粟内史' in the *Shǐjì*, claim that the Ministry of Finance was largely institutionalized when Emperor Wén 文 of Hàn came to power.[379]

> *Shǐjì*, "The Hereditary House of Chancellor Chén" (*Chén chéngxiàng shìjiā* 陳丞相世家): 陛下即問问決獄, 責廷尉; 問錢穀, 責治粟内史.[380]
>
> If Your Highness wants to ask about judging lawsuits, ask the Minister of Trials. If you want to ask about coinage and grain, ask the Minister of Finance.

There is little evidence in the *Lǐyē* manuscripts regarding this issue. However, a few texts may be instructive with regards to the status and position of the 'Ministry of Finance' (*nèishǐ* 内史) within the imperial hierarchy. Manuscript 8-657, for example, opens with the following lines:

> 8-657 I–II: ☒亥朔辛丑[38], 琅邪叚(假)【守】□敢告内史、屬邦、郡守主: 琅邪尉徒治即【默】☒┘ 琅邪守四百卌四里, [. . .][381]
>
> ☒ new moon on a [missing]*hài*, on a *xīnchǒu*, □【Governor】*ad Interim* of Lángyá dares to report to the Ministry of Finance, all the dependent states and the Commandery Gover-

---

376 '*Nèishǐ* 内史' was an office originally established in the Western Zhōu era (Li 2008: 75–76, 308).

377 Chén Wěi 2012: 63; Hulsewé 1978: 194–95; Yamada 1990: 2. The expression '*zhìsù nèishǐ* 治粟内史 'literally translates to "Secretary of the Interior governing millet."

378 Fù Jiāyí 2007: 63.

379 Some scholars argue that the 'Main Treasury' (*dànèi* 大内) was created and institutionalized during the reign of King Zhèng, either before or shortly after he became the First Emperor (Kudō 2010; Yamada 1990: 3; Yáng Zhènhóng 2014: 60).

380 Zhāng Dàkě 1990: 1267.

381 Chén Wěi 2012: 193.

nors: The Commandery Military Commandant of Lángyá moves to govern (military matters) in Jí【mò】☒┘ The Governor of Lángyá is 434 *lǐ* away, [. . .]

Comments: 'Dependent states' (*bāng* 邦), formerly known as 'states', referred to territories inhabited by indigenous ethnic groups after the 26th year of King Zhèng (221 BC). Based on a passage from the "Table of All Official Ranks and Offices" (*Bǎi guān biǎo* 百官表), a section in the *Book of Hàn* (*Hànshū* 漢書), and another passage on strip 201 of the "Qín lǜ shíbā zhǒng", Chén Wěi reads '*shǔ* 屬' as an organization responsible for administering indigenous groups in dependent states.[382]

Jímò 即默 (*tsəkmə̂k) could be a variant of Jímò 即墨 (*tsəkmə̂k), a county during the Hàn dynasty that belonged to Jiāodōng 膠東 Commandery in present-day Shāndōng Province.[383] We learn from this manuscript that the Governor's office in Lángyá Commandery was 434 'miles' (*lǐ* 里) (ca. 180 kilometers) away from the office of the Military Commandant (also in Lángyá Commandery). This indicates that offices did not need to be situated near each other. One mile measured approximately 415.80 meters in the Qín period.[384]

The sequence of terms listed in the first line of manuscript 8-657 could mirror the hierarchic positions of the respective offices and officials within the imperial structure. This ordering places the 'Ministry of Finance' (*nèishǐ* 内史) at the Imperial Court, followed by the subordinate "Governors of dependent states and commanderies" (*bāng jùn shǒu zhǔ* 邦郡守主).

Based on fragments from manuscripts 8-105, 8-1270 and 8-1845, the Ministry of Finance also appears to have dealt with 'military matters' (*jūn shì* 軍事) and 'statistics' (*jì* 計) submitted by county and commandery officials.

8-105: 彭陽 内史[385]

(From) Péngyáng (County to) the Ministry of Finance.

Comments: Péngyáng 彭陽 was a county in Āndìng 安定 Commandery (see Tab. 4.3).[386]

8-206 I, rI–II: 武關内史. └ 進書李季□足┘ 自發.[387]

(From) the frontier pass Wǔ (to) the Ministry of Finance. Make an entry in the document that (the person named) Lǐjì □ sufficient └Sent by himself.

382 Chén Wěi 2012: 193.
383 Chén Wěi 2012: 193. Phonological reconstructions are taken from Schuessler 2009: 111, 113.
384 Hulsewé 1961: 206–207.
385 Chén Wěi 2012: 63.
386 Chén Wěi 2012: 63.
387 Chén Wěi 2012: 113.

8-228 I–VI: □□内史守衷下: 縣以律令傳別□☒┘ 縣界中□□者縣各別下書焉□☒┘ □□地□□□□報【沅】陽, 言書到☒┘ □□□□□商丞□下報商, 書到☒┘ 十月丁巳[54], 南郡守恒下真書洞庭☒┘ □□□手.[388]

□□ Zhōng, Incumbent of the Ministry of Finance, (sends a document to) the lower authorities: The counties (should) use the statutes and ordinances to copy and send □☒┘ in county area □□ the counties (should) make a copy of each and (send) the document to the lower authorities in them □☒┘ □□ territories □□□□ reply to【Yuán】yáng, the protocol arrived ☒┘ □□□□□ Prefect of Shāng □ (sends) a reply to Shāng, the document arrived ☒┘ Month 10 on a *dīngsì*, Héng, Governor of Nán Commandery, sends the original document to the lower authorities in Dòngtíng ☒┘ By the hand of □□□.

8-1270: 内史軍事盡□☒[389]

The military matters of the Ministry of Finance exhaust □☒

The top part of this manuscript is stained with black ink. The lower section is missing.

8-1387: ☒蜀中内史☒ [390]

☒ in the middle of Shǔ to the Ministry of Finance ☒

This passage could also be translated as "the Ministry of Finance in the middle of Shǔ." Shǔ 蜀 was a commandery in the Qín empire (Tab. 4.1).

8-1510 I–III/rI–III: 廿七年三月丙午[43]朔己酉[46], 庫後敢言之: 兵當輸内史, 在貳春□□□□┘ 五石一鈞七斤, 度用船六丈以上者四樓(艘). 謁令司空遣吏、船徒取. 敢言┘ 之.☒ └三月辛亥[48], 遷陵守丞敦狐告司空主, 以律令從事.ノ. . .┘ 昭行┘ 三月己酉[46]水下下九, 佐赾以來.ノ釦手.[391]

In the 27th year (of King Zhèng) [220 BC], month 3, new moon on a *bǐngwǔ*, *jǐyǒu* [day 4], Hòu from the Warehouse dares to say: The weapons should be transferred to the Ministry of Finance, in Èrchūn □□□□┘ Five piculs, one pound and seven catties. To cross to the other side, use a boat. For more than six yards, (use) four boats. Request to order the Commissioner of Public Works to dispatch officials and boats to fetch the convicts. (He) dares to say┘ ☒ └In month 3 on a *xīnhài* [day 6], Dūnhú, Deputy Prefect of Qiānlíng, reports to the Commissioner of Public Works. Follow up on the matter in accordance with the statutes and ordinances. /. . . ┘ Dispatched by Zhāo.┘ In month 3 on a *jǐyǒu* [day 4], at water below (mark) 9, (the document) is brought by Assistant X [pronunciation unknown]. / By the hand of Kòu.

388 Chén Wěi 2012: 119.
389 Chén Wěi 2012: 303.
390 Chén Wěi 2012: 319.
391 Chén Wěi 2012: 341.

'*Zhàng* 丈' was a unit of length or distance equal to ten 'feet' (*chǐ* 尺), approximately 2.30 meters. A 'pound' (*jūn* 鈞) was a unit of weight equivalent to 30 'catties' (*jīn* 斤).[392]

8-1845 I–II: 卅二┛年遷陵內史計.┛ ☐[393]

32nd ┛ year (of King Zhèng) [215 BC]. The statistics of the Ministry of Finance in Qiānlíng. ┛ ☐

If '*jì* 計' is understood as a verb in the second line of this passage and used similarly to '*fā* 發' or '*xíng* 行', the sentence could also be translated as: "Qiānlíng (submits) the statistics to the Ministry of Finance." In this context, the term '*nèishǐ* 內史' may represent a branch unit of the Ministry of Finance at the commandery level that received and reviewed statistics directly from the County Offices.

Although evidence is scarce and not entirely conclusive, '*nèishǐ* 內史' seems to be an abbreviated form of the 'Ministry of Finance' (*zhìsù nèishǐ* 治粟內史). This institution directed the imperial finances and logistical operations of the military well into the reign of the First Emperor, and perhaps beyond, until around 208 BC. The full designation '*zhìsù nèishǐ* 治粟內史' does not appear on the *Lǐyē* documents from layers 5, 6 and 8. Instead, the term '*nèishǐ* 內史' is generally used to refer to an agency at the Imperial Court managed by 'Incumbents' (*shǒu* 守), with branch units at the commandery level.[394]

Following the "Statutes Concerning Granaries" (*cāng lǜ* 倉律) in the *Shuìhǔdì* documents – which mainly discuss the distribution rather than the storage of food – Qín counties reported directly to central institutions like the 'Grand Granary' (*dà cāng* 大倉 or *tài cāng* 太倉), terms nonexistent in the *Lǐyē* corpus and a practice discontinued in the Hàn dynasty.[395] Nevertheless, this might explain why some counties and even local 'border checkpoints' or 'frontier passes' (*guān* 關) maintained contact with the Ministry of Finance, a main division of the central government.

---

392 Lau and Staack 2016: 302.
393 Chén Wěi 2012: 399.
394 Lau and Staack, et al. suggest that, during Qín times, every holding of an office was designated as '*shǒu* 守', for example, in combination with terms like '*shǎonèi* 少內' (8-152), '*sīkōng* 司空' (8-135) and '*tiánguān* 田官' (9-981), all of which mean 'Officeholder' or 'Incumbent' (Chén Sōngcháng 2004: 19; Lau and Staack 2016: 86; Sūn Wénbó 2014a; Yáng Zōngbīng 2004: 11–14). This designation is often omitted in the *Lǐyē* texts.
395 Hulsewé 1978: 195–200; 1985: 41. According to the "Record of Official Titles" (*Bǎi guān zhì* 百官志) in the *Book of the Later Hàn* (*Hòu Hàn shū* 後漢書), counties, townships and 'marches' (*dào* 道) submitted reports to their respective commanderies. The commandery then gathered and forwarded the information to the Court (Bielenstein 1980: 43–47). In the "Statutes Concerning Granaries" (*Cāng lǜ* 倉律) on strips 49–51 of the *Shuìhǔdì* documents, the distribution of food supplies is strictly regulated and differentiated for officials, commoners and convicts (Hulsewé 1985: 31).

'Prosecutors' (*yùshǐ* 御史) were high-level officials appointed by the central government to address specific matters at the commandery and county levels. Manuscript 8-141+8-668 mentions a 'Crossbow Archer' (*fānǔ* 發弩), a county-level official, and reports that the "Court has sent the document of the Prosecutor to the lower authorities" (*tíng xià yùshǐ shū* 廷下御史書).[396] This document, specifying procedures for executing and reviewing lawsuits at the county level, might have been compiled by the central government. However, the Court forwarding it from the Imperial Court to the Military Commandant could have been, in this case, also the County Court.

A similar entry on manuscript 8-152 suggests that '*tíng* 廷' refers to the County Court. It also indicates that the document of the Prosecutor circulated among county government units before the requested information was submitted to the higher authorities in the capital by Dòngtíng Commandery.[397]

8-152 I, rI: 卅二年四月丙午[43]朔甲寅[51], 少内守是敢言之: 廷下御史書舉事可爲┛ 恒程者、洞庭上帬(裙)直, 書到言. 今書已到, 敢言之. └四月甲寅[51]日中, 佐處以來. ノ欣發. 處手.[398]

In the 32nd year (of King Zhèng) [215 BC], month 4, new moon on a *bǐngwǔ*, *jiǎyín* [day 9], Shì, Incumbent of the Lesser Treasury, dares to say: The (County) Court sent the document of the Prosecutor to the lower authorities. (They) follow up on the matter and approve that it is handled according to the standard procedures. Dòngtíng (then) submits (the document) concerning the value of the lower garments made of cotton to the higher authorities. Communicate orally upon delivery. Now, the document has already arrived. (He) dares to say. └ Month 4 on a *jiǎyín* [day 9], at midday, (the document) is brought by Assistant Chǔ. / Sent by Xīn. By the hand of Chǔ.

Comments: The 'Lesser Treasury' (*shǎonèi* 少内) acted as an intermediary between the Bureaus and Offices of the county and the County Court.

The verb-object collocation "*jǔ shì* 舉事" is understood by Chén Wěi as "*xíng shì* 行事" or "to follow up on the matter."[399] As outlined by Bǔ Xiànqún, '*jǔ'shū* 舉書' was a document particularly common in Hàn dynasty manuscripts that contained accusations made by superior officials against subordinates conducting reports.[400] Failure to respond to these accusations within a specified time frame would constitute a violation of the 'decrees' (*zhì* 制) and result in disciplinary action. This interpretation of '*jǔ* 舉' is fitting as it reflects the proce-

396 For a transcription and translation of *Lǐyē* manuscript 8-141+8-668, see chapter three.

397 Chén Wěi also supports the hypothesis that the 'County Court' (*xiàntíng* 縣廷) forwarded the document of the Prosecutor to the respective county government unit (Chén Wěi 2012: 97).

398 Chén Wěi 2012: 92.

399 Chén Wěi 2012: 92.

400 Bǔ Xiànqún 1997.

dure of following up on a binding request according to the prescribed standards issued by a higher authority.[401]

From manuscripts 8-153, 8-158 and 8-159, it is evident that "*luò (qún) qún* 络帬(裙)" is an abbreviated form of "*qún (qún)* 帬(裙)".[402] Additionally, manuscript 8-158 indicates that this was a "document containing information about the lower garments made of cotton" (*qún (qún) zhí shū* 帬(裙)直書).[403]

Lǐyē manuscript 8-159 might have been the actual "document of the Prosecutor" (*yùshǐ shū* 御史書) referenced in manuscript 8-152, in which the central government requested the government seat in Dòngtíng to verify the proper handling of the matter and assess the value of cotton garments. On manuscript 8-159, an ordinance issued by the 'Chief Minister' (*chéngxiàng* 丞相) prompts an investigation by the 'Deputy Prosecutor' (*yùshǐ chéng* 御史丞), for which the 'Document of Decrees' (*zhì shū* 制書) may have served as a regulatory preface:

8-159 I–III: 制書曰: 舉事可爲恒程者上丞相, 上洞庭絡帬(裙)程有□□□┘ 卅二年二月丁未[44]朔□亥, 御史丞去疾: 丞相令曰舉事可爲恒┘ 程者□上帬(裙)直. 即䧹(應)令, 弗䧹(應), 謹案致 . . . .┘[404]

The Document of Decrees states: Follow up on the matter, approve that it is handled according to the standard procedures and submit to the Chief Minister. The procedure concerning the value of the lower garments made of cotton, which has been submitted by Dòngtíng Commandery, contains □□□┘ In the 32nd year (of King Zhèng) [215 BC], month 2, new moon on a *dīngwèi*, □*hài*, Deputy Prosecutor Qùjí (reports): The ordinance of the Chief Minister says to follow up on the matter and approve that it is handled according to the ┘ standard procedures □ submit (the document containing) the value of the lower garments made of cotton. If [it] complies with the ordinance, or [if it] does not comply, investigate (the issue) carefully and established (the outcome) firmly . . . . . .┘

Comments: '*Chéngxiàng* 丞相' was the 'Chief Minister' and the highest-ranking administrative official in the imperial hierarchy.[405] The 'Document of Decrees' (*zhì shū* 制書) is mentioned only on this manuscript and the heavily damaged manuscript 8-1648, in connection with "sending bird feathers" (*fā yǔ* 發羽).[406]

401 '*Jǔ' shū* 舉書' appears three times in the *Lǐyē* corpus: once on manuscript 8-152 and twice on manuscript 8-159. Both instances are related to matters concerning the 'Prosecutor' (*yùshǐ* 御史) (8-152) or the 'Deputy Prosecutor' (*yùshǐ chéng* 御史丞) (8-159).

402 Chén Wěi 2012: 92.

403 For a translation of manuscript 8-158, see chapter 4.4.

404 Chén Wěi 2012: 96.

405 Chén Wěi 2012: 66.

406 Chén Wěi 2012: 373.

4.2 'Commanderies' (*jùn* 郡)

The spatial reconfiguration of the empire into smaller territorial units created a modular network of interrelations, from which new institutions and agents could emerge. There are very few records about the structure and organization of bureaucracy at the county level and below during the Qín imperial era.[407] This chapter pieces together a list of commanderies and officials who directed and oversaw these territorial units. The next chapter provides a brief description of Qín county officials and an overview of counties in Dòngtíng Commandery and adjacent regions that were in contact with government units in Dòngtíng Commandery. The analysis primarily focuses on the *Lǐyē* manuscripts from layers 5, 6, 8, 12, 14 and 15, while other epigraphic material from early imperial period is considered as needed.

Commanderies were large territorial units led by '(Superior) Governors' (*tài shǒufǔ* 泰守府 or *shǒu(fǔ)* 守(府)), 'Deputy Governors' (*shǒu chéng* 守丞) and 'Governors *ad Interim*' (*jiǎ(jiǎ)shǒu* 叚(假)守).[408] As transmitted sources indicate, the Qín kingdom had 12 commanderies by 246 BC and, in the course of two decades, expanded to include 36 by 221 BC.[409] Based on his readings of the *Lǐyē* corpus, Yóu Yìfēi argues that Dòngtíng comprised at least 14 counties in the Qín dynasty, one of which was Qiānlíng County.[410] This total does not account for Línhàn 臨漢 and Tuóyáng 陁陽 counties, which cannot be assigned to specific commanderies and appear on damaged or partially illegible documents.[411]

Compared with the "Treatise on Geography" (*Dìlǐ zhì* 地理志) in the *Book of Hàn* (*Hànshū* 漢書), Dòngtíng ranks first in terms of the absolute number of counties, even surpassing Wǔlíng 武陵, which at that time encompassed 13 counties and was considered one of the largest commanderies. Based on my analysis of the *Lǐyē* corpus, Dòngtíng emerges as one of 15 commanderies documented in the texts and was by no means a small region limited to Dòngtíng Lake and Mount Xiāng 湘.[412] Dòngtíng was bordered by Nán 南 Commandery to the north, with ten confirmed counties; Bā 巴 Commandery to the west, with seven confirmed counties; and Cāngwú 蒼梧 Commandery to the east, with one confirmed and one unconfirmed county. This

---

407 Zhāng Chūnlóng 2019.

408 Manuscript 8-1523 is an interesting example, where the Governor of Dòngtíng is referred to as "Governor of Dòngtíng" (*Dòngtíng shǒu* 洞庭守) in the first line (month 7), and as "Governor of Dòngtíng *ad Interim*" (*Dòngtíng jiǎ(jiǎ)shǒu* 叚(假)守) in lines II and III (month 8).

409 Kiser and Cài Yǒng 2003: 530.

410 Yóu Yìfēi 2015.

411 This observation refers to manuscripts 8-1555 and 8-2188, respectively.

412 For an alternate view on Dòngtíng Commandery, see Sanft 2014a: 92–93.

suggests that communication with Nán Commandery and Bā Commandery was likely more frequent.[413] An overview of Qín commanderies recorded in the *Lǐyē* documents is provided below (Tab. 4.1).

Tab. 4.1: List of imperial Qín 'commanderies' (*jùn* 郡) on *Lǐyē* manuscripts from layers 5, 6, 8 and 14 (sorted by alphabetical order).

No.	Commandery name	Officials in charge	Occurrence: Manuscript number (s)
1	Bā 巴	'Governor *ad Interim*' (*jiǎ(jiǎ)shǒu* 叚(假)守)	**2:** 8-61+8-293+8-2012, 8-207
2	Cāngwú 蒼梧	'Military Commandant' (*wèi* 尉)	**3:** 8-376, 8-657, 8-758
3	Dài 代[414]	–	**1:** 8-528+8-532+8-674
4	Dòngtíng 洞庭	'(Superior) Governor' (*(tài)shǒu* (泰)守, *shǒu(fǔ)* 守(府), 'Governor *ad Interim*' (*jiǎ(jiǎ)shǒu* 叚(假)守), 'Military Commandant' (*wèi* 尉)	**>120**
5	Hándān 邯鄲	–	**1:** 8-894
6	Hénèi 河内	–	**1:** 14-169[415]
7	Héngshān 衡山	'Governor' (*shǒu* 守)	**2:** 8-1234
8	Lángyá 琅邪	'Governor' (*shǒu* 守), 'Governor *ad Interim*' (*jiǎ(jiǎ) shǒu* 叚(假)守), 'Military Commandant' (*wèi* 尉)	**1:** 8-657
9	Lújiāng 廬江	–	**1:** 8-1873
10	Nán 南	'(Superior) Governor' (*tàishǒu* 泰守, *shǒu* 守), 'Governor *ad Interim*' (*jiǎ(jiǎ)shǒu* 叚(假)守)	**3:** 8-228, 8-772, 8-974
11	Sānchuān 叁川	–	**1:** 14-638[416]
12	Shǔ 蜀	–	**1:** 8-1387
13	Tàishān 泰山	–	**1:** 8-462+8-685
14	Tàiyuán 泰原	–	**1:** 8-2040
15	Yànmén 鴈(雁)門	'(Superior) Governor' (*tàishǒu* 泰守)	**1:** 8-410

413 Mǎ Yí 2005.

414 The government seat of Dài 代 Commandery was located in the southwest of present-day Yù 蔚 County, Héběi Province (Chén Wěi 2012: 174).

415 Yóu Yìfēi 2015: 36.

416 See Hé Yǒuzǔ 2016a.

Concerning official titles and responsibilities, the "Wéi yù děng zhuàng" and records in the "Zòuyànshū", excavated from a Western Hàn tomb at Zhāngjiāshān 張家山 (186 BC), list four titles at the commandery level:[417]

1. '(Commandery) Governor' (*(jùn)shǒu* (郡)守 or *tàishǒu* 太守): Appointed by the central government to administer a commandery.[418]
2. 'Military Commandant' (*(dū)wèi* (都)尉): Appointed by the central government to oversee military matters and policies in the absence of the Commandery Governor.[419] This title was either combined with the name of a commandery or abbreviated simply as '*wèi* 尉'. Beginning in Early Hàn times, the 'Military Commandant' was occasionally referred to as '*dūwèi* 都尉'.[420]
3. 'Commandery Secretaries' or 'Provincial Secretaries' (*zúshǐ* 卒史): Tasked with administrative duties and the initial evaluation of 'judicial reviews' (*fù yù* 覆獄) regarding legal violations committed by commandery officials.[421] As indicated in a letter of recommendation from the "Wéi yù děng zhuàng", aspiring Secretaries were expected to embody specific virtuous traits.

> *Yuèlù*, "Wéi yù děng zhuàng", strip 168: 皆請(清)絜(潔)毋(無)害, 敦殼(慤)守吏, 心平端禮.[422]
>
> They are all officeholders who have integrity and moral purity and are impeccable, honest and faithful. In their hearts, they are impartial, upright and proceed from ritual propriety.[423]

417 Lau and Staack 2016: 310.

418 Refers to strips 25, 40 and 63 of the "Wéi yù děng zhuàng", and to strips 125, 129 and 154 of the "Zòuyànshū" (Lau and Staack 2016: 310).

419 Military matters and policies could include the conscription and training of recruits, annual military exercises held during month 8, the inspection of fortifications, the prevention of organized crime, and related activities.

420 Refers to strips 1 and 129 of the "Zòuyànshū" (Lau and Staack 2016: 176–177); See also Ān Zuòzhāng and Xióng Tiějī 2007: 574–581; Bielenstein 1980: 94; Lǐ Yùfú 2002: 322–324; Loewe 1967: 60–61, 74, 76.

421 Refers to strips 149 and 170 of the "Wéi yù děng zhuàng"; and to strips 124 and 128 of the "Zòuyànshū" (Lau and Staack 2016: 225–226). For more information on the office of the 'Commandery Secretary' (*zúshǐ* 卒史), see Lau and Lüdke 2012, footnote 923. For a detailed investigation of judicial review, see Mizuma 2012. For a reconstruction of criminal proceedings and sentencing in the Early Hàn dynasty, as reflected in the "Zòuyànshū", see Lau 2002: 343–395.

422 Lau and Staack 2016: 244.

423 Translation is taken from Lau 2014: 179, with a minor modification in the second sentence, where "*duān lǐ* 端禮" was originally translated as "in conformity with the rituals."

4. 'Supervisory Prosecutors' (*jiānyùshǐ* 監御史): Appointed by the central government to inspect preassigned commanderies and review pending criminal offenses committed by commandery officials, including the Governor. In the Hàn dynasty, Supervisory Prosecutors were re-designated 'Regional Inspectors' (*cìshǐ* 刺史) and had a considerably narrower scope of responsibility and institutional authority.[424]

The *Lǐyē* corpus conveys a similar image, with the addition of more hierarchical positions. Tab. 4.2 summarizes the titles and main responsibilities of commandery-level officials as reflected in the texts.

Tab. 4.2: List of titles and main responsibilities of commandery level officials on the *Lǐyē* manuscripts from layers 5, 6 and 8.

No.	Title	Main responsibilities	Occurrence: Manuscript number (s)
		'Governors' (*shǒu* 守)	
1.1	'Superior Governor' (*dà(tài)shǒu* 大(泰)守)	– Maintain communication with high-ranking officials from other commanderies and counties – Issue 'ordinances' (*lìng* 令) – Send documents relating to household registration to the 'Household Bureaus' (*hùcǎo* 戶曹) – Send documents relating to public finances on one's own behalf to the 'Lesser Treasuries' (*shǎonèi* 少內) – Receive and reply to general reports from 'Prefects' (*chéng* 丞) – Receive documents from the 'Bureau of the (County) Military Commandant' (*wèicáo* 尉曹) – Regularly receive 'statistics' (*jì* 計) from the 'Prefects' (*chéng* 丞) on the new moon day of month 4	**52:** 5-23, 8-67+8-652, 8-122, 8-137, 8-140, 8-155, 8-165, 8-175, 8-247, 8-273+8-520, 8-359, 8-410, 8-433, 8-434, 8-653, 8-663, 8-60+8-656+8-665+8-748, 8-762, 8-681, 8-704+8-706, 8-728+8-1474, 8-756, 8-768, 8-768, 8-772, 8-8772, 8-806, 8-908, 8-978, 8-959+8-1291, 8-1103, 8-1119, 8-1155, 8-1225, 8-1229, 8-1258, 8-1404, 8-1477, 8-1481, 8-1525, 8-1560, 8-1586, 8-1624, 8-1695, 8-1756, 8-1829, 8-2125, 8-2138, 8-2144, 8-2353, 8-2381, 8-2502

424 Refers to strips 14 and 170 of the "Wéi yù děng zhuàng" (Lau and Staack 2016: 101–117). See also Ān Zuòzhāng and Xióng Tiějī 2007: 57–58.

Tab. 4.2 (continued)

No.	Title	Main responsibilities	Occurrence: Manuscript number (s)
		– Receive documents from the 'Judiciary Bureau of the East' (*yùdōngcáo* 獄東曹) and the 'Judiciary Bureau of the South' (*yùnáncáo* 獄南曹) – Receive harvest reports on day 15 of month 9 from the 'Office of Agricultural Fields' (*tiánguān* 田官)	
1.2	'Governor' (*shǒu(fǔ)* 守(府)	– Coordinate military matters – Communicate with high-ranking officials from other counties – Send documents relating to military matters on one's own behalf to the 'Bureau of the (County) Military Commandant' (*wèicáo* 尉曹)	**8:** 8-61+8-293+8-2012, 8-71, 8-228, 8-657, 8-755, 8-1523, 8-2159, 8-2164
1.3	'Governor *ad Interim*' (*jiǎ(jiǎ)shǒu* 叚(假)守)	– Communicate with high-ranking county officials – Report to the Governor – Report to the 'Ministry of Finance' (*nèishǐ* 内史) and representatives of 'dependent states' (*shǔ bāng* 屬邦)	**6:** 8-61+8-293+8-2012, 8-657, 8-759, 8-974, 8-1523, 8-2115
		'Military Commandants' (*wèi* 尉)	
2.1	'Commandery Military Commandant' (*(jùn)wèi* 郡尉)	– Enforce 'ordinances' (*lìng* 令) issued by the 'Governor' (*tài (tài) shǒufǔ* 泰(太)守府) – Oversee the 'Bureaus of the (County) Military Commandants' (*wèicáo* 尉曹) – Receive documents from the 'Judiciary Bureau of the South' (*yùnáncáo* 獄南曹) and the 'Bureaus of the (County) Military Commandants' (*wèicáo* 尉曹) – Exchange documents relating to food supplies with the 'Granaries' (*cāng* 倉) – Award military ranks at local level	**18:** 8-69, 8-71, 8-98, 8-167+8-194+8-472+8-1011, 8-247, 8-376, 8-461, 8-657 (4x), 8-728+8-1474, 8-1225, 8-1311, 8-1477, 8-1517, 8-1563, 8-1823

Tab. 4.2 (continued)

No.	Title	Main responsibilities	Occurrence: Manuscript number (s)
2.2	'Military Commandant of a Dependent State' (*bāngwèi* 邦尉)[425]	– Coordinate military matters of a 'dependent state' (*bāng* 邦)	**2:** 8-461, 8-649
2.3	'Commandery Cavalry Commandant' (*qí[jùn] wèi* 騎[郡]尉)[426]		**1:** 8-461
2.4	'Cavalry Commandant of a Dependent State' (*qíbāngwèi* 騎邦尉)[427]		**1:** 8-461
		'Supervisors' (*wèijiān* 尉監)	
3.1	'Supervisors' (*wèijiān* 尉監)		**2:** 8-1032, 8-1429
		'Commandery Secretaries' (*zúshǐ* 卒史)	
4.1	'Commandery Secretaries' (*zúshǐ* 卒史)	– Create maps of a border area between adjacent commanderies and counties and have them approved by a 'Prosecutor' (*yùshǐ* 御史) – Conduct 'judicial review' (*fù yù* 覆獄) of criminal offenses	**4:** 8-135, 8-167+8-194+8-472+8-1011, 8-224+8-412+8-1415, 8-247
4.2	'Commandery Secretaries *ad Interim*' (*jiǎ (jiǎ)zúshǐ* 叚(假)卒史)	– Communicate with the 'Prefects' (*chéng* 丞)	**1:** 8-78

425 This title was officially changed to 'Commandery Military Commandant' (*jùnwèi* 郡尉) in the 26th year of King Zhèng (221 BC) (see *Lǐyē* manuscript 8-461, column B, line XXV – translation in chapter 4.1).
426 Due to the analogous structure of manuscript 8-461, I assume the missing character here is 'commandery' (*jùn* 尉).
427 This title was officially changed to 'Commandery Cavalry Commandant' (*qíjùnwèi* 騎郡尉)' in the 26th year of King *Zhèng* (221 BC) (see manuscript 8-461, column B, line XXIV – translation in chapter 4.1).

Tab. 4.2 (continued)

No.	Title		Main responsibilities	Occurrence: Manuscript number (s)
	'Official in Charge of the Commissioners of Public Works' (*zhǔ sīkōng* 主司空)			
5.1	'Official in Charge of the Commissioners of Public Works' (*zhǔ sīkōng* 主司空)	–	Receive the "documents of the Bureaus of the (County) Military Commandants" (*wèicáo shū* 尉曹書)	**1:** 8-1616

The *Lǐyē* manuscripts likely refer to three types of Governors:

1) 'Superior Governors' (*dà(tài)shǒu* 大(泰)守),
2) 'Governors' (*shǒu(fǔ)* 守(府)),
3) 'Governors *ad Interim*' (*jiǎ(jiǎ)shǒu* 叚(假)守).

Similarly, at least four types of Military Commandants are mentioned:

1) 'Commandery Military Commandants' (*(jùn)wèi* (郡)尉),
2) 'Military Commandants of a Dependent State' (*bāngwèi* 邦尉),
3) 'Commandery Cavalry Commandants' (*qí[jùn]wèi* 騎[郡]尉),
4) 'Cavalry Commandants of a Dependent State' (*qíbāngwèi* 騎邦尉).

The texts further distinguish between two types of Commandery Secretaries:

1) 'Commandery Secretaries' (*zúshǐ* 卒史),
2) 'Commandery Secretaries *ad Interim*' (*jiǎ(jiǎ)zúshǐ* 叚(假)卒史)).

An 'Official in Charge of the Commissioners of Public Works' (*Dòngtíng zhǔ sīkōng* 洞庭主司空) also seems to have served Dòngtíng Commandery. Unlike the respectful address "*sīkōng zhǔ* 司空主" used for the Commissioners of Public Works, this title is prefixed with '*zhǔ* 主'. This incongruence appears exclusively on manuscript 8-1616, hence the title is provisional.

The findings suggest that the official title 'Supervisory Prosecutor' (*jiānyùshǐ* 監御史) on strip 14 of the "Wéi yù děng zhuàng" is divided into 'Prosecutor' (*yùshǐ* 御史) and 'Supervisor' (*wèijiān* 尉監) in the *Lǐyē* corpus. At the very least, it is possible that the Prosecutor was directly appointed by, and accountable to, the central government.[428] According to *Lǐyē* manuscript 8-224+8-412+8-1415, the Pros-

428 For a translation of the "Wéi yù děng zhuàng", see Lau and Staack 2016. The *Lǐyē* manuscripts further categorize 'Prosecutor' (*yùshǐ* 御史) into four types: 1) 'Prosecutor of the Fifth Rank' (*yùshǐ dàifu* 御史大夫) (8-528+8-532+8-674), 2) 'Prosecutor' (*yùshǐ* 御史) (8-153, 8-528+8-532

ecutor was also responsible for reviewing 'maps' (*(dì) tú* (地)圖), usually drawn up by the 'Commandery Secretaries' (*zúshǐ* 卒史). Each commandery appears to have independently followed a set of unified technical standards and commissioned the production of maps for its own territories. Any deviation from the actual terrain or failing to proofread the initial draft made by the Commandery Secretary could have resulted in the punishment of the '(Bureau) Incumbent' or 'Governor' (*shǒu* 守) and every official below them:

8-224+8-412+8-1415 I–II: 其旁郡縣與椄(接)界者毋下二縣, 以□爲審, 即令卒史主者操圖詣┘御史, 御史案雠更并, 定爲輿地圖. 有不雠、非實者, 自守以下主者[429]

If there are not less than two counties with a common border to an adjacent commandery and county, by □ establish firmly. Then order the Commandery Secretary to conduct (the drawing of the) map and pay a visit to┘ the Prosecutor. The Prosecutor investigates, proofreads, further combines and finalizes the map. (If) there is (a Prosecutor) who does not proofread, or (if the map) does not correspond to reality, everyone, beginning with the Governor down to his subordinates, (will be punished).

Comments: The official title '*shǒu* 守' is absent in this case and may stand for either the '(Bureau) Incumbent' or the 'Governor'. Chén Wěi suggests that the term likely refers to the Governor.[430] As the latter part of the manuscript is missing, the consequences of failing to follow procedure cannot be fully reconstructed.

Manuscript 8-543+8-667 seems to deal with the production of maps for the Yǒu 酉 River and Yǒuyáng 酉陽 County region. However, due to significant damage, it lacks detailed information and does not allow any conclusive insights.[431]

Manuscript 8-1429 is heavily damaged and contains many illegible graphs. This makes it difficult to ascertain whether the term '*wèijiān* 尉監' referred to a position at the commandery level, in which case it should be glossed as 'Supervisor', or whether it was an official title at all. Together with the "administrative unit of the Supervisor" (*wèijiān fǔ* 尉監府) on manuscript 8-1032, this is the only occurrence of '*wèijiān* 尉監' in the texts from layers 5, 6 and 8.[432] The character '*jiān* 監'

+8-674, 8-632, 8-1514), 3) 'Prosecutor *ad Interim*' (*jiǎ (jiǎ) yùshǐ* 叚(假)御史) (8-528+8-532+8-674), and 4) 'Deputy Prosecutor' (*yùshǐ chéng* 御史丞) (8-159).

429 Chén Wěi 2012: 118.

430 Chén Wěi 2012: 118

431 For a transcription of manuscript 8-543+8-667, see Chén Wěi 2012: 177.

432 Similar occurrences include the "documents of the Supervisor's administrative unit" (*jiānfǔ shū* 監府書) on manuscript 8-1644, and the "matters of the Supervisor's administrative unit" (*jiānfǔ shì* 監府事) on manuscript 8-1006. The exact meaning of '*fǔ* 府' is unclear in early imperial manuscripts. It might refer to 'government storehouses' engaged in commercial dealings or 'workshops' (*guānfǔ* 官府), also labeled 'detention places' (*xìsuǒ* 繫所), where individuals were likely held in custody to work off government debts (Hulsewé 1985: 244). According to an interis also understood as ‘to supervise’ on manuscripts 8-907+8-923+8-1422, 8-984, 8-992 and 8-1751+8-2207, for example.

> 8-1429 I–II, rI: 卌一□□□□□□□□□□日灌會□□一封尉監□□□一□□┘ □□□. . . . . . . └百[433]
>
> 41 □□□□□□□□□□ day irrigate assemble □□ one letter to the Commandery Supervisor □□□ one □□┘ □□□. . . . . . . └one-hundred

Whenever post required the ‘seal’ (*yìn* 印) of a higher authority to be sent to a government seat or other location, it was occasionally specified which seal was used, along with details of the destination, the date of dispatch and the number of documents included in the message. Manuscript 8-1429 appears to parallel these types of texts, one of which is found on manuscript 8-1155:

> 8-1155: 獄東曹書一封, 丞印, 詣泰守府. 廿八年九月己亥[36]水下四刻, 隸臣申以來.[434]
>
> There is one document from the Judiciary Bureau of the East with the Prefect’s seal. (It is) addressed to the Governor. In the 28th year (of King Zhèng) [219 BC], month 9 on a *jǐhài*, at water below mark 4, (the document) is brought by Bondservant Shēn.

Given the relatively high occurrence and comparable structure of such texts in the *Lǐyē* corpus, paired with the fact that ‘*fǔ* 府’ typically designates higher administrative units at the commandery level, it is plausible that ‘*wèijiān* 尉監’ was an official title at the commandery level.

4.3 ‘Counties’ (*xiàn* 縣)

‘Counties’ (*xiàn* 縣) were territorial subdivisions of ‘commanderies’ (*jùn* 郡) during the late pre-imperial and early imperial dynasties. Modern scholars have long assumed that some counties in present-day Húnán Province were established during the reign of Emperor Gāozǔ 高祖 (r. 202–195 BC), the founder and first Emperor of the Hàn dynasty.[435] However, new information from the *Lǐyē* corpus reveals that Dòngtíng Commandery and Cāngwú Commandery, whose counties covered the areas of modern-day Húběi and Húnán Provinces, as well as the

view with Prof. Dr. Chén Wěi at Wǔhàn University, May 15, 2019, ‘*fǔ* 府’ may have designated a higher administrative unit, potentially at the commandery level.

433 Chén Wěi 2012: 323.

434 Chén Wěi 2012: 285.

435 Zhāng Chūnlóng 2019.

Chóngqìng municipality, were already in existence before that period. Tab. 4.3 shows a list of counties and their respective commanderies recorded in the *Lǐyē* cache.

Tab. 4.3: List of imperial Qín 'commanderies' (*jùn* 郡) and their respective 'counties' (*xiàn* 縣) on the *Lǐyē* manuscripts from layers 5, 6, 8, 10, 12, 14 and 15. The list only includes counties whose commanderies are explicitly mentioned in the texts. Counties known merely from transmitted sources that are not assigned to one of these commanderies are absent (sorted by commandery, county, official(s) in charge, occurrence(s) and manuscript number(s)).

No.	County name	Officials in charge	Occurrence(s): Manuscript number(s)
1		**Bā Commandery (*Bā jùn* 巴郡)**	
1.1	Dàngqú 宕渠[436]	–	**1:** 8-657
1.2	Fúlíng 涪陵	–	**3:** 8-650+8-1462, 8-1094, 8-1206
1.3	Jiāngzhōu 江州	–	**1:** 8-61+8-293+8-2012
1.4	Lángzhōng 閬中	–	**1:** 8-2191
1.5	Qúrěn 朐忍[437]	–	**9:** 8-63, 8-373, 8-445, 8-988, 8-1469, 8-1563, 8-1574+8-1787, 8-1732, 8-1958
1.6	Zhǐ 枳[438]	–	**1:** 8-197
1.7	Zī(Zī)zhōng 粢(資)中	–	**1:** 8-2014
2		**Cāngwú Commandery (*Cāngwú jùn* 蒼梧郡)**	
2.1	Yìyáng 益陽	–	**2:** 8-151, 8-1494
3		**Dài Commandery (*Dài jùn* 代郡)**	
–			

436 See Chén Wěi 2012: 194. Dàngqú 宕渠 was a 'county march' (*xiàn dào* 縣道) under Bā 巴 Commandery. Its government seat was located in the northeast of present-day Qú 渠 County, Sìchuān Province.

437 The government seat of Qúrěn 朐忍 was located west of present-day Yúnyáng 云陽 County, Sìchuān Province (Chén Wěi 2012: 51).

438 This county also appears on strip 453 of the "Èrnián lǜlìng" (Chén Wěi 2012: 109). Zhǐ 枳 was a county in Bā 巴 Commandery, with its government seat in the west of present-day Fúlíng 涪陵 suburban area, Chóngqìng municipality (Chén Wěi 2012: 109). According to manuscript 8-746+8-1588, Zhǐ 枳 was a 'district' (*xiāng* 鄉) (for a transcription of the manuscript text, see Chén Wěi 2012: 215).

Tab. 4.3 (continued)

No.	County name	Officials in charge	Occurrence(s): Manuscript number(s)
4		**Dòngtíng Commandery (*Dòngtíng jùn* 洞庭郡)**	
4.1	Chényáng 辰陽	–	**1:** 8-373
4.2	Chōng 充	–	**7:** 8-242, 8-468, 8-632, 8-903, 8-987, 8-1624, 8-2430
4.3	Línyuán 臨沅[439]	'Prefect' (*chéng zhǔ* 丞主)	**16:** 8-50+8-422, 8-57, 8-66+8-208, 8-151, 8-547, 8-560, 8-695, 8-855, 8-970, 8-1032, 8-1432, 8-1445, 8-1460, 8-1722, 8-1911, 8-2412
4.4	Língyáng 零陽	–	**6:** 5-1, 8-159, 8-375, 8-439+8-519+8-537, 8-1886, 8-2430
4.5	Ménqiǎn 門淺[440]	–	**3:** 8-66+8-208, 8-159, 8-1184
4.6	Péng 蓬	'Prefect' (*chéng* 丞)	**2:** 8-109+8-386, 8-1558
4.7	Qiānlíng 遷陵	'Prefect' (*chéng (zhǔ)* 丞 (主)), 'Deputy Prefect' (*shǒuchéng* 守丞)	**> 290**
4.8	Shàngyǎn 上衍	–	**4:** 8-159, 8-1450, 8-1450, 8-2414
4.9	Shàngzhù 上軴	'Deputy Prefect' (*shǒuchéng* 守丞)	**2:** 8-1219, 15-259[441]
4.10	Tánchéng 鐔成	–	**1:** 8-1373
4.11	Wúyáng 無陽	–	**3:** 5-22, 8-1555, 10-1170[442]
4.12	Xīn Wǔlíng 新武陵	'Prefect' (*chéng* 丞)	**5:** 8-649, 8-657, 8-994, 8-1349, 8-1677

439 Línyuán 臨沅 belonged to Dòngtíng during the Qín dynasty and was renamed to Wǔlíng 武陵 Commandery in the succeeding Hàn dynasty (Chén Wěi 2012: 41).

440 On manuscript 8-159, Ménqiǎn 門淺 appears alongside Suǒ 索, Shàngyǎn 上衍 and Língyáng 零陽, all of which were Qín dynasty counties (Chén Wěi 2012: 52).

441 See Hé Yǒuzǔ 2016b.

442 Transcription taken from Zhāng Chūnlóng 2012: 453–464.

Tab. 4.3 (continued)

No.	County name	Officials in charge	Occurrence(s): Manuscript number(s)
4.13	Yǒuyáng 酉陽	'Prefect' (*chéng (zhǔ)* 丞(主)), 'Deputy Prefect' (*shǒuchéng* 守丞)	**27:** 5-34, 8-50+8-422, 8-65, 8-133, 8-145, 8-158, 8-159, 8-167+8-194+8-472+8-1011, 8-201, 8-226, 8-543+8-667, 8-647, 8-650+8-1462, 8-713, 8-747, 8-1130, 8-1131, 8-1174, 8-1295, 8-1432, 8-1448, 8-1465, 8-1565, 8-1669, 8-1886, 8-2120, 8-2443
4.14	Yuánlíng 沅陵	–	**17:** 6-24, 8-145, 8-186, 8-244, 8-255, 8-265, 8-167+8-194+8-8-472+8-1011, 8-492, 8-647, 8-940, 8-1058, 8-1426, 8-1618, 8-1729, 8-1897, 8-2221, 8-2436
4.15	Yuányáng 沅陽	–	**6:** 8-228, 8-759, 8-830+8-1010, 8-1523, 8-1523, 8-1626
5		**Hándān Commandery (*Hándān jùn* 邯鄲郡)**	
–			
6		**Hénèi Commandery (*Hénèi jùn* 河内郡)**	
6.1	Zhǐ 軹	–	**1:** 14-169[443]
7		**Héngshān Commandery (*Héngshān jùn* 衡山郡)**	
–			
8		**Lángyá Commandery (*Lángyá jùn* 琅邪郡)**	
8.1	Gāolíng 高陵	–	**1:** 8-1533
8.2	Wēi(Wèi)jī 巍(魏)箕	–	**2:** 8-2098, 8-2133
9		**Lújiāng Commandery (*Lújiāng jùn* 廬江郡)**	
–			
10		**Nán Commandery (*Nán jùn* 南郡)**	
10.1	Dāngyáng 當陽	–	**2:** 8-2235, 8-2430
10.2	Jiānglíng 江陵	–	**2:** 8-1328, 8-1444
10.3	Jìnglíng 競陵	–	**1:** 8-896

443 Transcription taken from Yóu Yìfēi 2015.

Tab. 4.3 (continued)

No.	County name	Officials in charge	Occurrence(s): Manuscript number(s)
10.4	Línjŭ 臨沮[444]	'Prefect' (*chéng* 丞)	**1:** 8-140
10.5	Lǐyáng 醴陽	'Prefect' (*chéng* 丞)	**2:** 8-761, 8-2319
10.6	Xiāo 銷	–	**1:** 8-453
10.7	Yān 鄢	–	**1:** 8-807
10.8	Yílíng 夷陵	–	**3:** 8-160, 8-1250, 8-1452
10.9	Zhōulíng 州陵	–	**1:** 14-948[445]
10.10	Zǐguī 秭歸	–	**1:** 8-1516
11		**Sānchuān Commandery (*Sānchuān jùn* 叁川郡)**	
11.1	Luòyáng 雒陽[446]	–	**1:** 8-232
11.2	Píng 平	–	**1:** 8-754+8-1007
11.3	Suŏ 索[447]	–	**3:** 8-1775, 8-1931, 14-638[448]
11.4	Yíyáng 宜陽	'Prefect' (*chéng* 丞)	**1:** 8-1831
12		**Shŭ Commandery (*Shŭ jùn* 蜀郡)**	
12.1	Bódào 僰道	–	**1:** 8-60+8-656+8-665+8-748
12.2	Chéngdū 成都	–	**4:** 6-8, 8-38, 8-961, 8-2276
12.3	Línqióng 臨邛	–	**1:** 12-2301[449]
12.4	Pí 郫	–	**3:** 8-1025, 8-1309, 8-1364

444 Línjŭ 臨沮 was a county with its government seat in the northwest of modern-day Yuǎn'ān 遠安 County, Húběi Province (Chén Wěi 2012: 81).
445 Transcription taken from Yóu Yìfēi 2015.
446 Luòyáng 雒陽 County was renamed to Hénán 河南 Commandery during the reign of Emperor Gāozǔ 高祖 of Hàn (r. 202–195 BC) (Chén Wěi 2012: 119).
447 Suŏ 索 County, which also appears in the "Èrnián lǜlìng" (Yàn Chānggùi 2006), was likely situated to the north of Qiānlíng County (Yóu Yìfēi 2015; Xiè Kūn 2015).
448 Transcription taken from Yóu Yìfēi 2015.
449 Sòng Shǎohuá, et al. 2013: 220.

Tab. 4.3 (continued)

No.	County name	Officials in charge	Occurrence(s): Manuscript number(s)
12.5	Qī 郪	'Prefect' (*chéng* 丞)	**2:** 8-75+8-166+8-485, 8-1023
12.6	Zǐtóng 梓潼[450]	–	**2:** 8-71, 8-1445
13		**Tàishān Commandery (*Tàishān jùn* 泰山郡)**	
–			
14		**Tàiyuán Commandery (*Tàiyuán jùn* 泰原郡)**	
–			
15		**Yànmén Commandery (*Yànmén jùn* 鴈(雁)門郡)**	
15.1	Píngchéng 平成	–	**1:** 8-2040

As shown in Tab. 4.3, the *Lǐyē* texts from layers 5, 6, 8, 12, 14 and 15 explicitly mention 47 counties in 15 commanderies. Unsurprisingly, Dòngtíng Commandery leads with 15 counties, followed by Nán 南 Commandery with ten, Bā 巴 Commandery with seven, Shǔ 蜀 Commandery with six, and Sānchuān 叁川 Commandery with four.

In general, 'Office Holders' or 'Incumbents' from court to county level are designated as '*shǒu* 守'. The title '*chéng* 丞' is almost exclusively reserved for 'Prefects'.[451] Nine counties, including Qiānlíng, appear to have been directed by 'Prefects' (*chéng (zhǔ)* 丞(主)), with an additional three counties overseen by 'Deputy Prefects' (*shǒuchéng* 守丞). The remaining counties are either associated with officials of lower rank or appear on manuscripts with illegible graphs. This does not exclude the possibility that all counties were governed by Prefects and/or Deputy Prefects.

Manuscript 8-158 records a logistical exchange within Dòngtíng Commandery between Qiānlíng and Yǒuyáng County. Even seemingly minor issues arising between Prefects or Deputy Prefects from different counties within the same commandery occasionally required intervention or formal approval from the 'Commandery Governor' (*shǒu* 守).

450 Zǐtóng 梓潼 was a county with its government seat in modern-day Sìchuān Province (Chén Wěi 2012: 54).

451 By contrast, '*shǒu* 守' is almost exclusively reserved for 'Prefects' in the "Wéi yù děng zhuàng". See strips 1, 3, 4, 13, 25, 31, 40, 61 and 68 (Lau and Staack 2016: 86, footnote 486).

8-158 I–II, rI: 丗二年四月丙午[43]朔甲寅[51], 遷陵守丞色敢告酉陽┘ 丞主: 令史下絡帬(裙)直書已到, 敢告主. └四月丙辰[53]旦, 守府快行旁. 欣手.[452]

In the 32nd year (of King Zhèng) [215 BC], month 4, new moon on a *bǐngwǔ*, *jiǎyín* [day 9], Sè, the Deputy Prefect of Qiānlíng, dares to say to the Prefect┘ of Yǒuyáng: The document sent by the Prefectural Secretary to the lower authorities concerning the value of the lower garments made of cotton has already arrived. (He) dares to say to the Prefect. └In month 4 in the morning of a *bǐngchén* [day 11], Governor Kuài dispatches (the document) to adjacent counties. By the hand of Xīn.

'County Courts' (*xiàntíng* 縣廷) served as government seats at the county level for Prefects and Deputy Prefects. Subordinate functionaries operated out of 'Bureaus' (*cáo* 曹) and 'Offices' (*guān* 官). The differentiation, mutual relationships and responsibilities of these two agencies were largely unknown until the discovery of the *Lǐyē* corpus.[453]

'Prefects' (*chéng* 丞) were responsible for administrative and military matters within their counties, although direct control over county military matters was managed by the 'Military Commandants' (*wèi(shǒu)* 尉(守) or *wèi(zhǔ)* 尉(主)).[454] Manuscripts 8-1225 and 8-1616 indicate that some documents addressed to the Governor, issued by the Bureau of the Military Commandant, required the county 'seal' (*yìn* 印) prior to delivery:[455]

8-1225 I–II: 尉曹書二封, 遷陵印, 一封詣洞庭泰(太)守府, 一封詣洞庭尉府.┘ 九月辛丑[38]水下二刻, 走☐以來.[456]

There are two documents from the Bureau of the Military Commandant with the seal of Qiānlíng. One is addressed to the Governor of Dòngtíng, and one is addressed to the Seat of the Military Commandant of Dòngtíng.┘ In month 9 on a *xīnchǒu*, at water below mark 2, (the document) is brought by Runner ☐.

452 Chén Wěi 2012: 95–96.

453 Sūn Wénbó 2014b.

454 This is corroborated by manuscripts 8-1225, 8-1616 and 8-1563. According to an interview with Prof. Dr. Chén Wěi at Wǔhàn University, May 15, 2019, the suffix '*zhǔ* 主' is a respectful address used for higher-ranking officials, such as Prefects or Military Commandants. For additional notes on the 'Bureaus of the Military Commandant' (*wèicáo* 尉曹) appearing on manuscripts 8-71, 8-253, 8-453, 8-1225, 8-1616 and 16-3, see also Chén Wěi 2012: 55.

455 Chén Mèngjiā discusses officials titled 'Commandery Governor' (*jùntàishǒu* 郡太守) and 'Commandant' (*jùnwèi* 郡尉) based on his analysis of the *Zhāngjiāshān* texts from the Early Hàn dynasty (Chén Mèngjiā 1985).

456 Chén Wěi 2012: 295.

8-1616: 尉曹書一封詣洞庭主司空. ノ□☒[457]

There is one document from the Bureau of the Military Commandant addressed to the Official in Charge of the Commissioners of Public Works in Dòngtíng Commandery. / □☒

These manuscripts reveal that the imperial administration employed Military Commandants at both the county and commandery levels. At the county level, the administrative units led by County Military Commandants were known as 'Bureaus of the Military Commandant' (*wèicáo* 尉曹). At the commandery level, these units, directed by the 'Commandery Military Commandants' (*jùnwèi* 郡尉), were referred to as '*wèifǔ* 尉府'. This term was often prefixed with the name of the respective commandery, as exemplified by '*Dòngtíng wèi* 洞庭尉' on manuscript 8-1563. This is further affirmed in line BXXV of manuscript 8-461, in which the "Military Commandant of a Commandery and State" becomes "Military Commandant of a Commandery" (*jùnbāngwèi wéi jùnwèi* 郡邦尉爲郡尉).[458]

Military Commandants also oversaw 'Offices' (*guān* 官) within their designated counties. Manuscript 8-657 is the only instance among the *Lǐyē* manuscripts from layers 5, 6 and 8 that references an 'Office of the Military Commandant' (*wèiguān* 尉官) (see Tab. 4.4). In contrast, 'Bureaus of the Military Commandant' (*wèicáo* 尉曹) appear in five manuscripts, namely on 8-71, 8-253, 8-453, 8-1225 and 8-1616.[459]

The seats of Commandery Governors and Military Commandants did not necessarily need to be in the same location. Prefects and Deputy Prefects typically forwarded decisions on military matters from higher authorities directly to their respective County Military Commandant:

8-657 I–V, rI–III: ☒亥朔辛丑[38], 琅邪叚(假)【守】□敢告內史、 屬邦、郡守主: 琅邪尉徙治即【默】☒┙ 琅邪守四百卌四里, 卒可令縣官有辟, 吏卒衣用及卒有物故當辟徵遝☒┙ 告琅邪尉, 毋告琅邪守. 告琅邪守固留費, 且輒御論吏當坐者. 它如律令. 敢□☒┙ □一書. · 以蒼梧尉印行事. ノ六月乙未[32], 洞庭守禮謂縣嗇夫聽書從事□┙ □軍吏在縣界中者各告之. 新武陵別四道, 以次傳. 別書寫上洞庭 └尉. 皆勿留. ノ葆手. ┙ ノ驕手. ノ八月甲戌[11], 遷陵守丞膻之敢告尉官主: 以律令從事. 傳別【書】┙ 貳春, 下卒長奢官. ノ□手. ノ丙子[13]旦食走印行. ☒┙ [. . .][460]

New moon on a ☒*hài, xīnchǒu*, □, 【Governor】 *ad Interim* of Lángyá dares to report to the Ministry of Finance and the Head Governors of dependent states and commanderies: The Military Commandant of Lángyá (Commandery) moves to govern (military matters) in 【Jímò】

457 Chén Wěi 2012: 369.

458 For a transcription and translation of manuscript 8-461, see chapter 4.1.

459 For a discussion of 'Bureaus' (*cáo* 曹) and 'Offices' (*guān* 官), see chapter 4.4.

460 Chén Wěi 2012: 193.

☒┘ The Governor of Lángyá is 434 miles away. The conscript soldiers could be ordered by the County Offices to be inflicted with punishments. (But as) the clothes of the officials and conscript soldiers are used, the conscript soldiers have things (to do), (strong) evidence for inflicting punishments should be taken into account and ☒┘ Report to the Commandery Military Commandant, do not report to the Governor of Lángyá. Report to the Governor of Lángyá to strongly save expenses, and to immediately hold back judgments on accused officials who should be prosecuted for criminal offenses. Everything else (should be handled) in accordance with the statutes and ordinances. (He) dares □☒┘ □ one document. • Use the seal of the Military Commandant of Cāngwú (Commandery) to follow up on the matter. / In month 6 on a *yǐwèi*, Lǐ, Governor of Dòngtíng, addresses the Overseers of the Counties to obey the documents and follow up on the matter.□┘ □ The army officials who are within the boundaries of the county (should) report it to one another. Xīn Wǔlíng (Commandery) is divided into four marches. Transmit in respective order. In addition, make copies of the document and submit them to the Military Commandant of └Dòngtíng (Commandery). No one should procrastinate. / By the hand of Bǎo.┘ / By the hand of Jiāo. / In month 8 on a *jiǎxū*, Shānzhī, Deputy Prefect of Qiānlíng (County), dares to report to the Head of the Military Commandant's Office: Use the statutes and ordinances to follow up on the matter. Make copies of the documents for┘ Èrchūn (District), and (send them to the) lower Office of Shē, Commander of the Soldiers in his section. / By the hand of □. / On a *bǐngzǐ*, during the morning meal, Runner Yìn delivers (the document). ☒┘ [. . .]

Comments: Jímò 即默 (*tsəkmə̂k) could be a variant of Jímò 即墨 (*tsəkmə̂k), a county during the Hàn dynasty that belonged to Jiāodōng 膠東 Commandery in present-day Shāndōng Province.[461] Zhōu Xiǎolù, Lù Dōngzhī and Xīn Déyǒng argue, based on their interpretations of the "Clay Seals of Qín" (*Qín fēngní* 秦封泥), that during the Warring States period, Jímò 即默 was one of the 'five capitals' (*wǔ dū* 五都) of the Qí 齊 state, each belonging to a different 'county' (*xiàn* 縣).[462] In the 26th year of his reign (221 BC), the First Emperor conquered Qí and transformed it into a 'commandery' (*jùn* 郡). Jímò 即默 does not appear on any other *Lǐyē* manuscript from layers 5, 6 or 8. On this manuscript, only the Military Commandant of Lángyá Commandery moves to Jímò – perhaps temporarily – while the Governor remains in Lángyá. As noted by Chén Wěi, the character '*mò* 默' cannot be deciphered with certainty. Therefore, the arguments provided by Zhōu Xiǎolù, Lù Dōngzhī and Xīn Déyǒng should be viewed with caution.

During this period, 434 'miles' (*lǐ* 里) is equal to approximately 182 kilometers.[463] The term '*bì* 辟' is synonymous with '*zuì* 罪' and means '(to be liable to) punishment' in legal manuscripts from early imperial times.[464] When collocated with the character '*yǒu* 有', the phrase '*yǒu bì* 有辟' means 'to be liable to punishment'. Subjects accused of criminal offenses could be forced to provide 'evidence' (*zhǐ* 徵) or undergo 'forensic inspections' (*zhěn* 診) during the investigation.

461 Chén Wěi 2012: 194. Phonological reconstructions are taken from Schuessler 2009: 111, 113.
462 Xīn Déyǒng 2009: 69; Zhōu Xiǎolù and Lù Dōngzhī 2000: 268.
463 Ten 'miles' (*lǐ* 里) corresponded to ca. 4.2 kilometers in the Qín dynasty (one '*lǐ* 里' to ca. 420 meters) (Lau and Staack 2016: 153).
464 Chén Wěi 2016; Lau and Staack 2016: 89, 322.

> '*Bié shū* 別書' was a method of transmitting written information and means 'to make a copy of a document'. On strip 8 of the 'Declaration Document' (*Yǔ shū* 語書) in the *Shuìhǔdì* texts, '*bié shū* 別書' is used in a similar context and also means 'to make a copy of a document'.[465]

Governors and Commandery Military Commandants operated at a distance from the central authority, under conditions that did not allow for immediate consultations during emergencies.[466] The decision-making process detailed on manuscript 8-657 spanned at least two months – from 'month 6' (*liù yuè* 六月) to 'month 8' (*bā yuè* 八月) – and could have taken even longer, as the initial section of the text is irreparably lost.

The geographical separation between Governors and Commandery Military Commandants, sometimes extending hundreds of kilometers, significantly lengthened the chain of command and increased the complexity of decision-making. This spatial distance often forced officials to wait for approval from higher authorities, risking long delays or necessitating independent actions, which could potentially lead to legal repercussions. Interestingly, an order is given on manuscript 8-657 that a particular issue should be reported to the Commandery Military Commandant rather than the Governor of Lángyá (*gào Lángyá wèi, wú gào Lángyá shǒu* 告琅邪尉, 毋告琅邪守). This deviation could be due to the absence of the Commandery Governor in the region or a strategic decision to circumvent legal liability. The Governor is instructed "to withhold judgments on accused officials who should be prosecuted for criminal offenses" (*zhé què lùn lì dāng zuò zhě* 輒卻論吏當坐者).[467]

The model of the Qín dynasty as a fully centralized regime emerging after centuries of disunion may not adequately capture the full range of paleographic evidence found in the *Lǐyē* corpus. On the contrary, the imperial administration seems to have granted more autonomy to local and regional governments than previously assumed.[468] Edgar Kiser and Cài Yǒng have elaborated on decentralized, collective rulership in Early China, arguing that such deliberate constraints on bureaucratization may have led to more efficient forms of administration. This perspective challenges the traditional view of the Qín dynasty being a highly

---

465 Zhī Qiáng 2012.

466 Loewe 2006: 88–89.

467 This observation partially aligns with excerpts from the "Zòuyànshū", which state that 'Military Commandants' ((*dū*)*wèi* (都)尉) were in charge of military matters during the absence of the Commandery Governor. The term '*dūwèi* 都尉' appears on strips 1 and 129 of the "Zòuyànshū" (Lau and Staack 2016: 176–177). See also Ān Zuòzhāng and Xióng Tiějī 2007: 574–581; Bielenstein 1980: 94; Lǐ Yùfú 2002: 322–324; Loewe 1967: 60–61, 74, 76.

468 For a discussion on the unification of the empire based on readings of early historical and philosophical texts, see Pines 2000.

centralized and rigidly controlled state, suggesting a more nuanced understanding of its administrative practices and the distribution of power.[469]

Decentralized and collective approaches were also adopted in the creation of 'maps' (*tú* 圖), which were carefully studied and utilized by Military Commandants for logistical and tactical purposes.[470] This is illustrated in the chapter "On Maps" (*Dìtú* 地圖) from the *Writings of Master Guǎn* (*Guǎnzǐ* 管子) and is further supported by military maps discovered in Western Hàn dynasty tombs at Mǎwángduī 馬王堆, located in modern-day Chángshā, Húnán Province (Fig. 4.3).[471]

> *Writings of Master Guǎn* (*Guǎnzǐ* 管子), "On Maps" (*Dìtú* 地圖): 凡兵主者必先审審知地圖轘轅之險. 濫車之水名山通谷經川陸丘阜之所在, 苴草林木蒲葦之所茂道里之遠近, 城郭之大小, 名邑廢邑困殖之地必盡知之. 地形之出入相錯者, 盡藏之.[472]
>
> All military commanders must first examine and come to know maps. They must know thoroughly the location of winding, gatelike, streams that may inundate their chariots, famous mountains, passable valleys, rivers, highlands and hills. They must also know where grasses, trees and rushes grow, the distances of roads, the size of the city and suburban walls, famous and deserted settlements and barren and fertile land. They should thoroughly store up (in their minds) the location of ways in and out of the terrain.[473]

Fragmentary ordinances reconstructed from manuscripts 8-224+8-412+8-1415 and 8-543+8-667 reveal that county officials were commissioned by higher authorities to record topographic, military and demographic information concerning regions under their control.[474] Manuscript 8-224+8-412+8-1415 substantiates the independent production of maps at the county and commandery level in accordance with official standards.[475] Such intelligence constituted a powerful tool, granting significant advantages that could be weaponized in the hands of rivals. Sīmǎ Qiān – the most prominent historian of the Western Hàn dynasty – credits Liú Bāng's 劉邦 conquest of the Qín and his ascension to the throne as the First Emperor of Hàn to his advisor Xiāo Hé 蕭何 (died 193 BC), who salvaged the Qín 'maps and documents' (*tú shū* 圖書) from the imperial palace before it was set ablaze.

---

469 See Kiser and Cài Yǒng 2003.

470 On the etymological development of '*tú* 圖' in early Chinese texts, see Behr 2007.

471 The *Writings of Master Guǎn* (*Guǎnzǐ* 管子) is a politico-philosophical text that received considerable attention during the Western Hàn dynasty. A concise introduction to the text *Writings of Master Guǎn* is provided by Rickett 1985: 3–7. For more information on early Chinese maps and cartography, see Cáo Wǎnrú 1983; Gutkind Bulling 1978; Hsu Mei-Ling 1978, 1993.

472 Zhào Shǒuzhèng 1987: 269.

473 Translation is taken from Rickett 1985: 389–390, with minor modifications.

474 For a transcription and translation of manuscript 8-224+8-412+8-1415, see chapter 4.2.

475 Chén Wěi 2012: 101. Strips 328–331 of the "Èrnián lǜlìng" also address cartography, but pertain to a later period of the Hàn dynasty.

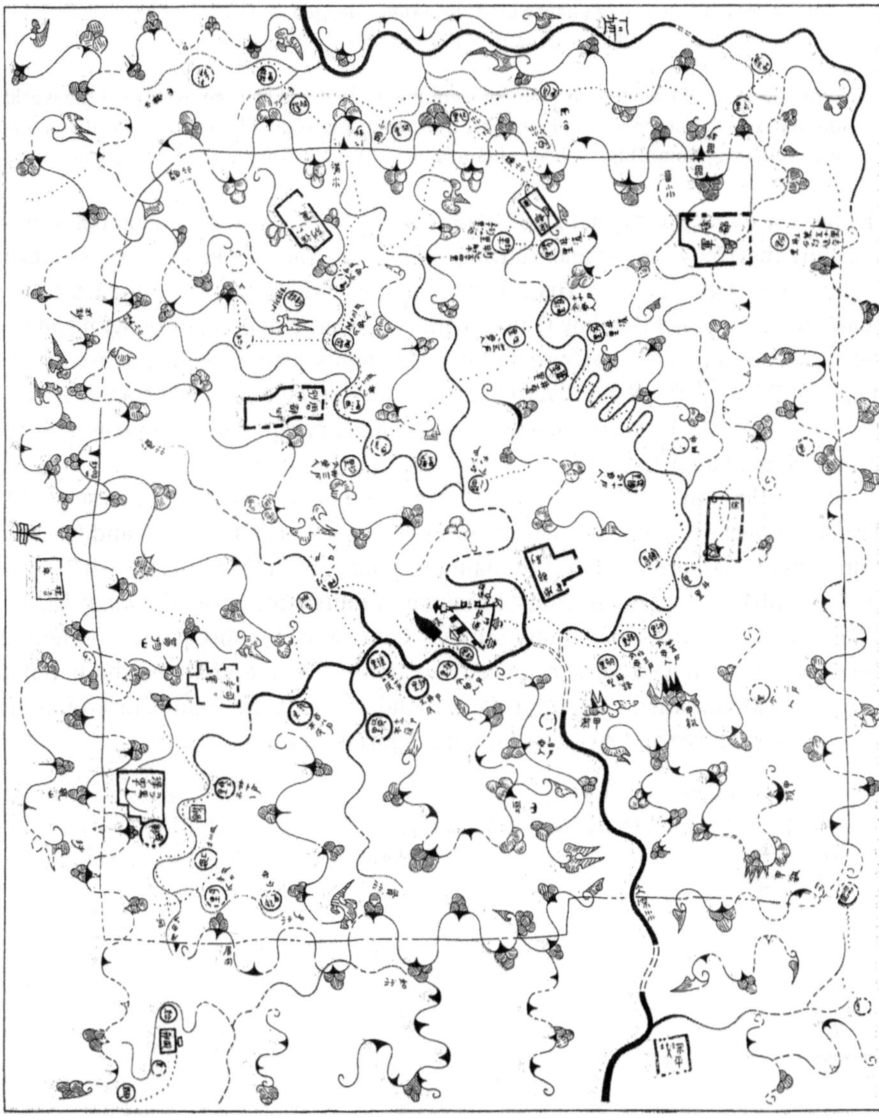

Fig. 4.3: Example of a reconstructed Western Hàn dynasty military map on silk, found in Tomb no. 3 at Mǎwángduī 馬王堆, modern-day Chángshā, Húnán Province. The original map is dated 168 BC. Its dimensions are 28 x 36 cm. The top of the map is facing south.[476] It portrays places of inhabitants,

476 Hsu Mei-Ling 1978: 48.

Shǐjì, “Hereditary House of Chancellor Xiāo” (*Xiāo xiāngguó shìjiā* 蕭相國世家): 漢王所以具知天下阸塞, 戶口多少, 彊弱之處, 民所疾苦者, 以何具得秦圖書也.[477]

What the King of Hàn needed to know were the narrow and blocked (passes) of the world, the number of households and persons, the strong and weak places and what the people were suffering from. This has been obtained by Hé from the Qín maps and documents.

The heavily damaged fragment 8-543+8-667 seems to detail the production of a map depicting the Yǒu 酉 River and Yǒuyáng 酉阳 area. The Yǒu River was a strategic water route flowing into the Yuán 沅 River and eventually into Dòngtíng Lake. The Yuán River connected Qiānlíng County with the government seat of Dòngtíng and is the earliest documented water route linking a county with a commandery in Early Imperial China.[478] The Yuán River enabled faster and more efficient transportation of people, essential resources, documents and military equipment compared to the arduous mountain paths of Qiānlíng.

Beyond military operations, Military Commandants collaborated closely with ‘Prefects’ (*chéng* 丞), ‘Commissioners of Public Works’ (*sīkōng* 司空) and ‘Section Commanders’ (*xiàozhǎng* 校長) to maintain public order. Manuscript 8-167+8-194+8-472+8-1011 records an incident in which bandits stole a ‘boat’ (*chuán* 船) intended for public use. The County Military Commandant subsequently requested the Prefect to initiate a mission to retrieve the boat and apprehend the bandits. The case was classified as ‘urgent’ (*jí* 急), underscoring the significance of boats as a scarce resource and rivers as vital transportation routes.

8-167+8-194+8-472+8-1011 I–IV, rI: 尉敬敢再捧(拜)謁丞公: 校長寬以遷陵船徙卒史┘【酉陽, 酉陽】□□【船】□元(沅)陵, 寬以船屬酉陽校長徐. 今司空┘ □□□□□□丞公令吏徒往取之, 及以書告酉陽令┘ 来歸之. 盜賊事急, 敬已遣寬與校長囚吾追求盜 └發田官不得者. 敢再捧(拜)謁之.[479]

Fig. 4.3 (continued)
military operations, transportation lines, roads and other topographic features. An extensive network of roads and transportation lines facilitated the movement of armies and the exchange of information.[480] The degree of accuracy is greater in the central regions than in the periphery. The original map is polychrome.

477 Zhāng Dàkě 1990: 1234.

478 *Húnán shěng wénwù kǎogǔ yánjiūsuǒ* 2007: 235.

479 Chén Wěi 2012: 101.

480 Wáng Zǐjīn 1994: 28–32.

Jìng, (County) Military Commandant, dares to confer another request on Prefect Gōng: Kuān, Section Commander, uses the boats of Qiānlíng to move Commandery Secretaries┘【to Yǒuyáng, Yǒuyáng】□□【boat】□ Yuánlíng (County). Kuān provides a boat to Xú, Section Commander in Yǒuyáng (County). Now, the Commissioner of Public Works┘ □□□□□□ Prefect Gōng orders officials and convicts to go and collect them [the boats], to report to Yǒuyáng in accordance with the documents and order to┘ return them. The matter of the robbers and bandits is urgent. Jìng has already dispatched Kuān and Section Commander Qiúwú to pursue and search for the bandits. └What has been sent to the Office of Agricultural Fields is not what we have received. (He) dares to confer upon him another request.

Comments: '*Zài bài* 再拜' is a formulaic expression of respect or courtesy, also found in the *Shuìhǔdì* documents, typically at the beginning or end of letters.[481] Ulrich Lau and Thies Staack translate '*lì tú* 吏徒' as 'officers and conscripts' and '*dào zéi* 盜賊' as 'robbers and bandits'.[482]

'*zhuī* 追' was used in early legal texts, when an official 'pursued a suspect' based on evidence found at the crime scene or information regarding the suspect's identity, appearance, whereabouts and alleged crime.[483] Convicts were organized into 'Units of Five People' (*wǔ* 伍) when employed 'to pursue or search for bandits' (*zhuī qiú dào* 追求盜).[484] An excerpt from the "Èrnián lǜlìng" states:

> *Zhāngjiāshān*, "Èrnián lǜlìng", strip 141: 吏將徒, 追求盜賊, 必伍之, 盜賊以短兵殺傷其將及伍人, 而弗能捕得, 皆戍邊二歲.[485]
>
> Officials leading conscripts to pursue and search for bandits must organise them into Units of Five. When they (the units) are not able to arrest the bandits, although they killed or injured their Commander or a member of their unit with short-range weapons, they are all to perform two years of (additional) military service at the frontier.[486]
>
> Comments: In stipulations from the "Èrnián lǜlìng", '*shù* 戍' refers to 'military service' lasting one to four years, often performed in border regions. '*Shù* 戍' was a form of "punishment for soldiers drafted to combat banditry for failing to arrest bandits who killed or injured their commander or other members of their military unit; for not daring to attack out of cowardice; for not detecting a bandit raid; or for failing to discover contraband gold while on border guard."[487]

Manuscript 8-140 describes a situation in which a member of a 'Unit of Five People' (*shìwǔ* 士五 or *wǔ* 伍) fails to return to Qiānlíng after being dispatched to a

481 Chén Wěi 2012: 101.
482 Lau and Staack 2016: 103, 317.
483 Lau and Staack 2016: 91–92, 317, 322.
484 Lau and Staack 2016: 124.
485 Lau and Staack 2016: 124.
486 Translation is taken from Lau and Staack 2016: 124.
487 Lau and Staack 2016: 101, footnote 542. These findings are based on strips 76, 96–97, 141, 143–144 and 226 of the "Èrnián lǜlìng".

'military outpost' (*shǔ* 署).[488] The County Military Commandant *ad Interim* reports the missing individual and requests a thorough investigation.

8-140 I–V, rI: ☒朔甲午[31], 尉守傰敢言之: 遷陵丞昌曰: 屯戍士五(伍)桑唐趙歸┛ ☒日已, 以迺十一月戊寅[15]遣之署. 遷陵曰: 趙不到, 具爲報・問: 審以卅┛ ☒【署】, 不智(知)趙不到故, 謁告遷陵以從事. 敢言之.ノ六月甲午[31]┛ 臨沮丞禿敢告遷陵丞主、令史, 可以律令從事. 敢告主.ノ胥手.┛ 九月庚戌[47]朔丁卯[4], 遷陵丞昌告尉主, 以律令從事.ノ氣手.ノ九月戊辰[5]旦, 守府快行. ┗☒俉手.[489]

☒ new moon on a *jiǎwǔ*, Péng, County Military Commandant *ad Interim*, dares to report: Chāng, Prefect of Qiānlíng, says: Zhào, member of a Unit of Five People, at the military outpost in Sāngtáng, returns┛ ☒ day ends. Previously, in month 11 on a *wùyín*, (he) was dispatched to a military outpost. Qiānlíng says: Zhào has not arrived. Prepare a response. • Asks: Establish the facts by 30┛ ☒【military outpost】, do not know the reason why Zhào has not arrived. Request to report to Qiānlíng to follow up on the matter. (He) dares to say. / In month 6 on a *jiǎwǔ*,┛ Tū, Prefect of Línjǔ, dares to report to the Prefect of Qiānlíng and the Prefectural Secretary to approve and use the statutes and ordinances to follow up on the matter. (He) dares to report to the Prefect. / By the hand of Xū.┛ In month 9, new moon on a *gēngxū*, *dīngmǎo*, Chāng, Prefect of Qiānlíng, reports to the (County) Military Commandant to approve the statutes and ordinances to follow up on the matter. / By the hand of Qì. / In month 9 in the morning of a *wùchén*, Governor Kuài sends (the document). ┗☒ by the hand of Wú.

Comments: Lǐ Xuéqín believes that '*shǔ* 署' stands for a 'defense sector' or 'station of a unit' (*fángdì* 防地).[490] Wolfgang Behr argues that, since early imperial times, '*shǔ* 署' was used as an "exoactive-directive verb" meaning 'to place *somewhere*' or 'to put *into* a position', and it "also has a nominal usage as 'post', common in bureaucratic contexts as 'government office'."[491] In the *Shuìhǔdì* texts, the term still designates 'military outposts' and 'emplacements' located in border regions. Zhāng Jùnmín advocates a more generic reading of '*shǔ* 署' as 'workplace' (*gōngzuò chǎngsuǒ* 工作場所).[492] The term '*shěn* 審' means 'to establish the facts (of the matter)' or 'to be firmly established'.[493]

The imperial government invested significant effort into controlling food and labor, redistributing people and livestock and loaning grain to workers cultivating farmland. To date, very little is known about these practices prior to the Hàn dynasty. According to manuscript 8-482, Military Commandants were required to submit three types of 'evaluation records' (*kè zhì* 課志).[494] These included:

488 For more on the title and social status of members of a 'Unit of Five People' (*shìwǔ* 士五 or *wǔ* 伍) or 'Commoners without Rank', as Robin Yates terms them, see Yates 1987: 201–203.

489 Chén Wěi 2012: 80.

490 Lǐ Xuéqín 2003.

491 Behr 2007: 125.

492 Zhāng Jùnmín 2003.

493 Lau and Staack 2016: 319.

494 See Yates 1990.

1) "Evaluation of conscript soldiers who died" (*zú sǐ wáng kè* 卒死亡課),
2) "Evaluation of Overseers of Bandits working the agricultural fields" (*sīkòu tián kè* 司寇田課),
3) "Evaluation of conscript soldiers working the agricultural fields" (*zú tián kè* 卒田課).

Unfortunately, the text does not specify which officials or government units received these records, although manuscript 8-132+8-334 seems to elaborate on the "evaluation of conscript soldiers who died" (*zú sǐ wáng kè* 卒死亡課).

8-482 AI–III, BI–II: A【尉】課志:┘ 卒死亡課,┘ 司寇田課,┘ B卒田課.┘ ·凡三課.[495]

A Evaluation records of the 【Military Commandant】:┘ Evaluation of conscript soldiers who died,┘ evaluation of the Overseers of Bandits working the agricultural fields,┘ B evaluation of conscript soldiers working the agricultural fields.┘ · In total, there are three evaluations.

Comment: This text shows that 'Overseers of Bandits' (*sīkòu* 司寇) and 'conscript soldiers' (*zú* 卒) were also employed in the cultivation of farmlands.

8-132+8-334 AI–III, BI–II: A ☒宂募群戍卒百卌三人.┘ ☒廿六人. · 死一人.┘ ☒六百廿六人而死者一人.┘ B 尉守抓課.┘ 十一月己酉[46]視事, 盡十二月辛未[8].[496]

A ☒ surplus recruitment of various conscript soldiers who were sentenced to perform military service, 143 people.┘ ☒ 26 people. · One person died.┘ ☒ 626 people, and one person died.┘ B Evaluated by Military Commandant Zhuā.┘ In month 11 on a *jǐyǒu*, (he) oversees the matter until month 12 on a *xīnwèi*.

Ordinances on tax revenues in the form of harvest yields were enforced by the Military Commandants, and any deviations from the norm or unaddressed issues were reported to higher authorities. Manuscript 8-67+8-652, dating back to the 26th year of King Zhèng (221 BC), specifies the exact time of year when officials were expected to report the annual harvest to the 'administrative unit' (*fǔ* 府) of the Governor:

8-67+8-652 I–III: 廿六年十二月癸丑[50]朔辛巳[18], 尉守蜀敢告之: 大(太)守令曰: 秦人□□□┘ 侯中秦吏自捕取, 歲上物數會九月朢(望)大(太)守府, 毋有亦言.┘ 問之尉, 毋當令者. 敢告之. [. . .][497]

In the 26th year of King Zhèng, month 12, new moon on a *guǐchǒu*, *xīnsì* [day 29], Military Commandant Shǔ dares to report: The ordinance of the Governor states: The people of Qín

495 Chén Wěi 2012: 165.
496 Chén Wěi 2012: 70.
497 Chén Wěi 2012: 52.

□□□┘ government officials (in the dependent) states catch and seize on their own. Annually send the things that are counted and collected in month 9 on day 15 to the administrative unit of the Governor. Even if there is nothing, do also report.┘ Clarify with the Military Commandant. There is no corresponding ordinance. (He) dares to report. [. . .]

Comments: The phrase *'jiǔ yuè wàng* 九月望' designates day 15 of month 9.[498] Chén Wěi interprets *'yán* 言' as 'to report', which aligns with Lau and Staack's reading of this character in certain early imperial texts.[499]

In a nutshell, the rank of 'Military Commandant' (*wèi(shǒu)* 尉(守)) should be understood as representing an entire division of Qín officials responsible for handling military matters and other tasks across different levels of administration. Tab. 4.4 illustrates that the documents from layer 8 mention two titles related to the 'County Military Commandant'. The character *'wèi* 尉' is not found in materials from layers 5 and 6.

Tab. 4.4: Titles related to the 'County Military Commandant' (*wèi(shǒu)* 尉(守)) appearing on the *Lǐyē* manuscripts from layer 8.

No.	Title	Responsibilities	Occurrence: Manuscript number(s)
1	'County Military Commandant' (*wèi (shǒu)* 尉(守))	– Communicate with 'Prefects' (*chéng* 丞) to ensure they follow the statutes and ordinances – Receive reports relating to military matters from 'Prefects' (*chéng* 丞) – Hand in 'evaluation records' (*kè zhì* 課志) of conscript soldiers who died, and of Overseers of Bandits and conscript soldiers working the 'agricultural fields' (*tián* 田)	*'wèizhǔ* 尉主' **6:** 8-69, 8-140, 8-201, 8-528+8-532+8-674, 8-1193, 8-2260 *'wèishǒu* 尉守' **7:** 8-67+8-652, 8-85, 8-132+8-334, 8-140, 8-671+8-721+8-2163 (2x), 8-1563
2	'Secretary to the County Military Commandant' (*wèishǐ* 尉史)	– Facilitate communication with the '(County) Court' (*tíng* 廷)	**8:** 8-356, 8-671+8-721+8-2163, 8-717, 8-761, 8-1008+8-1461+8-1532, 8-1128, 8-1364, 8-1562

498 Chén Wěi 2012: 53.
499 Chén Wěi 2012: 53; Lau and Staack 2016: 19–20, 75–77.

As the title suggests, 'Commissioners of Public Works' (*sīkōng* 司空) were responsible for directing public labor and coordinating human resources, as seen, for example, in the assignment of 'conscript soldiers' (*zú* 卒) or 'convict (servants)' (*tú(lì)* 徒(隸)) to specific tasks, officials or institutions. Positioned below the Prefects and Deputy Prefects, Commissioners of Public Works coordinated matters between the 'Courts' (*tíng* 廷), 'Bureaus' (*cáo* 曹) and 'Offices' (*guān* 官) of a county and frequently acted as intermediaries among these agencies. The incomplete list of duties on manuscript 8-486 paints a somewhat vague picture of the 'evaluations' (*kè* 課) conducted by these officials. These evaluations included detailed accounts of 'Grain Pounders' (*chōng* 舂) giving birth, public 'boats' (*chuán* 船) used for transporting people, material resources, postal documents and labor assignments, among other related activities.

> 8-486 AI-V, BI-V: A 司空課志:┘ □爲□□□┘ □課,┘ □□□□課,┘ 舂產子課,┘ B □船課, ┘ □□□課,┘ 作務□□┘ . . . ┘ . . .[500]
>
> A Evaluation records by the Commissioner of Public Works:┘ □ become □□□┘ □evaluation,┘ □□□□ evaluation┘ evaluation of Grain Pounders giving birth,┘ B □ evaluation of boats,┘ □□□ evaluation,┘ conducting labor assignments □□┘ . . . ┘ . . .

Manuscript 8-145 could represent a type of ledger or duty roster complementing the evaluations mentioned above. It details the duties of various convicts, commoners and officials, along with their personal names, health conditions and assignments to imperial institutions. The list is partially sorted by rank, age and sex and concludes with an account of '(male) Minor Wall Pounders' (*xiǎo chéngdàn* 小城旦) and '(female) Minor Grain Pounders' (*xiǎo chōng* 小舂). The 'minor' (*xiǎo* 小) status of Grain Pounders was determined not by the precise age but by body height, which was not to exceed six 'feet' (*chǐ* 尺) or 1.39 meters.[501]

> 8-145 A, BI–VII, CI–XI, DI–XII, EI–IX, FI–IXV, rI: A . . . ┘ B . . . 圂、叚、卻.┘ 七人市工用.┘ 八人與吏上計.┘ 一人爲舄: 劇┘ 九人上省.┘ 二人病: 復、卯.┘ 一人□徒酉陽.┘ C □□□人.┘ □□十三人.┘ 隸妻擊(繫)舂八人.┘ 隸妻居貲十一人.┘ 受倉隸妻七人.┘ • 凡八十七人.┘ 其二人付畜官.┘ 四人付貳舂.┘ 廿四人付田官.┘ 二人除道沅陵.┘ 四人徒養: 枼、痤、帶、復.┘ D 二人取芒: 阮、道.┘ 一人守船: 遏.┘ 三人司寇: 豨、猳、款.┘ 二人付都鄉 .┘ 三人付尉.┘ 一人付□.┘ 二人付少內.┘ 七人取□: 綰、林、嬈、粲、鮮、夜、喪.┘ 六人捕羽: 刻、婢、□、□、娃、變.┘ 二人付啓陵.┘ 三人付倉.┘ 二人付庫.┘ E 二人傳徙酉陽 .┘ 一人爲笥: 齊.┘ 一人爲席: 姱.┘ 三人治枲: 梜、茲、緣.┘ 五人墼: 婢、般、橐、南、儋.┘ 二人上眚(省).┘ 一人作廟.┘ 一人作務: 青.┘ 一人作園: 夕.┘ F • 小城旦九人:┘ 其一人付少 內.┘ 六

500 Chén Wěi 2012: 165–166.
501 Lau 1999: 45.

人付田官.┙ 一人捕羽: 强.┙ 一人與吏上計.┙ ·小舂五人.┙ 其三人付田官.┙ 一人徒養: 姊.┙ 一人病: □. └☑□圂敢言之, 寫上, 敢言之.ノ 痤手.[502]

A . . . ┙ B . . . Hùn, Jiă, Què.┙ Seven people trade with work utensils.┙ Eight people assist officials in submitting the statistics to the higher authorities.┙ One person makes wooden sandals: Jù.┙ Nine people are promoted to do autumn hunting.┙ Two people are ill: Fù, Măo.┙ One person □ convict in Yǒuyáng (County).┙ C □□□ people.┙ □□ 13 people.┙ Eight (female) Bondservants who are placed under detention as Grain Pounders.┙ 11 (female) Bondservants redeem their fines.┙ Seven (female) Bondservants received from the Granary.┙ · In total, there are 87 people.┙ Among these, two people are handed over to the Office of Domestic Animals.┙ Four people are handed over to Èrchūn (District).┙ 24 people are handed over to the Office of Agricultural Fields.┙ Two people clear the road in Yuánlíng (County).┙ Four people (work as) Provisioners for convicts: Yè, Cuó, Dài, Fù.┙ D Two people collect awns: Ruăn, Dào.┙ One person guards the boats: È.┙ Three people are Firewood Gatherers: Huò, Xù, Kuăn.┙ Two people are handed over to the Dū (District).┙ Three people are handed over to the (County) Military Commandant.┙ One person is handed over to □.┙ Two people are handed over to the Lesser Treasury.┙ Seven people collect □: X [pronunciation unknown], Lín, Ráo, Càn, Xiān, Yè, Sàng.┙ Six people catch bird feathers: Kè, Bì, □, □, Wá, Biàn.┙ Two people are handed over to Qīlíng (District).┙ Three people are handed over to the Granary.┙ Two people are handed over to the Warehouse.┙ E Two people transfer the convicts to Yǒuyáng (County).┙ One person makes bamboo baskets: Qí.┙ One person makes mats: Kuā.┙ Three people administer the (storage/distribution) of hemp-nettle: Jiá, Zī, Yuán.┙ Five people (make) unfired bricks: Bì, Bān, Tuó, Nán, Dān.┙ Two people are promoted to do autumn hunting.┙ One person constructs an ancestral temple.┙ One person conducts labor assignments: Qīng.┙ One person works in the gardens: Xī.┙ F · There are nine Minor Wall Pounders.┙ Among these, one person is handed over to the Lesser Treasury.┙ Six people are handed over to the Office of Agricultural Fields.┙ One person catches birds: Qiáng.┙ One person assists officials in submitting the statistics to the higher authorities.┙ · There are five Minor Grain Pounders.┙ Among these, three people are handed over to the Office of Agricultural Fields.┙ One person (works as) a cook for convicts: Zĭ.┙ One person is ill: □. └☑□ Hùn dares to say. Write to the higher authorities. (He) dares to say. / By the hand of Cuó.

Comments: A more recent study by a group at Wŭhàn University suggests combining this text with fragment 9-2294, which would date the entry to month 10 of the 32nd year of King Zhèng (215 BC): "In the 32nd year (of King Zhèng) (215 BC), month 10, new moon on a *jĭyǒu, yĭhài* [day 27]" (*shìèr nián shí yuè jĭyǒu shuò yĭhài* 卅二年十月己酉[46]朔乙亥[12]).[503]

In the transcription proposed by Chén Wĕi, the names 'Fù 復' and 'Măo 卯' in line BVI are not separated from one another.[504]

502 Chén Wĕi 2012: 85–86.

503 For more information on the revised reading of this manuscript text, see *Wŭhàn dàxué jiănbó yánjiū zhōngxīn* 2014.

504 Chén Wĕi 2012: 84.

'*Xì* 繫' is usually used as a verb and understood as 'to place (a suspect) under detention' in Early Qín and Hàn dynasty legal texts.[505] A statute on strip 185 of the *Yuèlù Statutes and Ordinances* (*Yuèlù lǜlìng* 嶽麓律令), detailing the consequences of absconding, reads: "(Absconding for) less than 12 months constitutes 'vagrancy' and is to be punished with detention among the Wall and Grain Pounders" (*bù yíng shíèr yuè wéi jiāng yáng, jī (xì) chéngdàn chōng* 不盈十二月爲將陽, 毄(繫)城旦舂).[506] Zhāng Jīnguāng understands '*xì* 繫' as a 'large rope' that similarly means 'to take into custody' or 'to tie up'.[507]

"Convicts who work off fines, redemption fees or debts" are referred to as "*jū zī shú zé (zhài) zhě* 居貲贖責(債)者" in the "Qín lǜ shíbā zhǒng".[508]

In this context, '*fù* 付' (to hand over) is an antonym of '*shòu* 受' (to receive) and conveys a similar meaning in Early Zhōu bronze inscriptions.[509]

Whereas certain convicts or individuals engaged in trade remain unnamed in this text, others performing manual work are explicitly named. For example, Jù 劇 is noted for making 'wooden sandals' (*xì* 舄) and others are engaged to 'catch bird feathers' (*bǔ yǔ* 捕羽) or make 'unfired bricks' (*jī* 墼), 'mats' (*xí* 席) or 'bamboo baskets' (*sì* 笥). The personal names of 'Provisioners' (*yǎng* 養), individuals working in the 'gardens' (*yuán* 園), and those who had fallen 'ill' (*bìng* 病) are also recorded. All these individuals were under the direct supervision of the Office of the Commissioner of Public Works. There does not seem to be a specific order or hierarchical sequence to the duties, titles or number of individuals employed in these tasks. This lack of order is similarly reflected in manuscript 8-663, which, like manuscript 8-145, appears to be a ledger or duty roster:

8-663 AI-VI, BI–VI, rI: A 二人付□□□.┘ 一人付田官.┘ 一人付司空: 枚.┘ 一人作務: 臣.┘ 一人求白翰羽: 章.┘ 一人廷守府: 快.┘ B 其廿六人付田官.┘ 一人守園: 壹孫.┘ 二人司寇守: 囚、嫭.┘ 二人付庫: 恬、擾.┘ 二人市工用: 餔、亥.┘ 二人付尉□□. ☒ └五月甲寅[51]倉是敢言之: 寫上. 敢言之. ☒[510]

A Two people are handed over to □□□.┘ One person is handed over to the Office of Agricultural Fields.┘ One person is handed over to the Commissioner of Public Works: Méi. ┘

505 Lau and Staack 2016: 320.

506 Lau and Staack 2016: 121.

507 Zhāng Jīnguāng 2004. Xú Shìhóng, relying on his analysis of the "Èrnián lǜlìng", differentiates between "tying up Wall and Grain Pounders" (*jì chéngdàn chōng* 繫城旦舂), "punishing Wall and Grain Pounders through mutilation" (*xíng wéi chéngdàn chōng* 刑爲城旦舂) and "punishing Wall and Grain Pounders without mutilation" (*wán wéi chéngdàn chōng* 完爲城旦舂) (Xú Shìhóng 2004). For more information on 'Wall and Grain Pounders' (*chéngdàn chōng* 城旦舂), based on Qín dynasty legal texts, see Zhāng Xīnchāo 2014.

508 Lau and Staack 2016: 288–289, footnote 1348. The term '*shú* 贖' is read as 'to pay a fee for redemption (from a certain punishment)' (Lau and Staack 2016: 319).

509 Lau and Staack 2016: 98.

510 Chén Wěi 2012: 196.

One person conducts labor assignments: Chén.┘ One person looks for white bird feathers that can be used as a pen: Zhāng.┘ One person is Governor at the Court: Kuài.┘ B Among these, 26 people are handed over to the Office of Agricultural Fields.┘ One person is in charge of the gardens: Yīsūn. ┘ Two people are Overseers of Bandits: Qiú, Hù. ┘ Two people are handed over to the Warehouse: Tián, Rǎo.┘ Two people trade with work utensils: X [pronunciation unknown], Hài.┘ Two people are handed over to the Military Commandant □□. ☒ └ In month 5 on a *jiǎyín*, Shì from the Granary dares to say: Write to the higher authorities. (He) dares to say. ☒

Comments: When the title of the Granary official is omitted, as in line rI of this manuscript, it is generally assumed that the individual referred to is the 'Granary Overseer' (*cāng sèfū* 倉嗇夫).[511] This interpretation is problematic. On manuscript 8-136+8-144, for example, '*cāng sèfū* 倉嗇夫' seems to refer instead to the 'Granary Incumbent' (*cāng shǒu* 倉守).[512] According to strip 161 of the "Statutes for the Installation of Officials" (*Zhì lì lǜ* 置吏律) from the *Shuìhǔdì* documents, substituting the 'Granary Overseer' (*cāng sèfū* 倉嗇夫) or 'Prefectural Secretary' (*lìngshǐ* 令史) with an official of a similar rank appears to have been common practice.[513]

The mention of artisans by name in connection with the manufacturing of goods is not unprecedented. Excavated objects from the Qín dynasty often bear short inscriptions or stamped seals that served to authenticate the quality of the work. These seals, along with the increased standardization of such items, also reinforced awareness of the empire and notified that the objects were public property.[514] A large portion of clay tiles and more than half of the restored terracotta warriors from the First Emperor's necropolis bear the signatures of potters or numerical symbols representing different stages of production.[515] None of these signatures belongs to soldiers, officials or generals, highlighting that only artisans or individuals working off fines, debts, redemption fees and other forms of punishment were relevant. The meticulous recording of commoners and low-level convicts by name in the *Lǐyē* manuscripts is an exceptional practice for Qín dynasty documents, reflecting a remarkable attention to detail in the administration of public labor.

Relying on the information compiled under the supervision of the Commissioners of Public Works, the 'Bureaus of the Commissioner of Public Works' (*sīkōngcáo* 司空曹) maintained comprehensive 'statistical registers' (*jì lù* 計録) documenting the number of convicts, boats, vessels, ransoms and legal acts of transgression within their respective counties. Manuscript 8-480 illustrates how

511 Chén Wěi 2012: 39–40.
512 For a transcription and translation of manuscript 8-136+8-144, see chapter five.
513 Kim 2016: 565.
514 On this topic, see also Barbieri-Low 2007; Sanft 2014a: 66.
515 For a discussion of the Qín dynasty terracotta figures unearthed near the First Emperor's tomb, see, for example, Kesner 1995.

these registers were structured and what type of information they contained (Fig. 4.4):

8-480 AI–III, BI–III, CI–II: A 司空曹計録:┛ 船計,┛ 器計,┛ B 贖計,┛ 貲責計,┛ 徒計.┛ C 凡五計.┛ 史尚主.[516]

A Statistical registers by the Bureau of the Commissioner of Public Works:┛ Statistics of boats,┛ statistics of vessels,┛ B statistics of ransoms,┛ statistics of fines and accountabilities (for transgressors),┛ statistics of convicts,┛ C In total, there are five statistics.┛ Secretary Shàng is in charge.

Comments: '*Shú* 贖' is understood as 'to pay a fee for redemption (from a certain punishment).'[517] In the *Lǐyē* and *Zhāngjiāshān* manuscripts, the term often appears in conjunction with the punishment of 'being shaved' (*nài* 耐), as in "(being punished by paying the fee for) redemption from being shaved" (*shú nài* 贖耐).[518]

The 'Statistics of fines and accountabilities (for transgressors)' (*zī zé jì* 貲責計) also appears on the heavily damaged manuscript 8-1686, dated 212 BC. The character '*zī* 貲' is translated in the "Wéi yù děng zhuàng" as either the verb 'to fine (a transgressor)' or the noun 'fine'.[519]

According to a passage on strip 187 of the *Shuìhǔdì* documents, the annual figures for 'vessels' (*qì* 器) were compiled by all Offices and forwarded to the 'Ministry of Finance' (*nèishǐ* 內史). Vessels reported as missing were replaced by month 9 of the lunar year:

Shuìhǔdì, "Qín lǜ shíbā zhǒng", strip 187: 都官歲上出器求補者數, 上會九月內史.[520]

All Offices annually submit the figures of the vessels that have been given out and for which they request replacement (to the higher authorities). In month 9, accounts are submitted to the Ministry of Finance.

Based on this information, it is conceivable that statistical registers, such as the one found on manuscript 8-480, were submitted to higher authorities by Bureaus and Offices in months 8 and 9, or in the intercalary month 9, of the lunar year.

Many *Lǐyē* documents that mention Commissioners of Public Works distributing food cite 'Wall and Grain Pounders' (*chéngdàn chōng* 城旦舂) as recipients.[521] In some cases, 'coins' (*qián* 錢) were also distributed as compensation for their services and physical labor.[522] The overall structure of these texts consists of:

---

516 Chén Wěi 2012: 164.
517 Lau and Staack 2016: 319.
518 See, for example, manuscripts 8-149+8-489 or 8-1008+8-1461+8-1532, both of which are transcribed and translated in chapter five.
519 Lau and Staack 2016: 322.
520 Hulsewé 1985: 90.
521 See Xiè Kūn 2015.
522 For more on money in Ancient China, see Scheidel 2015; Thierry 2017; Valmisa 2019.

Fig. 4.4: Monochrome photograph of *Lǐyē* manuscript 8-480. Ink on wood. Dimensions Unknown.[523] The text is divided into three columns: A (top three lines), B (middle three lines), and C (bottom two lines).

1) The precise date of the entry,
2) The name of the Granary, Commissioner of Public Works or officials involved,
3) The total amount of food distributed,
4) The compensation period,
5) The name of the Commissioner of Public Works and/or his Assistant,
6) The type of food distributed,

523 *Húnán shěng wénwù kǎogǔ yánjiūsuǒ* 2012a: 71.

7) The amount of food the convicts were entitled to receive per day,
8) The exact recipients of the food,
9) The name of the supervisor overseeing the distribution,
10) The official responsible for the document.[524]

Manuscript 8-212+8-426+8-1632 (Fig. 4.5, 4.6) describes such events at the Jìng 徑 'Grain Store' (*kuài* 廥) in month 1 of the 31st year of King Zhèng (216 BC):[525]

> 8-212+8-426+8-1632 I–II: 徑廥粟米一石九斗五升六分升五. 卅一年正月甲寅[51]朔丁巳[54], 司空守增、佐得出以食舂、小城旦渭等卌七人, 積卌七日, 日四升六分升一.┘ 令史□視平. 得手.[526]
>
> The Jìng Grain Store (distributes) one picul, nine bushels and 5.6 pints of millet and hulled rice. In the 1st month of the 31st year (of King Zhèng) [216 BC], new moon on a *jiǎyín, dīngsì* [day 4], Zēng, Commissioner of Public Works, and his Assistant Dé, give out food (rations) to Grain Pounders, Minor Wall Pounder Wèi and others. There are 47 people for an accumulated period of 47 days. Each day, (they are entitled to receive) 4.6 pints.┘ Prefectural Secretary □ supervises the weighing (of the food). By the hand of Dé.
>
> Comments: The total amount of food distributed to the convicts for an accumulated period of 47 days was 195 and 5/6 pints of millet and hulled rice. This amount is equivalent to 1 picul, 9 bushels and 5.6 pints.

Instances of officials misallocating food rations or funds were compensated at a loss by the overseeing government unit. Missing rations or faulty calculations were thoroughly investigated, and accused officials were pursued meticulously, even after relocating to new work assignments. If perpetrators were reassigned, their outstanding debts and infractions were transferred to the new government unit under the supervision of a different Commissioner of Public Works.

The following text records one such incident, beginning with a report made by a relatively low-ranking official named Dīng 丁. The report was initially re-

524 This structure is also seen on manuscripts 8-212+8-426+8-1632, 8-216+8-351 and 8-337.
525 Lau and Staack differentiate between 'Grain Stores' (*kuài* 廥) and 'Granaries' (*cāng* 倉) (Lau and Staack 2016: 148, footnote 725). Strip 195 of the "Qín lǜ shíbā zhǒng" states: "Only [the walls] of Grain Stores and Granaries covered with reed should be made high" (*Shuìhǔdì Qín mù zhújiǎn zhěnglǐ xiǎo zǔbiān* 1990). The term 'Grain Stores' (*kuài* 廥) appears on *Lǐyē* manuscripts 8-1239, 8-1257, 8-1321, 8-1545, 8-1590, 8-1647, 8-1690 and 8-1739.
526 Chén Wěi 2012: 115.

Fig. 4.5: Monochrome photographs of manuscript 8-212+8-426+8-1632 (from left to right). Ink on wood. Dimensions unknown.[527] Line II, despite containing only a few graphs, extends across all three manuscript fragments. The phrase "(the) Prefectural Secretary □ supervises the weighing (of the food)" (*lìngshǐ* □ *shì píng* 令史□視平) is in the middle left, and the signatory clause "by the hand of Dé" (*Dé shǒu* 得手) is in the bottom left. So-called 'notches' (*kèchǐ* 刻齒) are carved into the left side of each fragment, conveying additional administrative information not shown in the published photographs by *Húnán shěng wénwù kǎogǔ yánjiūsuǒ*. In this instance, the undamaged notches state: "One picul, nine bushels and three pints" (*yī dàn jiǔ dǒu sān shēng* 一石九斗三升).[528] These types of manuscripts simultaneously served as 'tallies' (*quàn* 券), consisting of two matching pieces.[529]

ceived by the Prefect of Xúnyáng 旬陽 and then forwarded to the County Court of Qiānlíng. From there, it was passed down from the Prefect of Qiānlíng to the Deputy Prefect of Qiānlíng, and finally delivered by a 'Runner' (*zǒu* 走) to the Commissioner of Public Works. This entire process, detailed on manuscript 8-63, took more than 1.5 years to complete, beginning with the initial report in month 3 of

527 *Húnán shěng wénwù kǎogǔ yánjiūsuǒ* 2012a: 49, 64, 214.

528 For more information on these 'notches' (*kèchǐ* 刻齒), see Zhāng, Ohkawa und Momiyama 2015: 53–69.

529 Ma Tsang Wing 2017: 325–328.

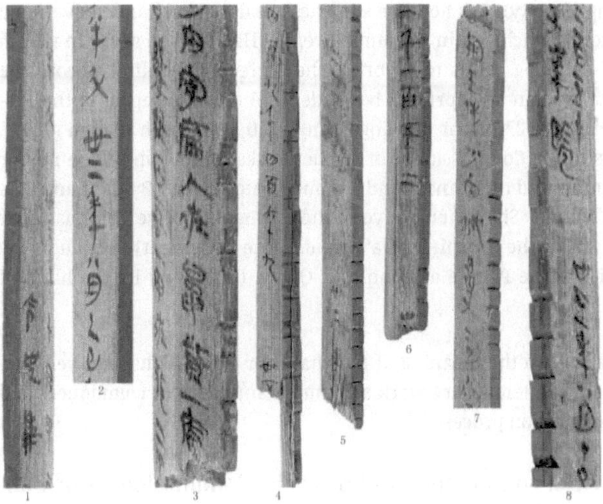

Fig. 4.6: Monochrome photographs of various *Lǐyē* 'tallies' (*quàn* 券) displaying carved 'notches' (*kèchǐ* 刻齒) on their sides. Ink on wood. Dimensions unknown.[530]

the 26th year of King Zhèng (221 BC) and concluding with the delivery of the amended document in month 10 of the 27th year of King Zhèng (220 BC).

> 8-63 I–V, rI: 廿六年三月壬午[19]朔癸卯[40], 左公田丁敢言之: 佐州里煩故爲公田吏, 徙屬. 事荅不備, 分┛ 負各十五石少半斗, 直錢三百一十四. 煩冗佐署遷陵. 今上責校券二, 謁告遷陵┛ 令官計者定, 以錢三百一十四受旬陽左公田錢計, 問可(何)計付, 署計年爲報. 敢言之.┛ 三月辛亥, 旬陽丞滂敢告遷陵丞主: 寫移, 移券, 可爲報. 敢告主. ノ兼手.┛ 廿七年十月庚子[37], 遷陵守丞敬告司空主, 以律令從事言. ノ𪏽手. 即走申行司空. └十月辛卯[28]旦, 朐忍𡩋秦士五(伍)狀以來. ノ慶半. 兵手.[531]
>
> In the 26th year (of King Zhèng) [221 BC], month 3, new moon on a *rénwǔ*, *guǐmǎo* [day 22], Dīng, Official in Charge of the Agricultural Fields, dares to say: Assistant Fán from the township Zhōu used to be an Official in Charge of the Agricultural Fields. He was transferred to another government unit. A matter was responded to in an unsatisfactory way. He divided┛ (a compensation) into two parts, with 15 piculs and less than half a bushel worth 314 coins each. Fán was temporarily recruited as an Assistant subordinate to Qiānlíng County. Now, the higher authorities demand a refund and a crosscheck of tally number two. Request to report to Qiānlíng┛ and order the office to compile the statistics and settle (the matter). Receive 314 coins from the Official in Charge of the Statistics regarding the agricultural

530 Ma Tsang Wing 2017: 328.
531 Chén Wěi 2012: 48–49.

fields and funds in Xúnyáng (County). Ask how the statistics are done and handed over, and which points of the statistics are included in the annual reply. (He) dares to say.┘ In month 3, day 30, Pāng, Prefect of Xúnyáng, dares to report to the Prefect of Qiānlíng: Record the transfer and transfer the tally, then a report can be made. (He) dares to report to the Prefect. / By the hand of Jiān.┘ In the 27th year (of King Zhèng) [220 BC], month 10 on a *gēngzǐ*, Deputy Prefect Jìng reports to the Commissioner of Public Works. Follow up on the matter in accordance with the statutes and ordinances and communicate orally. / By the hand of X [pronunciation unknown]. Runner Shēn then delivers (the document) to the Commissioner of Public Works. └Month 10, in the morning of a *xīnmǎo*, (the document) is brought by Zhuàng, member of a Unit of Five People at Rěnqín in Qúrěn (County). / By the hand of Qìng. By the hand of Bīng.

Comments: Scholarly consensus on the meaning of the character '*rǒng* 冗' has not yet been reached.[532] It appears to refer to a temporary work assignment for convicts, commoners and officials without a formal application process.[533]

Besides administering convicts, commoners and officials, Commissioners of Public Works coordinated with 'Prefects' (*chéng* 丞) and 'Section Commanders' (*xiàozhǎng* 校長) to assign 'conscript soldiers' (*zú* 卒) and 'Secretaries' (*shǐ* 史) to a wide range of tasks. In manuscript 8-167+8-194+8-472+8-1011, which deals with bandits stealing a public boat, the County Military Commandant forwards a request to the Prefect, asking for assistance in apprehending the perpetrators. Due to illegible characters in the text, the exact role of the Commissioner of Public Works cannot be determined. It is possible that he allocated a certain amount of labor to be used at the Prefect's disposal.[534]

Commissioners of Public Works held a status and privileges not accorded to other positions in the Qín dynasty. An official's career path was largely determined by internal structures and rewards specifically created within the framework of a civil meritocracy. Acts of valor, acquired skills and official endorsements provided individuals unprecedented opportunities to elevate their social standing. This process was ultimately facilitated by institutional mechanisms that shifted power away from the representatives of pre-imperial lineages.[535]

As verified by numerous *Lǐyē* documents, the traditional Zhōu dynastic order, consisting of 20 'nobility ranks' (*jué* 爵), remained an integral part of the Qín empire and was not merely replaced. Instead, it was merged and re-institutionalized to allow talented individuals to receive privileges once reserved for the aristocrats of

532 Lau and Staack 2016: 136–137.
533 Chén Wěi 2012: 43–44.
534 For a transcription and translation of manuscript 8-167+8-194+8-472+8-1011, refer to chapter 4.3.
535 Loewe 2006: 71–72.

the *ancien régime*. This transformation had a profound effect on social practices, greatly impacting Qín institutions and society at large.[536]

Manuscript 8-269 is a rare and excellent example of a 'Secretary's' (*shǐ* 史) professional trajectory, detailing the assessment he underwent for the head position at the 'Bureau of the Commissioner of Public Works' (*sīkōngcáo* 司空曹) at the age of 36, after more than 24 years of government service:

> 8-269 AI–V, BI–II, CI, DI: A 資中令史陽里釦伐閱:┙ 十一年九月隃爲史. ┙ 爲鄉史九歲一日.┙ 爲田部史四歲三月十一日.┙ 爲令史二月.┙ B □計.┙ 年卌六.┙ C 戶計.┙ D 可直司空曹.[537]
>
> A Achievements and experiences of Kòu, Prefectural Secretary of Zīzhōng County in the township of Yáng:┙ At eleven years and nine months, he was promoted to Secretary. He was a District Secretary for nine years and one day.┙ He was Secretary for the Department of Agricultural Fields for four years, three months and eleven days.┙ He was a Prefectural Secretary for two months.┙ B □ statistics.┙ 36 years old.┙ C Household statistics.┙ D He can assume [the position] at the Bureau of the Commissioner of Public Works.
>
> Comments: According to the "Treatise on Geography" (*Dìlǐ zhì* 地理志) in the *Book of Hàn* (*Hànshū* 漢書), Zīzhōng County (*Zīzhōng xiàn* 資中縣) belonged to Qiánwèi 犍爲 Commandery during the Hàn dynasty. Its government seat was situated in the 'city' or, more accurately, at the 'marketplace' (*shì* 市) of Zīyáng 資陽, in modern-day Sìchuān Province.[538]
>
> Strip 10 of the "Annalistic Record" (*Biān nián jì* 編年記) from the *Shuìhǔdì* documents contains the phrase '*yú shǐ* 揄史', meaning 'to move forward as an official'.[539]
>
> When referring to Kòu's age (columns AII, BII), the character '*nián* 年' is used, as opposed to '*suì* 歲', which stands for the years comprising a certain time period.

The figures provided on this manuscript tell us that Kòu served as an imperial Secretary for most of his life, starting at age eleven (and nine months). Before his promotion to the rank of 'Prefectural Secretary' (*lìngshǐ* 令史), he worked as a 'District Secretary' (*xiāngshǐ* 鄉史) for exactly nine years and one day, and as a "Secretary for the Department of Agricultural Fields" (*tiánbùshǐ* 田部史) for four years, three months and eleven days. Assuming that the 'household statistics' (*hù jì* 戶計) are relatively complete without large, unregistered gaps, we can conclude that Kòu served as a 'Secretary' (*shǐ* 史) for over 12 years, from ages 11 to 23.[540] For an indi-

---

536 For more on the concept of 'elevating the worthy' (*shàng xián* 尚賢), social status and 'nobility' through meritocracy, see Lewis 2011; Richter 2005; Yates 1987.

537 Chén Wěi 2012: 125–126.

538 Chén Wěi 2012: 126.

539 Chén Wěi 2012: 126.

540 The definition of adulthood or 'being registered' (*fù* 傅) in the Qín dynasty seems to have mainly depended on social status and body height. Registration to a 'household' (*hù* 戶) likely applied to individuals in a 'Unit of Five' (*shìwǔ* 士五 or *wǔ* 伍) at the age of 16 or 17, or to those measuring a minimum of 1.5 meters (Yates 1987: 231). For example, the status of 'Grain Pounders'

vidual of lower rank to ascend the Qín hierarchy, practical experience, age and sex were among the predominant factors considered. Manuscript 8-269 highlights the remarkable effort invested in creating, maintaining, archiving and retrieving detailed household records. These records were kept by local governments and played a pivotal role in granting personal endorsements and rewards within the institutionalized Qín system of civil meritocracy.

4.4 'Bureaus' (*cáo* 曹) and 'Offices' (*guān* 官)

The distinctions between Qín 'Bureaus' (*cáo* 曹) and 'Offices' (*guān* 官) were largely unclear before the discovery of the *Lǐyē* corpus and have since been a subject of scholarly debate. These two institutions seem to have differed primarily in terms of their purpose and function.[541] Bureaus coordinated military matters, household registration, the distribution of food and resources, public labor, financial affairs and legal investigations at the county level. They also handled various 'statistics' (*jì* 計) related to personnel, convicts and goods. While Bureaus relied on the logistical efforts the Offices – and vice versa – they appear to have been direct auxiliaries to the 'County Courts' (*xiàntíng* 縣廷), focusing on higher-level administrative duties. Postal documents reveal that the County Court of Qiānlíng regularly exchanged correspondence with the 'Household Bureau' (*hùcáo* 戶曹), the 'Granary Bureau' (*cāngcáo* 倉曹), the 'Bureau for Officials' (*lìcáo* 吏曹), the 'Bureau of Ordinances' (*lìngcáo* 令曹) and the 'Reinvestigation Bureau' (*fùcáo* 覆曹).[542]

Like Bureaus, Offices were also located within 'districts' (*xiāng* 鄉), headed by 'Overseers' (*sèfū* 嗇夫), and were possibly more numerous and geographically dispersed than Bureaus. Unlike Bureaus, Offices appear to have been exempt from the direct authority of the Courts.[543] To date, no evidence of correspondence be-

(*chōng* 舂) was determined not by age but by body height, with minors being defined as those whose height did not exceed six 'feet' (*chǐ* 尺) or 1.39 meters (Lau 1999: 45).

541 Sūn Wénbó 2014b.

542 These exchanges are documented, for example, on manuscripts 8-263 ('Household Bureau' (*hùcáo* 戶曹)), 8-1288 ('Granary Bureau' (*cāngcáo* 倉曹)), 8-2507 ('Bureau for Officials' (*lìcáo* 吏曹)) and 8-1859 ('Bureau of Ordinances' (*lìngcáo* 令曹)).

543 Qiú Xīguī 1992: 430–523. At the 'Judiciary Bureaus' (*yùcáo* 獄曹), we mainly see 'Judiciary Secretaries' (*yùshǐ* 獄史) (6-28, 8-133, 8-186, 8-683, 8-754+8-1007, 8-940, 8-987, 8-1007, 8-1058, 8-1232, 8-1441, 8-1448) and 'Judiciary Assistants' (*yùzuǒ* 獄佐) (5-1, 8-255, 8-265, 8-492, 8-877, 8-988, 8-1232, 8-1441, 8-1448, 8-1729) handling legal matters. Manuscript 8-255 suggests that Judiciary Assistants from counties in neighboring commanderies were sometimes ordered to reinvestigate legal cases from Dòngtíng Commandery: "For judicial review, Judiciary Assistant Jǐ from Yuánlíng (County) ˩ will prosecute cases in Qiānlíng. Transfer [him] to Dòngtíng (Commandery)" (*Fùyù Yuánlíng*

tween County Courts and Offices has been discovered, and no *Lǐyē* documents contain phrases like "from the (County) Court to Office X" (*tíng X guān* 廷X官) or "the (County) Court uses the postal system to deliver (documents) to Office X" (*tíng yǐ yóu xíng X guān* 廷以郵行X官).[544]

County Offices were mainly responsible for the cultivation of agricultural fields, handling military and financial matters as well as supervising personnel, convicts and livestock. In reference to manuscript 8-461, institutions formerly known as the 'Royal Courts' (*wáng shì* 王室) and 'Homes of Dukes' (*gōng shì* 公室) before the Qín unification were renamed 'County Offices' (*xiànguān* 縣官) during the First Emperor's reign (221–209 BC).[545] Tab. 4.5 collates all Bureaus and Offices appearing on the *Lǐyē* documents from layers 5, 6, 8, 9 and 16.

In his analysis of the *Lǐyē* materials, Wáng Yànhuī's concludes that Qiānlíng County consisted of nine 'Bureaus' (*cáo* 曹) in the Qín dynasty:

1) 'Lesser Treasury' (*shǎonèi* 少內),
2) 'Bureau of the (County) Military Commandant' (*wèicáo* 尉曹),
3) 'Bureau for Officials' (*lìcáo* 吏曹),
4) 'Household Bureau' (*hùcáo* 戶曹),
5) 'Granary Bureau' (*cāngcáo* 倉曹),
6) 'Armory Bureau' (*kùcáo* 庫曹),
7) 'Bureau of the Commissioner of Public Works' (*sīkōngcáo* 司空曹),
8) 'Judiciary Bureau' (*yùcáo* 獄曹),
9) 'Stables' (*jiù* 廄).[546]

According to Wáng Yànhuī, the 'Finance Department' (*jīnbù* 金布), responsible for the accounting and financial matters in designated regions, was not part of the Qín Bureau system, but rather fell under the authority of the 'Lesser Treasury' (*shǎonèi* 少內). Only 'Prefects' (*chéng* 丞) headed the 'Judiciary Bureaus' (*yùcáo* 獄曹), while 'Court Masters of Officials' (*tíngzhǔlì* 廷主吏) oversaw the 'Bureaus for Officials' (*lìcáo* 吏曹).

---

yùzuǒ Jǐ zhì suǒ Qiānlíng chuán Dòngtíng 覆獄沅陵獄佐⅃己治所遷陵傳洞庭) (transcription taken from Chén Wěi 2012: 123).

544 This conclusion is based on my analysis of the *Lǐyē* documents from layers 5, 6 and 8.

545 For a transcription and translation of manuscript 8-461, see chapter 4.1.

546 Wáng Yànhuī 2012.

Tab. 4.5: 'Bureaus' (*cáo* 曹) and 'Offices' (*guān* 官) on the *Lǐyē* manuscripts from layers 5, 6, 8, 9 and 16.

No.	Designation	Occurrence(s): Manuscript number(s)
		'Bureaus' (*cáo* 曹)
1.1	'Bureau of the (County) Military Commandant' (*wèicáo* 尉曹)	**5:** 8-71, 8-253, 8-453, 8-1225, 8-1616
1.2	'Household Bureau' (*hùcáo* 戶曹)	**7:** 8-263, 8-488, 8-978, 8-1072, 8-1318, 8-1489, 8-1533
1.3	'Bureau of the Commissioner of Public Works' (*sīkōngcáo* 司空曹)	**4:** 8-269, 8-375, 8-480, 8-1428
1.4	'Bureau for Officials' (*lìcáo* 吏曹)	**10:** 8-98, 8-241, 8-554, 8-699, 8-829, 8-1126, 8-1700, 8-2149, 8-2507, 9-982[547]
1.5	'Granary Bureau' (*cāngcáo* 倉曹)	**8:** 8-481, 8-496, 8-500, 8-776, 8-1201, 8-1288, 8-1463, 8-1777+8-1868
1.5	'Bureau for Carriages' (*chēcáo* 車曹)	**1:** 8-405
1.6	'Judiciary Bureau of the East' (*yùdōngcáo* 獄東曹)	**5:** 5-22, 8-273+8-520, 8-959+8-1291, 8-996, 8-1155
1.7	'Judiciary Bureau of the South' (*yùnáncáo* 獄南曹)	**3:** 8-728+8-1474, 8-1760, 8-1886
1.8	'Bureau of Ordinances' (*lìngcáo* 令曹)	**2:** 8-778, 8-1859
1.9	'Inquiry Bureau' (*juàncáo* 讂曹)[548]	**2:** 9-1701+8-389 + 8-40, 9-2326
1.10	'Reinvestigation Bureau' (*fùcáo* 覆曹)	**1:** 8-2550
1.11	'Bureaus of Adjacent Counties' (*pángcáo* 旁曹)	**1:** 8-174
1.12	'Bureau for Official Ranks' (*juécáo* 爵曹)	**1:** 8-247
1.13	'Central Bureau' (*zhōngcáo* 中曹)[549]	**1:** 8-61+8-293+8-2012

547 *Húnán shěng wénwù kǎogǔ yánjiū suǒ, Xiāngxī Tǔjiā miáozú zìzhìzhōu wénwù chǔ, Lóngshān xiàn wénwù guǎnlǐ suǒ* 2003: 33.

548 This information is taken from Yates 2022.

549 In the Hàn dynasty, the headquarters of the Commandery Governor included a 'Central Bureau' (*zhōngcáo* 中曹) (Chén Mèngjiā 1980b: 98).

Tab. 4.5 (continued)

No.	Designation	Occurrence(s): Manuscript number(s)
	'Offices' (*guān* 官)[550]	
2.1	"County Offices (of Dòngtíng)" (*(Dòngtíng) xiànguān* ((洞庭)縣官)	**10:** 5-1, 8-454, 8-461, 8-657, 8-705, 8-734, 8-1083, 8-793+8-1547, 8-2367, 8-2492
2.2	'Office of the (County) Military Commandant' (*wèiguān* 尉官)	**1:** 8-657
2.3	'Office of Agricultural Fields' (*tiánguān* 田官)	**22:** 8-16, 8-74, 8-145, 8-162, 8-167+8-194+8-472+8-1011, 8-444, 8-479, 8-481, 8-580, 8-594, 8-663, 8-672, 8-764, 8-781+8-1102, 8-900, 8-1114+8-1150, 8-1328, 8-1430, 8-1566, 8-1574+8-1787, 8-1608, 8-2246
2.4	'Office of Domestic Animals' (*xùguān* 畜官)	**11:** 8-50+8-422, 8-137, 8-145, 8-162, 8-199+8-688, 8-285, 8-481, 8-490+8-501, 8-919, 8-1114+8-1150, 8-1558
2.5	'Crime Office' (*gànguān* 榦官)	**1:** 8-1831
2.6	'Office for Officials' (*lìguān* 吏官)	**1:** 8-1645
2.7	'Office of Ordinances' (*lìngguān* 令官)	**3:** 8-63, 8-143, 8-673+8-2002
2.8	'Boat Office' (*chuánguān* 船官)	**1:** 6-4
2.9	'Office of Dū (District)' (*Dū guān* 都官)	**1:** 8-649
2.10	'District Office' (*xiāngguān* 鄉官)	**2:** 8-198+8-213+8-2013, 8-2225
2.11	'District Department Office' (*xiāngbùguān* 鄉部官)	**1:** 8-297+8-1600
2.12	'Hidden Offices' (*yǐnguān* 隱官)	**1:** 16-5

In my analysis of the *Lǐyē* corpus, I exclude the 'Lesser Treasury' (*shǎonèi* 少內), the 'Armory Bureau' (*kùcáo* 庫曹), the 'Finance Department' (*jīnbù* 金布), the 'Court Personnel Bureau' (*tínglìcáo* 廷吏曹) and the 'Stables' (*jiù* 廄) from the list of Qín Bureaus and Offices for the following reasons: 1) these terms are not referenced on the manuscripts from layers 5, 6 and 8, or 2) the Lesser Treasury, Fi-

550 For more insights into Bureaus and other local Offices in early imperial China, see Barbieri-Low and Yates 2015.

nance Department and Stables are not explicitly designated as Bureaus or Offices in these texts.[551]

In contrast to Wáng Yànhuī and Robin Yates' argument that the designation of Bureaus was often preceded by the character for 'Court' (*tíng* 廷), as seen in names like 'Court Household Bureau' (*tínghùcáo* 廷戶曹) or 'Court Personnel Bureau' (*tínglìcáo* 廷吏曹),[552] I contend that this practice was regularly employed to indicate that a delivery was sent from the '(County) Court' (*tíng* 廷) to the 'Household Bureau' (*hùcáo* 戶曹) or the 'Bureau for Officials' (*lìcáo* 吏曹). Phrases like "*tínghùcáo* 廷戶曹" or "*tínglìcáo* 廷吏曹" should be divided and translated as "from the (County) Court to the [respective Bureau]." These expressions, therefore, do not stand for singular administrative units but rather two distinct entities, namely the '(County) Court' (*tíng* 廷) and the Bureau under the Court's authority. This interpretation explains the phrase "*tíng yǐ yóu xíng hùcáo* 廷以郵行戶曹" found on manuscript 8-1318, which translates as "the (County) Court uses the postal system to deliver (documents) to the Household Bureau."[553]

As observed by Shěn Gāng, the Warring States and Hàn dynasties relied heavily on 'statistics' (*jì* 計). This practice also held true for the Qín dynasty, although Qín officials compiled both 'statistics' (*jì* 計) and 'evaluations' (*kè* 課). Shěn Gāng argues that, while these two systems coexisted within the Qín administration, evaluations were eventually phased out during the Hàn dynasty.[554] Based on evidence from the *Lǐyē* corpus, Shěn Gāng's argument regarding Qín bureaucracy and the distinction between statistics and evaluations is well-founded.

In the *Shuìhǔdì* documents, the term '*kè* 課' – a verb meaning 'to evaluate' – is used in relation to cattle, horses and iron obtained from mining.[555] In the *Lǐyē* corpus, '*kè* 課' is used similarly yet includes commoners, conscript soldiers, convicts, the newborns of convicts, other domestic animals and documents concerning the so-called 'Lacquer Gardens' (*xiū yuán* 髹園).

8-383+8-484 AI–II, BI: A 田課志.┘ 髹園課.┘ B・凡一課.[556]

551 For the sake of terminological coherence, I substitute the terms 'Armory Bureau' (*kùcáo* 庫曹), 'Finance Bureau' (*jīnbù* 金布) and 'Court Personnel Bureau' (*tínglìcáo* 廷吏曹), as proposed by Robin Yates, with 'Warehouse Bureau' (*kùcáo* 庫曹), 'Finance Department' (*jīnbù* 金布) and 'Bureau for Officials' (*lìcáo* 吏曹), respectively.

552 The 'Court Household Bureau' (*tíng hùcáo* 廷戶曹) appears, for example, on manuscripts 8-263, 8-1072 and 8-1489; the 'Court Personnel Bureau' (*tíng lìcáo* 廷吏曹) on manuscripts 8-241, 8-554 and 8-1126.

553 For more on this subject, see chapter 2.3.

554 Shěn Gāng 2013.

555 Lau and Staack 2016: 245.

556 Chén Wěi 2012: 141.

A Evaluation records of the (Office/Bureau) of Agricultural Fields.⅃ Evaluation of the Lacquer Gardens.⅃ B • In total, there is one evaluation.

Comments: It remains unclear whether these evaluations were compiled by the Office or Bureau of Agricultural Fields. The term 'Lacquer Garden' (*xiū yuán* 髹園) also appears on strips 20 and 21 of the "Miscellaneous Excerpts from Qín Statutes" (*Qín lǜ zá chāo* 秦律雜抄) in the *Shuìhǔdì* documents. The character '*xiū* 髹' is synonymous with '*xiū* 髤'.[557]

Some of these 'evaluations' (*kè* 課) may have been selected by the respective government units before being sent to the commandery-level authorities. After processing at the commandery level, they were possibly forwarded to the central government for further review and implementation.[558]

By contrast, 'statistics' (*jì* 計) were static syntheses of reports aimed at creating general inventories. In the *Lǐyē* documents, statistics are more inclusive than 'evaluations' (*kè* 課), although there is some overlap in the variables they cover. Statistics encompassed a wide range of information, like harvest yields, domestic animals, tools, vessels, weapons, coins, lacquer, carriages, boats, district-level household numbers, convicts, land taxes, loans, ransoms, promotional letters, legal transgressions and criminal investigations. Granary statistics, in particular, focused on annual revenues and provided meticulous inventories of budgeted food rations.[559] Of special importance to the Granaries was information on convicts and livestock, catalogued in terms of their overall fiscal value or potential for generating revenue.

Manuscripts 8-481 and 8-495 contain a detailed account of ten types of statistics and eight types of evaluations conducted in either 'month 9' (*jiǔ yuè* 九月) or 'intercalary month 9' (*hòu jiǔ yuè* 後九月) of the lunar year. These reports were

557 Chén Wěi 2012: 141.

558 See Yates 2021.

559 For a study on the discrepancies in food distribution within the Granary and Grain Store systems, as documented in statutes from Shuìhǔdì 睡虎地 and Zhāngjiāshān 張家山, as well as administrative texts from Lǐyē and Xuánquán 懸泉, see Kim 2016. Kim examines how the statutes from Shuìhǔdì and Zhāngjiāshān imposed strict limits on the maximum amount of food allocated to convicts by Granaries and Grain Stores. In some cases, actual distributions often fell below these statutory limits. For example, '(female) Bondservants' (*lìqiè* 隸妾) generally received 2.5 bushels, or five-sixth of a bushel less 'millet' (*sù* 粟) than the amount cited in the *Shuìhǔdì* statutes (see manuscripts 8-760, 8-762, 8-763, 8-766, 8-1557 and 8-2249) (Kim 2016: 586).

compiled by the County Granaries and subsequently forwarded to higher authorities for further processing.

8-481 AI–IV, BI–IV, CI–V: A 倉曹計録:⅃ 禾稼計,⅃ 貸計,⅃ 畜計,⅃ B 器計,⅃ 錢計,⅃ 徒計;⅃ 畜官牛計,⅃ C 馬計,⅃ 羊計;⅃ 田官計.⅃ 凡十計.⅃ 史尚主.[560]

A Statistical registers of the Granary Bureau:⅃ Statistics of crops,⅃ statistics of loans,⅃ statistics of animals,⅃ B statistics of vessels,⅃ statistics of coins,⅃ statistics of convicts;⅃ statistics of oxen under the Office of Domestic Animals,⅃ C statistics of horses,⅃ statistics of sheep;⅃ statistics of the Office of Agricultural Fields.⅃ In total, there are ten statistics.⅃ Official Shàng is in charge.

8-495 AI–AV, BI–IV, C: 倉課志:⅃ 畜彘雞狗產子課,⅃ 畜彘雞狗死亡課, ⅃ 徒隸死亡課,⅃ 徒隸產子課,⅃ 作務產錢課,⅃ 徒隸行繇(徭)課,⅃ 畜鴈死亡課,⅃ 畜鴈產子課.⅃ · 凡☒[561]

Evaluation records of the Granary:⅃ Evaluation of domestic pigs, chickens and dogs that gave birth,⅃ evaluation of domestic pigs, chickens and dogs that died,⅃ evaluation of convict servants who died,⅃ evaluation of convict servants who gave birth⅃ evaluation of revenue (earned by convicts),⅃ evaluation of convict servants conducting labor assignments,⅃ evaluation of domestic geese that died,⅃ evaluation of domestic geese that gave birth.⅃ · In total ☒

Comments: Xiè Kūn argues that the material and textual characteristics of this manuscript fit the missing upper section of manuscript 8-150:[562]

8-150 I–II: ☒【年】課⅃ ☒□課

☒【year】evaluation⅃ ☒□ evaluation

Even if this is the case, manuscript 8-150 offers little additional information, aside from reinforcing the assumption that these statistics were part of the evaluations conducted at annual intervals.

In a study by Sūn Wénbó, the term 'statistical register' (*jì lù* 計録) is often associated with 'Bureau' (*cáo* 曹), while 'evaluation record' (*kè zhì* 課志) is consistently linked to 'Office' (*guān* 官).[563] In other words, Bureaus compiled statistical registers, and Offices produced evaluation records. According to this study and subse-

560 Chén Wěi 2012: 164.
561 Chén Wěi 2012: 169.
562 Xiè Kūn 2016.
563 Sūn Wénbó 2014b; Sūn Yànhuī 2014; See also Shěn Gāng 2013.

quent textual sources, Sūn Wénbó classifies the 'Finance Department' (*jīnbù* 金布) as a 'Bureau' (*cáo* 曹), and the 'Granary' (*cāng* 倉) as an 'Office' (*guān* 官).[564]

However, a closer examination of the documents from layers 5, 6 and 8 complicates this clear-cut distinction. Firstly, the 'Granary Office' (*cāngguān* 倉官) and 'Finance Bureau' (*jīnbùcáo* 金布曹) are nonexistent in these texts. Secondly, on manuscript 8-481, both the 'Office of Domestic Animals' (*chùguān* 畜官) and the 'Office of Agricultural Fields' (*tiánguān* 田官) submit their statistics to the 'Granary Bureau' (*cāngcáo* 倉曹) for inclusion in their 'statistical registers' (*jì lù* 計錄). As shown in Tab. 4.6, designations for government units sometimes omit the suffix 'Office' (*guān* 官) or 'Bureau' (*cáo* 曹), as seen with terms like 'Agricultural Fields' (*tián* 田), 'Military Commandant' (*wèi* 尉), 'District' (*xiāng* 鄉), 'Commissioner of Public Works' (*sīkōng* 司空) and 'Granary' (*cāng* 倉). Despite this Sūn Wénbó classifies all these institutions as 'Offices' due to their role in conducting 'evaluations' (kè 課).

Other administrative units at the Bureau and Office levels in the *Lǐyē* corpus exhibit similar designations. For example, institutions run by the Commissioner of Public Works or the Granaries are also referred to as Bureaus (*sīkōngcáo* 司空曹 or *cāngcáo* 倉曹). An auxiliary to the aforementioned 'Bureau for Officials' (*lìcáo* 吏曹) was the 'Office for Officials' (*lìguān* 吏官), which may have been staffed by personnel of different ranks and statuses. These findings suggest that attributing specific responsibilities, such as compiling statistics or evaluations, to individual government units is more intricate than previously assumed. Hence, Sūn Wénbó's classification of these units requires further scrutiny and should be considered cautiously until more evidence emerges.

The summary line of manuscript 8-490+8-501 reports a total of eight submitted evaluations, despite seven being expected. This discrepancy may be attributed to a clerical error, as the calligraphic and structural properties of fragments 8-490 and 8-501 seem to align perfectly (Fig. 4.7).

564 The 'Finance Department' (*jīnbù* 金布) compiles a 'statistical register' (*jì lù* 計錄) on manuscript 8-493, whereas an 'evaluation record' (*kè zhì* 課志) is issued by the 'Granary' (*cāng* 倉) on manuscript 8-495 (Chén Wěi 2012: 169).

Tab. 4.6: *Lǐyē* manuscripts from layers 5, 6 and 8 containing the terms 'evaluations' (*kè* 課) or 'statistics' (*jì* 計). The total number of manuscripts contained in one evaluation or statistic is usually summarized in the last line of the text (in cases it has survived), which is useful if preceding parts of the text are illegible or missing.

Manuscript number(s)	Government unit	Content of document	Total number of evaluations/ statistics
		'Evaluation records' (*kè zhì* 課志)	
		'Offices' (*guān* 官)	
8-479	'Office of Agricultural Fields' (*tiánguān* 田官)	Characters illegible	1
8-490+8-501	'Office of Domestic Animals' (*xùguān* 畜官)	– Evaluations of payments for domestic animals that died under the husbandry of convict servants; including butchering and selling (*túlì mùxù sǐ fù, bō mài kè* 徒隸牧畜死負、剝賣課) – Evaluations of untold (cases of) domestic animals that died under the husbandry of convict servants (*túlì mùxù xù sǐ bù qǐng kè* 徒隸牧畜畜死不請課) – Evaluations of horses that gave birth (*mǎ chǎn zǐ kè* 馬產子課) – Evaluations of domestic oxen that died (*xù niǔ sǐ wáng kè* 畜牛死亡課) – Evaluations of domestic oxen that gave birth (*xù niǔ chǎn zǐ kè* 畜牛產子課) – Evaluations of domestic sheep that died (*xù yáng sǐ wáng kè* 畜羊死亡課) – Evaluations of domestic sheep that gave birth (*xù yáng chǎn zǐ kè* 畜羊產子課)	7 (the summary line at the end of the text states 8)
8-1677	–	– Evaluations of captives (*lǔ kè* 虜課)	1

Tab. 4.6 (continued)

Manuscript number(s)	**Government unit**	**Content of document**	**Total number of evaluations/ statistics**
		'Registers of statistics' (*jì lù* 計録)	
		'Bureaus' (*cáo* 曹)	
8-480	'Bureau of the Commissioner of Public Works' (*sīkōngcáo* 司空曹)	– Statistics of boats (*chuán jì* 船計) – Statistics of vessels (*qì jì* 器計) – Statistics of ransoms (*shú jì* 贖計) – Statistics of fines and accountabilities (for transgressors) (*zī zé jì* 貲責計) – Statistics of convicts (*tú jì* 徒計)	5
8-481	'Granary Bureau' (*cāngcáo* 倉曹)	– Statistics of crops (*hé jià jì* 禾稼計) – Statistics of loans (*dài jì* 貸計) – Statistics of domestic animals (*xù jì* 畜計) – Statistics of vessels (*qì jì* 器計) – Statistics of coins (*qián jì* 錢計) – Statistics of convicts (*tú jì* 徒計) – Statistics of oxen under the Office of Domestic Animals (*xùguān niú jì* 畜官牛計) – Statistics of horses (*mǎ jì* 馬計) – Statistics of sheep (*yáng jì* 羊計) – Statistics of the Office of Agricultural Fields (*tiánguān jì* 田官計)	10

Tab. 4.6 (continued)

Manuscript number(s)	Government unit	Content of document	Total number of evaluations/ statistics
8-488	'Household Bureau' (*hùcáo* 戶曹)	– Statistics of district households (*xiāng hù jì* 鄉戶計) – Statistics of work assignments (*yáo (yáo) jì* 繇(徭)計) – Statistics of vessels (*qì jì* 器計) – Statistics of grain taxes and loans (*zū zhì jì* 租質計) – Statistics of promotional letters for agricultural land (*tián tí fēng jì* 田提封計) – Statistics of lacquer (*xiū jì* 髹計) – Statistics of criminal investigations (*jū jì* 鞫計)	7
8-493	'Finance Department' (*jīnbù* 金布)	– Statistics of Warehouse weaponry (*kù bīng jì* 庫兵計) – Statistics of carriages (*chē jì* 車計) – Statistics of tools for public use (*gōng yòng jì* 工用計) – Statistics of vessels for public use (*gōng yòng qì jì* 工用器計) – Statistics of vessels by the Lesser Treasury (*shǎonèi qì jì* 少內器計) – Statistics of 【metal】 and coins (【*jīn*】 *qián jì* 【金】錢計)	6
Unidentified whether 'Offices' (*guān* 官) or 'Bureaus' (*cáo* 曹)			
8-383+8-484	'Agricultural fields' (*tián* 田)	– Evaluations of Lacquer Gardens (*xiū yuán kè* 髹園課)	1
8-482	'Military Commandant' (*wèi* 尉)	– Evaluations of conscript soldiers who died (*zú sǐ wáng kè* 卒死亡課) – Evaluations of Overseers of Bandits working the fields (*sīkòu tián kè* 司寇田課) – Evaluations of conscripts working the fields (*zú tián kè* 卒田課)	3

Tab. 4.6 (continued)

Manuscript number(s)	Government unit	Content of document	Total number of evaluations/ statistics
8-483	'District' (*xiāng* 鄉)	– Evaluations of the counting of the black-headed (people) (*qiánshǒu lì kè* 黔首曆課) – Evaluations of children and children without fathers (*guǎ zǐ kè zǐ kè* 寡子課子課)	4 (of which 2 are illegible)
8-486	'Commissioner of Public Works' (*sīkōng* 司空)	– Evaluations of Grain Pounders who gave birth (*chōng chǎn zǐ kè* 舂產子課)	7 at least; there are 4 lines in column A, and 5 lines in column B. The last two lines of column B are missing.
8-495	'Granary' (*cāng* 倉)	– Evaluations of domestic pigs, chickens and dogs that gave birth (*xù zhì jī gǒu chǎn zǐ kè* 畜彘雞狗產子課) – Evaluations of domestic pigs, chickens and dogs that died (*xù zhì jī gǒu sǐ wáng kè* 畜彘雞狗死亡課) – Evaluations of convict servants who died (*túlì sǐ wáng kè* 徒隸死亡課) – Evaluations of convict servants who gave birth (*túlì chǎn zǐ kè* 徒隸產子課) – Evaluations of revenue (earned by convicts) (*zuò wù chán qián kè* 作務產錢課) – Evaluations of convict servants conducting labor assignments (*túlì xíng yáo (yáo) kè* 徒隸行繇(徭)課) – Evaluations of domestic geese that died (*xù yàn sǐ wáng kè* 畜鴈死亡課) – Evaluations of domestic geese that gave birth (*xù yàn chǎn zǐ kè* 畜鴈產子課)	8 (the summary line at the end of the text is incomplete. This figure is derived from summing the columns and lines in the document that are undamaged).

Fig. 4.7: Monochrome photographs of manuscripts 8-490 (l) and 8-501 (r). Ink on wood. Dimensions unknown.[565]

Although some Bureaus, like the 'Bureau of the Commissioner of Public Works' (*sīkōngcáo* 司空曹), were later abandoned, the general distinction between Bureaus and Offices had already been introduced in the early imperial Qín dynasty. This organizational structure continued into the Hàn dynasty, as confirmed by Western Hàn sources.[566]

565 *Húnán shěng wénwù kǎogǔ yánjiūsuǒ* 2012a: 73, 75.
566 Sūn Wénbó 2014b.

5 *Corvée* Labor

The Qín dynasty excelled at reinforcing social hierarchies inherited from earlier periods, as evidenced by legal codes that coerced of legal offenders into slavery or *corvée* labor.[567] In view of this, it is unsurprising that much of the empire's accumulated wealth and cultural achievements stemmed from the exploitation of conquered territories and populations. Under Qín law, fugitives and rebels from distant regions were tracked down and punished.[568] Many legal reforms during this period were designed to convert slaves, convicts and conscripts into laborers, making this demographic a key economic and agricultural force within the empire. To sustain large-scale construction projects, ensure efficient working conditions and maintain a taxation system that facilitated the redistribution of food and resources both during and between agricultural seasons, the regulation of land and labor was critical.[569]

The 'Bureau of the Commissioner of Public Works' (*sīkōngcáo* 司空曹), along with the 'Granary' (*cāng* 倉), 'Warehouse' (*kù* 庫) and 'Office of Domestic Animals' (*chùguān* 畜官), compiled statistics that factored in the availability of convicts.[570] Manuscript 8-480 is a 'statistical register' (*jì lù* 計録) created and submitted by the Bureau of the Commissioner of Public Works:

> 8-480 AI–III, BI–III, CI–II: A 司空曹計録:┘ 船計,┘ 器計,┘ B 贖計,┘ 貲責計,┘ 徒計.┘ C 凡五計.┘ 史尚主.[571]
>
> A Statistical registers by the Bureau of the Commissioner of Public Works:┘ Statistics of boats,┘ statistics of vessels,┘ B statistics of ransoms,┘ statistics of fines and accountabilities (for transgressors),┘ statistics of convicts,┘ C In total, there are five statistics.┘ Secretary Shàng is in charge.

---

567 *Corvée* labor, distinct from slavery, was a compulsory form of work imposed on non-slaves (i.e. officials or commoners) for a set period of time.

568 See Yates 2022.

569 A study on mitochondrial DNA from 19 human bone remains, excavated from a Qín dynasty tomb near the site of the terracotta army, shows that laborers who built the mausoleum of the First Emperor were genetically diverse. These findings suggest that many of these laborers migrated thousands of kilometers from southern regions of the empire to the Qín capital of Xiányáng 咸陽 or its surrounding areas (Xu Zhi, et al. 2008). Yóu Yìfēi argues that all Qín officials in Qiānlíng County, including conscript soldiers and commoners, were migrants ordered by the central government to relocate from outside of Dòngtíng (Yóu Yìfēi 2015). I would treat this claim with caution until more evidence is available, as some personal names in the *Lǐyē* corpus suggest local origins.

570 For more information on government units at the commandery and county level, see Tab. 4.3.

571 Chén Wěi 2012: 164.

https://doi.org/10.1515/9783110791853-005

Comment: The "statistics of fines and holding [transgressors] responsible" (*zī zé jì* 貲責計) also appears on the heavily damaged manuscript 8-1686, dated 212 BC.[572]

8-1686 I–II: 出莞席十. 卅五年八月丁巳[54]□☒┘ 其貲責(債)計☒ [573]

Give out mats made of club rush, ten (pieces). In the 35th year (of King Zhèng) [212 BC], month 8 on a *dīngsì*, □☒┘ Its statistics of fines and accountabilities (for transgressors) ☒

'Convict (servants)' (*tú(lì)* 徒(隸)) were individuals with a criminal record who received basic provisions and a fixed state income in order to repay fines, redemption fees or debts through compulsory labor.[574] Manuscript 8-1335, although damaged, details how the 'District Incumbent' (*xiāngshǒu* 鄉守) of Èrchūn 貳春 and his 'Assistant' (*zuǒ* 佐) allocated food rations to convicts under the oversights of a 'Prefectural Secretary' (*lìngshǐ* 令史):

8-1335 I–II: 粟米八升少半升. 令史逐視平.☒┘ 卅一年四月辛卯[28], 貳春鄉守氐夫、佐吾出食舂、白粲□等.☒[575]

(There are) 8.5 pints of millet and hulled rice. Prefectural Secretary Zhú supervises the weighing (of the food). ☒┘ In the 31st year (of King Zhèng) [216 BC], month 4 on a *xīnmǎo* [day 9], Dǐfū, Incumbent of Èrchūn District, and his Assistant Wú, give out food (rations) to Grain Pounders, Rice Sifters □ and others. ☒

Comments: According to the "Atlas of Ancient Chinese Weights and Measures" (*Zhōngguó gǔdài dùliànghéng tújí* 中國古代度量衡圖集), one 'pint' (*shēng* 升) was equivalent to 200 milliliters by modern standards, or 119.9 milliliters of edible grains after being peeled.[576]

On manuscript 8-736, the new moon day in month 4 of the 31st year (of King Zhèng) (216 BC) is a '*guǐwèi* 癸未', making '*xīnmǎo* 辛卯' fall on 'day 9'.[577]

For a discussion of the term '*píng* 平', see the comments on manuscript 8-217 in chapter six. The distribution of food rations in the *Lǐyē* corpus is explored in chapter six.

Ranks within the imperial system of *corvée* labor further intensified the stratification of the lower classes, reducing some individuals to tradeable commodities, as evidenced by manuscript 8-154:

572 For more comments on manuscript 8-480, refer to chapter 4.3.

573 Chén Wěi 2012: 378.

574 Cáo Lǚnìng 2015; Gāo Héng 1983; Lǐ Lì 2009.

575 Chén Wěi 2012: 312.

576 The amount of edible grains after peeling is based on a calculation introduced in Hú Píngshēng 2012b, which states that 834 milliliters of grain is equivalent to 500 mililiter of edible rice.

577 Compare with Chén Wěi 2012: 212; Xǔ Míngqiāng 2013b.

8-154 I–II, rI: 卅三年二月壬寅[39]朔朔日, 遷陵守丞都敢言之: 令曰恒以┘ 朔日上所買徒隸數. • 問之, 毋當令者, 敢言之. └二月壬寅[39]水十一刻刻下二, 郵人得行. 圂手.[578]

In the 33rd year (of King Zhèng) [214 BC], month 2, new moon on a *rényín*, new moon day [day 1], Dū, Deputy Governor of Qiānlíng, dares to say: The ordinance states to regularly submit a document┘ to the higher authorities on the 1st day (of the lunar month) with the number of convict servants that have been bought. • Clarify whether there is no corresponding ordinance. (He) dares to say. └In month 2 on a *rényín* [day 1], at water below mark 2 of 11 marks, Postman Dé delivers (the document). By the hand of Hùn.

In his 2003 analysis, Lǐ Xuéqín defines 'convicts' to include 'Bondservants' (*lìchén qiè* 隸臣妾), 'Wall and Grain Pounders' (*chéngdàn chōng* 城旦舂), and 'Firewood Gatherers and Rice Sifters' (*guǐxīn báicàn* 鬼薪白粲).[579] Readings of the *Yuèlù* manuscripts have confirmed these categories, with the proposed addition of the 'Overseers of Bandits' (*sīkòu* 司寇).[580] The *Lǐyē* texts group convicts into at least five different categories, based on factors such as social status, official rank, skills, age, sex and the compensation they received for their sentences:

1) 'Overseers of Bandits' (*sīkòu* 司寇),
2) 'Released Convicts' (*yǐnguān* 隱官),
3) 'Firewood Gatherers and Rice Sifters' (*guǐxīn báicàn* 鬼薪白粲),
4) 'Bondservants' (*lìchén qiè* 隸臣妾),
5) 'Wall and Grain Pounders' (*chéngdàn chōng* 城旦舂).[581]

578 Chén Wěi 2012: 93.

579 Lǐ Xuéqín 2003.

580 *Yüèlù* strip 640 lists 'Bondservants' (*lìchén qiè* 隸臣妾), 'Wall Pounders' (*chéngdàn* 城旦), 'Wall and Grain Pounders' (*chéngdàn chōng* 城旦舂), "Overseers of Bandits watching over Wall and Grain Pounders" (*chéngdàn chōng sīkòu* 城旦舂司寇) and 'Firewood Gatherers and Rice Sifters' (*guǐxīn báicàn* 鬼薪白粲) as convicts (individuals carrying out manual labor in exchange for food provided by the authorities) (Cáo Lǚnìng 2015; Yates 1995: 345–346). Convicts of a certain age or those unable to work could be redeemed by relatives or friends from their native region. If they lacked relatives or friends, or could not be identified, they were given food rations and sent to modern-day Sìchuān Province, where the population was still sparse. The *Yüèlù* legal texts do not clarify whether convicts who could not work and had no one to redeem them received government support. In the "Models for Sealing and Investigating" (*Fēng zhěn shì* 封診式) from the *Shuìhǔdì* texts, convicts were exchanged between private individuals and officials, particularly if they refused to work or follow orders (Qiú Xīguī 1981). For an overview captive hierarchies during the Qín dynasty, see Yates 2022; Zhāng Jīnguāng 2004: 533. Some scholars maintain that all soldiers captured by Qín forces were enslaved and made Bondservants (Sūn Míng 2018). Nonetheless, soldiers who surrendered and became Bondservants could improve their social status by performing feats on the battlefield (Yates 2022: 6). Social mobility was a privilege generally denied to slaves.

581 The term '*túlì* 徒隸', meaning 'convict servant', is often used interchangeably with '*tú* 徒' in scholarly discourse (Cáo Lǚnìng 2015). The following English designations of various convicts are

A manuscript excavated from layer 16 gives an overview of the imperial Qín convict and conscript system, outlining its various hierarchical subdivisions:

> 16-5 I–VII: 廿七年二月丙子[13]朔庚寅[27], 洞庭守禮謂縣嗇夫、卒史嘉、叚(假)卒史穀、屬尉, 令曰: 傳送委輸, 必先悉行┛ 城旦舂、隸臣妾、居資贖責(債). 急事不可留, 乃興繇(徭). 今洞庭兵輸內史及巴、南郡、蒼┛ 梧, 輸甲兵當傳者多. 節(即)傳之, 必先悉行乘城卒、隸臣妾、城旦舂、鬼薪白粲、居資贖┛ 責(債)、司寇、隱官、踐更縣者. 田時殹(也), 不欲興黔首. 嘉、穀、尉各謹案所部縣卒、徒隸、居┛ 資贖責(債), 司寇、隱官、踐更縣者簿, 有可令傳甲兵, 縣弗令傳之而興黔首, 可┛ 省少. 弗省少而多興者, 輒劾移縣, (縣)亟以律令具論, 當坐者言名史泰守府. 嘉、┛ 穀、尉在所縣上書, 嘉、穀、尉令人日夜枼(牒)行. 它如律令.[582]
>
> In the 27th year (of King Zhèng) [220 BC], month 2, new moon on a *bǐngzǐ*, *gēngyín* [day 15], Governor Lǐ of Dòngtíng addresses the County Overseer, the Commandery Secretary Jiā, the Commandery Secretary *ad Interim* Gǔ and the Commandery Military Commandant in an ordinance, stating: (If there are things that need) to be transported, one must first send┛ the Wall and Grain Pounders or Bondservants who work off fines, redemption fees or debts. If urgent matters cannot be postponed, then have them carried out by *corvée* labor. Now, the weapons of Dòngtíng are transported to the Ministry of Finance as well as to the commanderies Bā, Nán and Cāng┛ wú. Transporting suits of armor and weapons requires numerous people. In order to transport them, one must first send the Garrison Soldiers as well as the Bondservants, Wall and Grain Pounders, Firewood Gatherers and Rice Sifters who work off ┛ fines, redemption fees or debts, and (then) the Overseers of Bandits, (convicts from the) Hidden Offices and those who perform obligations in other counties. When the time to work the agricultural fields is delayed, one does not want to mobilize the black-headed (people). Jiā, Gǔ and the Commandery Military Commandant should each examine in their (respective) Departments or counties the registers of conscript soldiers, convicts, those who work off fines, redemption┛ fees or debts as well as the Overseers of Bandits, (convicts from the) Hidden Offices and (those who) perform obligations in other counties. If there are (people in these registers), who could be ordered to transport suits of armor and weapons, but the county authorities have not ordered (these people) to transport them and mobilized the black-headed (people) instead, they should order at least┛ as few as possible. If they mobilized more (than needed), immediately file an *ex-officio* charge (against the responsible person) and transfer (it) to the county. The county should immediately compile (the records) and pass judgment according to the statutes and ordinances on those who by law are prosecuted for a criminal offense and should (then) report their names and the decisions to the Commandery Governor. Jiā, Gǔ and the Commandery Military Commandant should submit (the documents) to the counties they are responsible for. Jiā,┛ Gǔ and the Commandery Military Commandant should order the people to deliver the tablets day and night. Everything else (should be handled) in accordance with the statutes and ordinances.[583]

---

taken from Lau and Staack 2016, with one modification: I have replaced the term 'Earth and Grain Pounders' (*chéngdàn chōng* 城旦舂) with 'Wall and Grain Pounders'.

582 *Húnán shěng wénwù kǎogǔ yánjiūsuǒ* 2007: 192–94.

583 Some parts of the English translation of line VI are taken from Lau and Staack 2016, footnote 550.

Comments: The office of 'Commandery Secretaries' or 'Provincial Secretaries' (*zúshǐ* 卒史) is discussed in Lau and Lüdke 2012, footnote 923. 'Hidden Offices' (*yǐnguān* 隱官) employed individuals who had been unjustly subjected to mutilation punishment due to judicial error or intentional perversion of justice, later corrected.[584]

As argued by Katsuya Onishi, the absence of the sentence-final predicate marker or copula '*yě* 也' in Qín dynasty textual sources might be attributed to the Qín authors' use of their dialect word, written as '*yì* 殹' (*qə) in place of '*yě* 也'.[585]

Following legal and administrative reforms after the Qín unification in 221 BC, the term '*mín* 民', meaning 'commoner', was largely substituted for 'black-headed (people)' (*qiánshǒu* 黔首).[586] Interestingly, the term '*mín* 民' only appears once in the *Lǐyē* corpus, on manuscript 9-1411.[587]

Tab. 5.1 lists all categories of convicts found on the *Lǐyē* documents from layers 5, 6 and 8, sorted by age, privileges and sex.

'Overseers of Bandits' (*sīkòu* 司寇) and convicts assigned to 'Hidden Offices' (*yǐnguān* 隱官) were penalized for minor offenses and enjoyed more privileges than other convicts. Classified under the punishment of 'being shaved' (*nài* 耐), these individuals engaged in lighter forms of penal labor and were responsible for supervising other convicts sentenced for more serious crimes.[588] In the *Shuìhǔdì* documents, Overseers of Bandits served two years on average and were occasionally tasked with searching for criminals or suspects. In addition, they were granted residency in township 'households' (*hù* 戶) with slightly larger plots of land compared to convicts from the Hidden Offices.[589] As illustrated in the following *Shuìhǔdì* text,

584 Lau and Staack 2016, footnote 400. For a study on manuscript materiality, textual composition and delivery methods, based on an analysis of manuscripts 9-2283, 16-5 and 16-6, see Ma Tsang Wing 2020b. For a textual analysis that includes the making and transmission of manuscripts 16-5 and 16-6, see Yáng Zhènhóng and Dān Yìnfēi 2014.

585 Onishi 1998.

586 Ma Tsang Wing 2020a. For a study on Qín reforms affecting official terminology, see Chén Kǎnlǐ 2014.

587 Chén Zhèngxián 2020.

588 Lau and Staack 2016: 111, 234, 318. A common disciplinary measure for minor offenses during the Qín dynasty was called '*nài* 耐' and involved shaving off a criminal's beard or hair (perhaps also the underlying skin) as a form of humiliation (see also Lau 1999: 49). Archeological evidence and transmitted texts do not always align regarding this punishment (Lüdke 2016: 221).

589 This information comes from strip 207 of the "Statutes Concerning the Commissioner of Public Works" in the "Qín lǜ shíbā zhǒng" (*Sīkōng lǜ* 司空律). Overseers of Bandits were reportedly categorized as "two-year hard labor convicts" (Yates 1995: 347). I would caution against generalizing the degrees of punishment associated with these terms. Numerous studies on the Qín system of punishments have sparked controversy, indicating that the nature and severity of punishments reveal significant variability depending on the specific time and location within the Qín empire. For further discussion, see Liú Hǎinián 1985; Shāng Qìngfū 1985.

Tab. 5.1: Categories of 'convict (servants)' (*tú(lì)* 徒(隸)) on the *Lǐyē* manuscripts from layers 5, 6 and 8. Convicts from the 'Hidden Offices' (*yǐnguān* 隱官) appear on manuscript 16-5 only (sorted by rank, age and sex).

No.	Type of 'convict (servants)' (*tú(lì)* 徒(隸))	Occurrence(s): Manuscript number(s)
	'Overseers of Bandits' (*sīkòu* 司寇)	
1.1	'Overseers of Bandits' (*sīkòu* 司寇)	**11:** 8-19, 8-145, 8-482, 8-533, 8-567, 8-663, 8-756, 8-1027, 8-1946, 8-2151, 8-2156
	'Released Convicts' (*yǐnguān* 隱官)	
2.1	'Released Convicts' (*yǐnguān* 隱官)	**1:** 16-5
	'Firewood Gatherers and Rice Sifters' (*guǐxīn báicàn* 鬼薪白粲)	
3.1	'Firewood Gatherers' (*guǐxīn* 鬼薪)	**10:** 8-533, 8-683, 8-805, 8-1069+8-1434+8-1520, 8-1161, 8-1279, 8-1515, 8-1566, 8-1143+8-1631, 8-2156
3.2	'Rice Sifters' (*báicàn* 白粲)	**7:** 8-805, 8-1143+8-1631, 8-1279, 8-1207+8-1255+8-1323, 8-1335, 8-1340, 8-1742+8-1956
	'Bondservants' (*lìchén qiè* 隸臣妾)	
4.1	'(male) (Adult) Bondservant' (*(dà) lìchén* (大)隸臣)	**7:** 8-686+8-973, 8-736, 8-1008+8-1461+8-1532, 8-1069+8-1434+8-1520, 8-1515, 8-1524, 8-1542
4.2	'(female) (Adult) Bondservant' (*(dà) lìqiè* (大)隸妾)	**14:** 8-196+8-1521, 8-199+8-688, 8-647, 8-651, 8-904+8-1343, 8-1143+8-1631, 8-1278+8-1757, 8-1279, 8-1524, 8-1538, 8-1557, 8-1566, 8-1584, 8-1641
4.3	'(male) Minor Bondservant' (*xiǎo lìchén* 小隸臣)	**4:** 8-448+8-1360, 8-1153+8-1342, 8-1551, 8-1580
4.4	'(female) Minor Bondservant' (*xiǎo lìqiè* 小隸妾)	**1:** 8-444
	'Wall and Grain Pounders' (*chéngdàn chōng* 城旦舂)	
5.1	'Elderly Wall Pounder' (*zhàng (zhàng) chéngdàn* (丈)仗城旦)	**5:** 8-686+8-973, 8-801, 8-1143+8-1631, 8-1278+8-1757, 8-1279
5.2	'(Adult) Wall Pounder' (*chéngdàn* 城旦)	**11:** 8-119, 8-196+8-1521, 8-533, 8-686+8-973, 8-904+8-1343, 8-1069+8-1434+8-1520, 8-1143+8-1631, 8-1279, 8-1566, 8-1894, 8-2423
5.3	'(Adult) Grain Pounder' (*chōng* 舂)	**14:** 8-59, 8-145, 8-212+8-426+8-1632, 8-216+8-351, 8-337, 8-486, 8-686+8-973, 8-1069+8-1434+8-1520, 8-1143+8-1631, 8-1279, 8-1335, 8-1531, 8-1566, 8-1576

Tab. 5.1 (continued)

No.	Type of 'convict (servants)' (*tú(lì)* 徒(隸))	Occurrence(s): Manuscript number(s)
5.4	'Minor Wall Pounder' (*xiǎo chéngdàn* 小城旦)	**7:** 8-145, 8-162, 8-212+8-426+8-1632, 8-216+8-351, 8-337, 8-1515, 8-1566
5.5	'Minor Grain Pounder' (*xiǎo (nǚzǐ) chōng* 小(女子)舂)	**3:** 8-145, 8-239, 8-1566

Overseers of Bandits were exempt from becoming 'Servants' (*pú* 僕) or 'Provisioners' (*yǎng* 養):

> *Shuìhǔdì*, "Qín lǜ shíbā zhǒng", strip 150: 司寇勿以爲僕、養.[590]
>
> Overseers of Bandits should not be used as Servants or Provisioners.

Aside from providing an incomplete summary of various household types in an unidentified region, *Lǐyē* manuscript 8-19 – damaged and missing both its top and bottom sections – lists five of the 20 noble 'ranks' (*jué* 爵) in hierarchic order, from the fifth lowest to the lowest rank. Notably, Overseers of Bandits are positioned immediately after members of a 'Unit of Five People' (*shìwǔ* 士五 or *wǔ* 伍), implying that they might have been the highest-ranking convicts, tiered just below commoners in the social hierarchy.

> 8-19 AI–VI, BI–V: A ☒□二戶.┘ 大夫一戶.┘ 大夫寡三戶.┘ 不更一戶.┘ 小上造三戶.┘ 小公士一戶.┘ B 士五(伍)七戶.☒┘ 司寇一【戶】.┘ 小男子□☒┘ 大女子□☒┘ · 凡廿五☒[591]
>
> A ☒□ two households.┘ Holders of the Fifth Rank, one household.┘ Widows of Holders of the Fifth Rank, three households.┘ Holders of the Fourth Rank, one household.┘ Minor Holders of the Second Rank, three households.┘ Minor Holders of the First Rank, one household.┘ B Members of a Unit of Five People, seven households.☒┘ Overseers of Bandits, one 【household】.┘ Minor males □☒┘ Minor females □☒┘ · In total, there are 25 ☒
>
> Comments: The 'Fifth Rank' (*dàifu* 大夫), 'Fourth Rank' (*búgèng* 不更), 'Second Rank' (*shàngzào* 上造) and 'First Rank' (*gōngshì* 公士) belonged to the 20 official 'ranks' (*jué* 爵) in the Qín and Hàn dynasties.[592] The First Rank was the lowest tier in this hierarchy. Officials, con-

590 Lau and Staack 2016: 222–223.

591 Chén Wěi 2012: 32–33.

592 For a complete list of the 20 'ranks' (*jué* 爵) from the Qín and Hàn imperial era, see Lau and Lüdke 2012; Nishijima 1961.

victs and commoners were further distinguished as either 'minor' (*xiǎo* 小) or 'adult' (*dà* 大).[593]

As recorded in the *Zhāngjiāshān* texts, Overseers of Bandits and convicts from the Hidden Offices may have been reintegrated into society after completing sentences by serving as 'Public Soldiers' (*gōngzú* 公卒), members of a 'Unit of Five People' (*shìwǔ* 士五 or *wǔ* 伍) or 'Freedmen' (*shùrén* 庶人).[594] However, the *Lǐyē* manuscripts from layers 5, 6 and 8 only mention the first two categories, namely 'Public Soldiers' and members of a 'Unit of Five People'.[595] Freedmen do not appear in these texts.

According to the *Yuèlù* statutes, privileged 'Holders of the Second Rank' (*shàng zào* 上造) or higher, who "committed an offense punishable by the most severe forms of penal labor as 'tattooed (or branded) Wall Pounders' (*qíng chéngdàn* 黥城旦)," could be reassigned to a mitigated class of 'Firewood Gatherers' (*guǐxīn* 鬼薪).[596] This provision extended to their wives, who, instead of being sentenced as 'Grain Pounders' (*chōng* 舂), would serve as 'Rice Sifters' (*báicàn* 白粲) if they were also Holders of the Second Rank or higher.

In the *Lǐyē* corpus, Overseers of Bandits, Firewood Gatherers and Rice Sifters often oversaw 'Wall and Grain Pounders' (*chéngdàn chōng* 城旦舂), who were among the most severely punished convicts. 'Wall Pounders' (*chéngdàn* 城旦), a male-only subdivision of Wall and Grain Pounders,[597] may account for the absence

593 For the distinction between 'minor' (*xiǎo* 小) and 'adult' (*dà* 大) officials and commoners, see Liú Mǐn 2004, 2009; Chén Pán 1975.

594 This information is taken from strips 312, 316 and 354–357 of the "Èrnián lǜlìng" (Lau and Staack 2016: 88). In terms of welfare, hereditary status and land or residency allotment, 'Public Soldiers' (*gōngzú* 公卒) – whether as members of a 'Unit of Five People' (*shìwǔ* 士五 or *wǔ* 伍) or 'Freedmen' (*shùrén* 庶人) – held no official rank and were of equal status. Anthony Barbieri-Low and Robin Yates translate '*shùrén* 庶人' as 'Freedman', describing them as individuals occupying a status between slaves, conscripts and unranked commoners. As they argue, '*shùrén* 庶人' "[signifies] slaves, bondservants, impounded persons or convict-laborers [. . .] [either] manumitted, redeemed or otherwise 'freed'" (Barbieri–Low and Yates 2015: 439, footnote no. 136 to Section 3.1 "Statutes on Assault" (*Zeilü* 賊律)).

595 The term 'Public Soldier' (*gōngzú* 公卒) specifically appears on manuscripts 8-113, 8-430, 8-445, 8-1563 and 8-2246. The term 'Unit of Five People' (*shìwǔ* 士五 or *wǔ* 伍) is mentioned 72 times.

596 Lau and Staack 2016: 133; '*Qíng* 黥' was a form of mutilation punishment and involved tattooing (or branding) the convict's forehead. The expressions '*qíng* 黥' or '*qíng chéngdàn* 黥城旦' do not appear in the *Lǐyē* manuscripts from layers 5, 6 and 8.

597 Legal cases 9 and 10 on strips 54–56 of the "Zòuyànshū" deal with the illicit 'private employment' (*sī shǐ* 私使) of Wall and Grain Pounders by officials (Lau and Staack 2016: 222–223). 'Wall Pounders' (*chéngdàn* 城旦) were a class of convicts dating back to at least the end of the Warring States period. In the Western Hàn dynasty, they were sentenced to four or five years of penal labor, often in remote border regions, and subjected to mutilations such as branding or the am-

of evidence showing Overseers of Bandits or Firewood Gatherers supervising Grain Pounders in the texts.[598] Grain Pounders were instead supervised by same-sex 'Rice Sifters' (*báicàn* 白粲).[599]

Wall Pounders and Firewood Gatherers are usually mentioned in the same entry, just as Grain Pounders appear alongside Rice Sifters. Firewood Gatherers do not share entries with Grain Pounders, nor are Wall Pounders listed with Rice Sifters. Manuscripts 8-2151 and 8-2156 record instances of Overseers of Bandits supervising Wall Pounders. No entry exists where an Overseer of Bandits oversees Grain Pounders or a group comprising both Wall and Grain Pounders.

8-2151 I–IV: ☒一年四月癸未[20]朔己□, ☒┘ ☒城旦司寇一人, ☒┘ ☒□薪廿人, ☒┘ ☒□□四人, ☒[600]

☒ 1st year (of King Zhèng), month 4, new moon on a *guǐwèi*, *jǐ* □, ☒┘ ☒ one Overseer of Bandits supervises ☒ Wall Pounders, ☒┘ ☒□ 20 Firewood,☒┘ ☒□□ four people, ☒

Comments: The year of this entry is most likely the 31st year of King Zhèng (216 BC). This conclusion can be drawn from manuscripts 8-736, 8-1278+8-1757 and 8-1759, where the new moon day *guǐwèi* in month 4 coincides with the 31st year of King Zhèng. The character preceding '*xīn* 薪' could be interpreted as '*guǐ* 鬼', resulting in the term 'Firewood Gatherers' (*guǐxīn* 鬼薪), as seen on manuscript 8-2156:

8-2156 I–III: ☒三月癸丑[50]朔壬戌[59], 【司空】□☒┘ ☒城旦司寇一人. ☒┘ ☒鬼薪十九人.☒[601]

☒ month 3, new moon on a *guǐchǒu*, *rénxū* [day 10], 【Commissioner of Public Works】□☒┘ ☒ one Overseer of Bandits supervises ☒ Wall Pounders. ☒┘ 19 Firewood Gatherers. ☒

8-1279 I–V: 【卅】年八月丙戌[23]朔癸卯[40]□☒┘ 城旦、鬼薪三 人.☒┘ 仗城旦一人.☒┘ 舂、白粲二人.☒┘ 隸妾三人.☒[602]

putation of the nose or left foot. Female 'Grain Pounders' (*chōng* 舂), their counterparts, were distinguished by shaven heads, red clothing, headwear and restraints, including neck irons and bindings on their hands and feet (Lau 1999: 53). Daily food rations for Grain Pounders, Minor Wall Pounders and possibly other convicts were set at four and one-sixth pints, or twenty-five sixths of a pint. For a transcription and translation of manuscript 8-216+8-351, see chapter six.

598 The "evaluation record by the Commissioner of Public Works" (*sīkōng kè zhì* 司空課志) in line AV of manuscript 8-486 also lists "Grain Pounders who gave birth" (*chōng chǎn zǐ* 舂產子) (Chén Wěi 2012: 166).

599 On the category of 'Rice Sifters' (*báicàn* 白粲), see also Lau and Staack 2016: 192.

600 Chén Wěi 2012: 438.

601 Chén Wěi 2012: 438.

602 Chén Wěi 2012: 305.

In the【30th】year (of King Zhèng) [217 BC], month 8, new moon on a *bǐngxū*, *guǐmǎo* [day 18] □☒┘ there are three Wall Pounders and Firewood Gatherers. ☒┘ One Elderly Wall Pounder. ☒┘ Two Grain Pounders and Rife Sifters. ☒┘ Three (female) Bondservants. ☒

'*Zhàng chéngdàn* 仗城旦' are 'Elderly Wall Pounders', a designation that is interchangeable with '*zhàng chéngdàn* 丈城旦', found in Early Qín and Hàn dynasty texts. The *Lǐyē* corpus identifies three types of Wall Pounders: 1) 'Minor Wall Pounders' (*xiǎo chéngdàn* 小城旦), 2) '(adult) Wall Pounders' (*(dà) chéngdàn* (大)城旦), and 3) 'Elderly Wall Pounders' (*zhàng (zhàng) chéngdàn* 仗(丈)城旦).[603]

The imperial institutions also differentiated between adult and nonadult convicts, with obligations and responsibilities varying with age. Manuscript 8-1566 distinguishes between adult and 'Minor Grain Pounders' (*xiǎo (nǚzǐ) chōng* 小(女子)舂), as well as adult and 'Minor Wall Pounders' (*xiǎo chéngdàn* 小城旦). Minor Firewood Gatherers and Minor Rice Sifters do not appear in the texts from layers 5, 6 and 8.

8-1566 I–II, rAI–III, rBI–III: 卅年六月丁亥[24]朔甲辰[41], 田官守敬敢言之: 疏書日食牘北(背)上.┘ 敢言之. ┘ A 城旦、鬼薪十八人.┘ 小城旦十人.┘ 舂廿二人.┘ B 小舂三人.┘ 隸妾居貲三人. └戊申[45], 水下五刻, 佐壬以來.ノ尚手. 逐手.[604]

In the 30th year (of King Zhèng) [217 BC], month 6, new moon on a *dīnghài*, *jiǎchén* [day 18], Jìng, Incumbent of the Office of Agricultural Fields, dares to say: The document containing (a list of) daily food rations is on the back side of the tablet.┘ (He) dares to say. └A 18 Wall Pounders and Firewood Gatherers.┘ Ten Minor Wall Pounders.┘ Two Grain Pounders.┘ B Three Minor Grain Pounders.┘ Three (female) Bondservants working off their fines.┘ On a *wùshēn* [day 22], at water below mark 5, (the document) is brought by Assistant Rén. / By the hand of Shàng. By the hand of Zhú.

'Elderly Wall Pounders' (*zhàng (zhàng) chéngdàn* 仗(丈)城旦) were assigned a separate status within Qín society, as reflected in the following manuscript, which contains a monthly record of convict laborers employed in Èrchūn District. 'Elderly' (*zhàng (zhàng)* 仗(丈)) was applied exclusively to male convicts and did not extend to female Grain Pounders.

8-1143+8-1631 AI–V, BI–V: A 卅年八月貳春鄉作徒薄(簿).┘ 城旦、鬼薪積九十人.┘ 仗城旦積卅人.┘ 舂、白粲積六十人.┘ 隸妾積百一十二人.┘ B・凡積二百九十二人.☒┘ 卅人甄.☒┘ 六人佐甄.☒┘ 廿二人負土.☒┘ 二人□瓦.☒[605]

603 Chén Wěi 2012: 203.
604 Chén Wěi 2012: 362.
605 Chén Wěi 2012: 283.

A In the 30th year (of King Zhèng) [217 BC], month 8, Èrchūn District compiles the register of convicts.⌐ Wall Pounders and Firewood Gatherers count 90 people.⌐ Elderly Wall Pounders count 30 people.⌐ Grain Pounders and Rice Sifters count 60 people.⌐ (Female) Bondservants count 112 people.⌐ B · In total, they count 292 people. ⧄⌐ 30 people make pottery. ⧄⌐ Six people assist in making pottery. ⧄⌐ 22 people carry earth. ⧄⌐ Two people □ tiles ⧄.

Comments: In addition to monthly reports, the total number of convicts employed in Èrchūn District was also tracked daily. Manuscript 9-18, for example, notes that seven convicts were counted in a single day. From this entry we learn that *'jī* 積' does not refer to the total number of convicts in the District, but rather the number of convicts dispatched on a daily, monthly or similar interval.[606]

Strict segregation by sex was also enforced among 'Bondservants' (*lìchén qiè* 隸臣妾).[607] Whereas *'lìchén qiè* 隸臣妾' is a unisex designation, *'chén* 臣' referred to male Bondservants and *'qiè* 妾' to female Bondservants or concubines. These Bondservants were allowed to raise families and perform official tasks, such as delivering documents or investigating minor offenses. Unlike Overseers of Bandits, Bondservants could become 'artisans' (*gōng* 工) if they had the requisite skills.[608] This allowed them to avoid becoming 'Servants' (*pú* 僕) or 'Provisioners' (*yǎng* 養).

Shuìhǔdì, "Qín lǜ shíbā zhǒng", strip 113: 隸臣有巧可以爲工者, 勿以爲人僕、養.[609]

When (male) Bondservants have skills and could be made artisans, do not make them Servants or Provisioners.

Some *Lǐyē* texts not only reveal figures of convicts but also specify their assigned tasks and personal names. Manuscripts 8-686+8-973 and 8-1069+8-1434-8-1520 are prime examples of well-preserved records from a 'Warehouse' (*kù* 庫) that meticulously compiled daily registers of convicts by name. These documents show how convicts were either 'given away' (*fù* 付) or 'received' (*shòu* 受) in terms of the total number.[610] Manuscript 8-1027 even identifies the township households to which various Overseers of Bandits were officially registered.

8-686+8-973 AI–IV, BI–III, rCIII: A 廿九年八月乙酉[22], 庫守悍作徒薄(簿): 受司空城旦四人、丈城旦一人、舂五人、受倉隸臣一人.・凡十一人.⌐ 城旦二人繕甲□□.⌐ 城旦一人治輸□□.⌐ 城旦一人約車: 登.⌐ B 丈城旦一人約車: 缶.⌐ 隸臣一人門: 負劇.⌐ 舂三人級: 娉、□、娃.⌐ C 廿

606 Zhāng Xīnchāo 2014.
607 For more information on the identity of Bondservants, see Lǐ Lì 2007.
608 Cáo Lǚnìng 2015.
609 Lau and Staack 2016: 222–223.
610 There are also monthly and annual registers of convicts in the *Lǐyē* texts. For a list of these records, see Liáng Wěijié 2013.

廿年上之☑ └八月乙酉[22], 庫守悍敢言之: 疏書作徒薄(簿)牘北(背)上, 敢言之. 逐手.┘ 乙酉[22]旦, 隸臣負劇行廷.[611]

A In the 29th year (of King Zhèng) [218 BC], month 8 on a *yǐyǒu*, Warehouse Incumbent Hàn compiles the register of convicts: (We) received from the Commissioner of Public Works four Wall Pounders, one Elderly Wall Pounder, five Grain Pounders, and from the Granary, one (male) Bondservant. • In total, (we received) 11 people.┘ B Two Wall Pounders repair suits or armor □□.┘ One Wall Pounder handles transportation □□.┘ One Wall Pounder equips the carriages: Fǒu.┘ One (male) Bondservant stands at the gate: Fùjù.┘ Three Grain Pounders order the silk threads: Kuā, □, Wá.┘ C 20, 20 years submit it (to the higher authorities). ☑ └In month 8 on a *yǐyǒu*, Warehouse Incumbent Hàn dares to say: The document containing the register of convicts is on the back side of the tablet. (He) dares to say. By the hand of Zhú.┘ In the morning of a *yǐyǒu*, (male) Bondservant Fùjù delivers (the document) to the (County) Court.

Comments: According to Chén Wěi, '*jí* 級' means the "quality of the silk" (*sī zhī cìdì* 絲之次第).[612] Kuā 姱 and Wá 娃 are female names, which reinforces that 'Grain Pounders' (*chōng* 舂) were female convicts. The sentence "20, 20 years submit it (to the higher authorities)" (*niàn niàn nián shàng zhī* 廿廿年上之) is written in noticeably large characters, possibly by someone other than Zhú. This inscription appears unrelated to the main body of the text (Fig. 5.1).

8-1069+8-1434-8-1520 I–IV, rI–II: 卅二年五月丙子[13]朔庚子[37], 庫武作徒薄(簿): 受司空城旦九人、鬼薪一人、舂三人; 受倉隸臣二人. • 凡十五人.┘ 其十二人爲萁: 奬、慶忌、䵋、䵋、船、何、冣、交、頡、徐、娃、聚; ┘ 一人紺: 竄. ┘ 二人捕羽: 亥、羅. └卅二年五月丙子[13]朔庚子[37], 庫武敢言之: 疏書作徒日薄(簿)一牒. 敢言之. 横手.┘ 五月庚子[37]日中時, 佐横以來. ノ圂發.[613]

In the 32nd year (of King Zhèng) [215 BC], month 5, new moon on a *bǐngzǐ*, *gēngzǐ* [day 25], Wǔ from the Warehouse compiles the register of convicts: (We) received from the Commissioner of Public Works nine Wall Pounders, one Firewood Gatherer and three Grain Pounders; (We) received from the Granary two (male) Bondservants. • In total, (we received) 15 people.┘ Among them, 12 people carry heavy objects: Jiǎng, Qìngjì, X [pronunciation unknown], X [pronunciation unknown], Chuán, Hé, Jù, Jiāo, Xié, Xú, Wá, Jù;┘ One person weaves: Cuàn.┘ Two people catch bird feathers: Hài, Luó. └ In the 32nd year (of King Zhèng) [215 BC], month 5, new moon on a *bǐngzǐ*, *gēngzǐ* [day 25], Wǔ from the Warehouse dares to say: The document with the daily register of convicts is written on one tablet. (He) dares to say. By the hand of Héng.┘ Month 5 on a *gēngzǐ* [day 25], at midday time, (the document) is brought by Assistant Héng. / Sent by Hùn.

611 Chén Wěi 2012: 203.
612 Chén Wěi 2012: 203–204.
613 Chén Wěi 2012: 272–273.

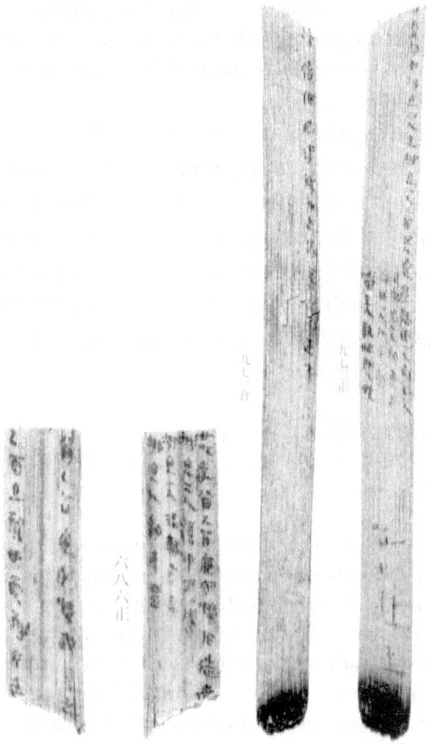

Fig. 5.1: Monochrome photographs of manuscripts 8-686 (l) and 8-973 (r), *verso* and *recto*. Ink on wood. Dimensions unknown.[614] Traces of fire damage are visible at the bottom of manuscript 8-973.

'*Kù* 庫' is an abbreviation for 'Overseer of the Warehouse' (*kùsèfū* 庫嗇夫) in this and similar contexts.[615] Following the transcriptions of comparable texts and the first line on the *recto* of this manuscript, I have included the character '*bù* 簿' in parentheses in the initial sentence. Chén Wěi reads '*xù* 萸' (*laʔ) as the verb '*yú* 輿' (*laʔ), which, in this instance, means 'to carry heavy objects on one's shoulder'.[616]

8-1027 I–II: 成里戶人司寇宜. ☐┘ 下妻𠚕.☐[617]

Yí, Overseer of Bandits, is a person from a household in the township of Chéng. ☐┘ The lower wife named X [pronunciation unknown]. ☐

---

614 *Húnán shěng wénwù kǎogǔ yánjiūsuǒ* 2012a: 98, 133.
615 Chén Wěi 2012: 90.
616 Chén Wěi 2012: 273. Phonological reconstructions are taken from Schuessler 2009: 55.
617 Chén Wěi 2012: 264.

The social status of 'lower wives' (*xià qī* 下妻) was comparable to that of 'unmarried wives', 'demoted wives' or 'concubines' (*qiè* 妾).[618] This designation is already known from strip 368 of the "Èrnián lǜlìng". 'Unmarried wives' were also referred to as '*xiǎo qī* 小妻' or '*páng qī* 旁妻'.

On manuscript 8-686+8-973 (translated above), the male Bondservant Fùjù 負劇 is tasked with guarding the 'gate' (*mén* 門) and 'dispatching' (*xíng* 行) the document to the County Court. Bondservants and other convicts, much like officials, had multiple responsibilities and were deployed according to their skills, aptitudes and the administrative needs. When specific tasks required certain abilities, these abilities were prioritized over other factors such as age or the severity of punishment. On manuscripts 8-1515 and 8-1524, for instance, both a male and a female Bondservant are assigned to work as messengers:

8-1515 I–III, rI: 卅年十月辛卯[28]朔乙未[32], 貳春鄉守綽敢告司空主: 主┛ 令鬼薪軫、小城旦乾人爲貳春鄉捕鳥及羽. 羽皆已┛ 備, 今已以甲午屬司空佐田, 可定薄(簿). 敢告主. ┗十月辛丑[38]旦, 隸臣良朱以來. ノ死手. 邛手.[619]

In the 30th year (of King Zhèng) [217 BC], month 10, new moon on a *xīnmǎo*, *yǐwèi* [day 5], Chuò, Incumbent of Èrchūn District, dares to say to the Commissioner of Public Works: He [the Commissioner of Public Works]┛ ordered Firewood Gatherer Zhěn and Minor Wall Pounder Qián to become bird and feather catchers in Èrchūn District. All bird feathers (that were needed) are already┛ completed. Now, already by day 4, these have been entrusted to Tián, Assistant to the Commissioner of Public Works. It can be recorded in the register of convicts. (He) dares to say to him [the Commissioner of Public Works]. ┗In month 10 in the morning of a *xīnchǒu* [day 11], (the document) is brought by (male) Bondservant Liángzhū. / By the hand of Sǐ. By the hand of Qióng.

Comments: In line with transcription conventions of other *Lǐyē* documents, I have substituted the comma in the first line of Chén Wěi's version with a colon, after "dares to say to the Head Commissioner of Public Works" (*gǎn gào sīkōng zhǔ* 敢告司空主). For a study on the documented history of featherwork in Early and Medieval China, see Milburn 2020.

8-1524 I–II, rI: 廿九年十二月丙寅[3]朔己卯[16], 司空色敢言之: 廷令隸臣□行┛ 書十六封, 曰傳言. 今已傳者, 敢言之. ┗己卯[16]水下六刻, 隸妾畜以來. ノ綽手. 郤手.[620]

In the 29th year (of King Zhèng) [218 BC], month 12, new moon on a *bǐngyín*, *jǐmǎo* [day 14], Sè, Commissioner of Public Works, dares to say: The County Court ordered (male) Bondservant □ to deliver┛ 16 documents. It states to transmit orally (what has been decided). Now, it has already been transmitted. (He) dares to say. ┗On a *jǐmǎo* [day 14], at water below mark

618 Chén Wěi 2012: 264. For a study on kinship, marriage and the influence of legal or administrative regulations on the status of women during the Qín and Hàn dynasty, see, for example, Hinsch 2002: 33–57, 79–91.

619 Chén Wěi 2012: 343.

620 Chén Wěi 2012: 348.

6, (the document) is brought by (female) Bondservant Xù. / By the hand of Chuò. By the hand of X [pronunciation unknown].

Compulsory labor as a Bondservant was a punishment inflicted not only on legal offenders but also on their close relatives. Other circumstances leading to this sentence included surrendering as an enemy combatant to the Qín army or, in particular cases, escaping as a 'captive' (*lǔ* 虜) and voluntarily returning after a certain period of time.[621]

Torture was deemed an adequate and acceptable method of ensuring accurate testimony, verifying confessions during an inconsistent defense, or extracting supplementary evidence from suspects. This practice is illustrated on manuscript 8-439+8-519+8-537, an arrest warrant for the absconded "convict in a Unit of Five People" (*tú wǔ* 徒伍), Liáokě 繚可. Dated to the 25th year of King Zhèng (222 BC), one year before the proclamation of the Qín empire, it reads:

8-439+8-519+8-537 I–II: 廿五年九月己丑[2], 將奔命校長周爰書: 敦長買、什長嘉皆告曰: 徒士五(伍)右里繚可, 行到零陽廡谿橋亡, 不智(知)□□☐ˉ 繚可年可廿五歲, 長可六尺八寸, 赤色, 多髮, 未產須, 衣絡袍一、絡單胡衣一, 操具弩二、絲弦四、矢二百、鉅劍一、米一石. ☐[622]

In the 25th year (of King Zhèng) [222 BC], month 9 on a *jǐchǒu*, Zhōu, Section Commander with an Emergency Deployment Troop, writes a protocol: Mǎi, Commanding Official of the Emergency Deployment Troop, and Jiā, Commander of a Unit of Ten People, unanimously report: Liáokě, a convict in a Unit of Five People from the township Yòu, went to the bridge of Wǔxī in Língyáng and absconded. We do not know □□☐ˉ Liáokě is about 25 years old, about six feet and eight inches tall, [his skin is of] red color, very hairy, no whiskers, (wears) one long cotton robe and one round collar cotton robe. (He) seized tools, two crossbows, four silk strings, 200 arrows, one iron sword and one picul of rice. ☐

Comments: This wooden strip contains text only on the *verso*, with the lower section being heavily damaged. Arrest warrants for absconded convicts are relatively rare in the *Lǐyē* corpus. Only manuscripts 8-534, 8-894, 8-1863+8-1866, 8-2098, 12-140 and 15-259 offer similar content.[623]

Chén Wěi interprets the compound noun '*bēn mìng* 奔命' as referring to 'troops appointed to fight in emergency situations'. He justifies this argument with excerpts from strips 399 and 1252 of the "Èrnián lǜlìng".[624] I agree with him on this point.

Língyáng 零陽, a county in Dòngtíng Commandery, also appears on manuscripts 5-1, 8-159, 8-375, 8-439+8-519+8-537, 8-1886 and 8-2430. The exact locations of Yòu 右 and the bridge at Wǔxī (or Wǔxīqiáo 廡谿橋) remain unknown.

621 For further information on the status of 'captives' (*lǔ* 虜), see Yates 2022.

622 Chén Wěi 2012: 149.

623 For a discussion on arrest warrants in the *Lǐyē* texts, see Hé Yǒuzǔ 2017. On becoming a fugitive, based on a study of manuscript 8-439+8-519+8-537, see Xiè Kūn 2017.

624 Chén Wěi 2009, 2012: 149.

The second and third occurrence of the character '*kě* 可' in line two should be read as an adverb, meaning 'about', 'around' or 'approximately'.[625] Earlier attested uses of this term appear, for example, in the "Outer Compendium of Explanations" (*Wài chǔ shuō* 外儲說) of the *Hánfēizǐ* 韓非子.[626] This particular usage is also found in later "Zòuyànshū" texts.

The measurement "six feet and eight inches" (*liù chǐ bā cùn* 六尺八寸) equates to 156.4 centimeters, and a 'picul' (*dàn* 石) was approximately 30 kilograms in the Qín dynasty.[627]

'*Yuán shū* 爰書' in the first line of the manuscript was a type of judicial protocol used for recording confessions obtained through torture. A commentary by the Táng dynasty historian Yán Shīgǔ 顏師古 (581–645 AD) in the *Book of Hàn* (*Hànshū* 漢書) states the following:

> *Book of Hàn* (*Hànshū* 漢書), "Biography of Zhāng Tāng" (*Zhāng Tāng zhuàn* 張湯傳): 爰,換也,以文書代替其口辭也.[628]
>
> '*Yuán*' is to exchange; namely to use a written document for what someone has spoken.

'*Yuán shū* 爰書' takes on additional meanings in the *Shuìhǔdì* documents, where it frequently serves as an introductory phrase to judicial texts. These documents address various aspects of legal procedures, including:

1) The conduct of officials during trials and forensic investigations,
2) The handling of confiscated property,
3) Procedures for hearings and confessions made by convicts,
4) Testimonies provided by individuals related to the case,
5) The issuance of final verdicts.[629]

Documents titled "Models for Sealing and Investigating" (*Fēng zhěn shì* 封診式), excavated from *Shuìhǔdì* tomb 11, are an insightful resource for understanding the term's usage in early judicial proceedings:[630]

> *Shuìhǔdì*, "Models for Sealing and Investigating" (*Fēng zhěn shì* 封診式), strips 2–5: 凡訊獄, 必先盡聽其言而書之, 各展其辭, 雖智(知)其訑, 勿庸輒詰. 其辭已盡書而毋(無)解, 乃以詰者詰之. 詰之有(又)盡聽書其解辭, 有(又)視其它毋(無)解者以復詰之. 詰之極而數訑, 更言不服, 其律當治(笞)諒(掠)者, 乃治(笞)諒(掠). 治(笞)諒(掠)之必書曰: 爰書: 以某數更言, 毋(無)解辭, 治(笞)訊某.[631]

625 Chén Wěi 2012: 150.
626 Chén Wěi 2012: 150.
627 Qiū Guāngmíng and Lóng Yángpíng 2001: 447.
628 Bān Gù and Yán Shīgǔ 1962: 2637.
629 For more details on judicial practice in the Qín and Hàn dynasties, see Andreini 2015; Lüdke 2016.
630 Hulsewé 1985: 183–207.
631 *Shuìhǔdì Qín mù zhújiǎn zhěnglǐ xiǎo zǔbiān* 1990: 148.

As a general rule for conducting interrogations in criminal cases, you first have to hear out the statements of those concerned and record these in writing, (allowing) each of the persons (interrogated) to present their statement (without interruption). Even if you know that the person concerned lies, there is no need to confront him immediately. If someone's statement has been recorded completely in writing, and there is still no (full) explanation, confront him with what is confrontational. When confronting him, again, fully hear him out and write down the statements he offers in explanation. Then look again at further points for which he has no (full) explanation and again confront him with these points. If he has been confronted (with inconsistencies) to the utmost and still repeatedly lies, changes his statements and does not succumb, then, where the respective statutes warrant torture, use torture. If you torture him by caning, it is mandatory to write down: "Protocol: Because X repeatedly has changed his statement and has not offered a statement which would provide a (full) explanation, I interrogated X with the application of the cane."[632]

Comment: The characters '*zhì* 治' (*r-lə, r-ləh) and '*chī* 笞' (*r-lhə), as well as '*liàng* 諒' (*raŋh) and '*liàng* 掠' (*raŋh), are phonetic loans.[633]

In more lenient cases, transgressions were often handled autonomously by prefectural or provincial authorities. This is illustrated on manuscript 8-1008+8-1461 +8-1532, where the Deputy Prefect of Qiānlíng personally delivers the verdict:

8-1008+8-1461+8-1532 I–III, rI: 令佐華自言: 故爲尉史, 養大隸臣豎負華補錢五百, 有約券. 豎捕戍卒□□事贖耐罪賜, 購千百五十┘ 二. 華謁出五百以自償.┘ 卅五年六月戊午[55]朔戊寅[15], 遷陵守丞銜告少内問: 如辤(辭), 次豎購當出界華, 及告豎令智(知)之. ノ華手. └ 華☑[634]

Prefectural Assistant Huà himself submitted a request (containing an appeal to the higher authorities): In the past, (when I was) Secretary to the Military Commandant, Provisioner and (male) Adult Bondservant Shù owed (me), Huà, an additional 500 coins. There is a tally concerning the agreement. Shù caught the Convict Soldier □□ matter (punished by paying the fee for) redemption from being shaved. He is entitled to receive a reward of┘ 1,152 (coins). I, Huà, request to give out 500 (coins) in order to recompense myself.┘ In the 35th year (of King Zhèng) [212 BC], month 6, new moon on a *wùwǔ*, *wùyín* [day 21], Xián, Deputy Prefect of Qiānlíng, reports to the Lesser Treasury with the enquiry: According to the complaint, Shù's reward should be given to Huà, and report to Shù to make this known to him. / By the hand of Huà. └ Huà ☑

Comments: The official title 'Prefectural Assistant' (*lìngzuǒ* 令佐) only appears in documents from Lǐyē and has not been found elsewhere.[635]

---

632 Translation taken from Lau and Staack 2016: 62, with minor modifications.

633 Phonological reconstructions are taken from Schuessler 2009: 76, 98–99.

634 Chén Wěi 2012: 261.

635 Lau and Staack 2016: 91. This title appears on manuscripts 6-5, 8-149, 8-197, 8-433, 8-460, 8-647, 8-919, 8-967, 8-1532 and 8-1583, for instance.

'*Zì yán* 自言' is a general term used when an official addresses the authorities voluntarily regarding a personal matter.[636] It may have originally been associated with '*zì jiàn* 自荐', meaning 'to recommend oneself' or 'to offer one's services'.[637]

In legal documents from the Qín and Hàn dynasties, 'tallies' (*quàn* 券) were certificates of written contracts divided into two or three parts used to record transactions of raw materials, cash and commodities between multiple parties.[638] The term '*quàn* 券' also signifies 'to make a contract on a tally' or 'to certify by a tally'.

Chén Wěi posits that '*cì* 賜' is a person's name.[639] I concur with Lau and Staack, identifying this term as 'reward' stipulated in units of gold or 'coins' (*qián* 錢).[640] '*Cí* 辭' carries multiple meanings in Qín and Hàn legal texts, such as 'statement (during interrogation)', 'defense', 'accusation' and 'to lodge a complaint'.[641]

Based on readings of a Qín mathematical manual in the *Yuèlù* manuscripts, one 'suit of armor' (*jiǎ* 甲) was worth two 'ounces' (*liǎng* 兩) and one 'mace' (*chuí* 錘) of gold – equivalent to 1,344 coins.[642] 1,152 coins equate to two piculs, hence one picul corresponds to 576 coins.[643] Line number two on the *verso* of *Lǐyē* manuscript 8-60+8-656+8-665+8-748 reflects the findings in the *Yuèlù* documents.

8-60+8-656+8-665+8-748 II: [. . .] 貲三甲, 爲錢四千卌二.[. . .][644]

A fine of three suits of armor equals 4,032 coins.

More complex investigations were likely carried out within the guidelines established by the 'Prosecutors' (*yùshǐ* 御史). Depending on the context, severity and unique circumstances of each case, some outcomes were reported back to the central administration via the County Courts, Commandery Military Commandants and Governors, ensuring that judicial investigations were properly executed by subordinate authorities.

8-728+8-1474 I, rI: ☐獄南曹書二封, 遷陵印: 一洞庭泰守府, 一洞庭尉府. • 九月☐ ㇗☐己亥[36] 餔時, 牢人誤以來.☐[645]

636 Lau and Staack 2016: 154–155.
637 Bǔ Xiànqún and Liú Yáng 2013: 81.
638 Lau and Staack 2016: 98, 318; For a discussion on 'tallies' (*quàn* 券), see also Ma Tsang Wing 2017.
639 Chén Wěi 2012: 261.
640 Lau and Staack 2016: 106.
641 Lau and Staack 2016: 156–157.
642 Chén Wěi 2014a.
643 For a study on the fixed rates and fluctuations of the price of gold in the Qín dynasty, see Qiū Guāngmíng 1992; Yú Zhènbō 2013.
644 Chén Wěi 2012: 43.
645 Chén Wěi 2012: 211.

> ☒ There are two documents from the Judiciary Bureau of the South with the seal of Qiānlíng: One (is addressed to) the Governor of Dòngtíng, and one (is addressed to) the Military Commandant of Dòngtíng. • Month 9 ☒ └☒ on a *jǐhài*, at dinner time, (the document) is brought by Prisoner Wù. ☒

Manuscript 8-644 recounts a case where a convict was assigned to guard a 'vessel' (*qì* 器). During his shift, the vessel goes missing, and the convict later dies. Since this rare incident was not regulated by existing statutes and ordinances, the legal prosecutor had to investigate whether the official responsible for the convict's supervision should be held accountable for the loss:

> 8-644 I–III, rI–II: 敬問之: 吏令徒守器而亡之, 徒┘ 當獨負. • 日足以責, 吏弗責, 負者死┘ 亡, 吏代負償. └ 徒守者往戍可(何)? 敬訊而負之, 可不可?┘ 其律令云何? 謁報.[646]

> Jìng clarifies the following: An official orders a convict to guard the vessel but he loses it. The convict┘ should pay the compensation himself. • One day's (worth of salary) is enough to pay the compensation, but the official has not yet (received) the payment. The one who is held accountable has died, and the official is instead to compensate. How about (cases) in which the convict guarding (the vessel) dies while working? Jìng is conducting an interrogation into who is responsible, is this permissible?┘ What do the statutes and ordinances say? Request a response (containing a decision taken by the higher authorities).

The absence of personal names and the use of generic terms such as 'official' (*lì* 吏) and 'convict' (*tú* 徒) suggest that the text was written for, what appears to be, educational purposes, like many texts in the "Zòuyànshū".

With regard to legal proceedings, one might expect the term '*fǎ* 法' – which Derk Bodde declares "by far the most important word in the Chinese legal vocabulary" – to be highly recurrent in the *Lǐyē* corpus.[647] However, the term is conspicuously absent from the manuscripts of layers 5, 6 and 8. Recent scholarship understands '*fǎ* 法' to mean '(penal) law code', 'precedent', 'class of offenses' and 'legal principle'. Miranda Brown and Charles Sanft propose a more inclusive, technical interpretation, suggesting that in Qín and Early Hàn legal texts, the term refers to a "legal category" associated with "heuristic, paradigmatic cases or prototypes" used by officials in judicial decision-making.[648] Arguments presented by Liáng Zhìpíng and Cáo Lǚnìng emphasize the interchangeability of the terms '*fǎ* 法', '*lǜ* 律' (statute) and '*xíng* 刑' (punishment or penalty). Yet, these claims

---

646 Chén Wěi 2012: 188.

647 Bodde 1963: 379.

648 Brown and Sanft 2011; 285, 292, 301–302. See also Xíng Yìtián 1983: 52–54; Xú Shìhóng 1997: 66–67.

have limitations and may only apply in certain instances within documented jurisprudence.[649] An excerpt from strip 174 of the *Zhāngjiāshān* texts (186 BC) reads:

> *Zhāngjiāshān*, "Zòuyànshū", strip 174: 異時魯灋(法)【曰】盜一錢到廿: 罰金一兩. [. . .][650]
>
> In a different age, the *fǎ* of the Lǔ state【mandated】that if between one and twenty coins were stolen, (the thief would) be fined one ounce of gold. [. . .][651]

In another passage from the same text, '*fǎ*' goes beyond the scope of criminal investigations:

> *Zhāngjiāshān*, "Zòuyànshū", strip 185: 律, 死置後之次, 妻次父母; 妻死歸寧. 與父母同法.[652]
>
> According to the statutes, the sequence to be followed in establishing heirs is that the wife is secondary to the father and mother. If the wife dies, the husband is granted a leave of absence for burial according to the same *fǎ* that is used for parents.[653]

While the terms '*fǎ* 法' and '*xíng* 刑' are absent in the *Lǐyē* corpus, the noun 'statutes' (*lǜ* 律) appears six times, and the fixed nominal compound 'statutes and ordinances' (*lǜlìng* 律令) 43 times.[654] '*Lǜlìng* 律令' frequently appears in fixed idioms; for example, 13 times in "to follow up on the matter in accordance with the statutes and ordinances" (*yǐ lǜlìng cóng shì* 以律令從事) and eight times in "everything else (should be handled) in accordance with the statutes and ordinances" (*tā rú lǜlìng* 它如律令). Together, these two expressions account for half of the total occurrences of the term in the *Lǐyē* texts, excluding eight instances found on damaged or illegible manuscripts.[655] Other occurrences of '*lǜlìng* 律令' are listed below:

---

649 Cáo Lǚnìng 2002: 151; Liáng Zhìpíng 1989: 61. For a semantic comparison between '*fǎ* 法' and '*lǜ* 律', see Tomiya Itaru 2010: 42.

650 Yates and Barbieri-Low 2015: 1370.

651 Translation is taken from Brown and Sanft 2011: 286. In this instance, '*fǎ* 法' could be an archaic statute or regulation, i.e. a rule dictating the proper course of action in a given circumstance.

652 Xíng Yìtián 2008.

653 Translation is taken from Brown and Sanft 2011: 287. My *pīnyīn* transcription. For more information on bereavement leave for officials, see Brown 2007: 26.

654 The term 'statutes' (*lǜ* 律) appears on manuscripts 8-110+8-669, 8-463, 8-508, 8-1198, 8-1517 and 8-1454+8-1629.

655 The fixed idiom "to follow up on the matter in accordance with the statutes and ordinances" (*yǐ lǜlìng cóng shì* 以律令從事) appears on manuscripts 8-21, 8-63, 8-131, 8-140 (twice), 8-229, 8-657, 8-904+8-1343, 8-1510, 8-1525, 8-1538, 8-1563 and 8-2115. "Everything else (should be handled) in accordance with the statutes and ordinances" (*tā rú lǜlìng* 它如律令) is found on manuscripts 5-6, 8-155, 8-657, 8-830+8-1010, 8-1560, 8-1668, 8-1901 and 8-2166.

8-157 rI–II: [. . .] 遷陵丞昌卻之啓陵: 廿七戶已有一典, 今有(又)除成爲典, 何律令┛ 𪚥(應)? 尉已除成、匄爲啓陵郵人, 其以律令. [. . .][656]

[. . .] Chāng, Prefect of Qiānlíng, declined to go to Qǐlíng. 27 households already have a Township Administrator. Now, Chú and Chéng are promoted to Township Administrators. Which statutes and ordinances┛ apply? The (County) Military Commandant has already (promoted) Chú, Chéng and Gài to Postmen. He used the statutes and ordinances. [. . .]

Comments: According to strip 201 of the "Èrnián lǜlìng", strip 329 of the "Statutes Concerning Households" (*hù lǜ* 戶律), and strip 390 of the "Statutes on the Establishment of Heirs" (*Zhìhòu lǜ* 置後律) from the *Zhāngjiāshān* texts, an 'Administrator' (*diǎn* 典) ranked just below the 'Principal' (*zhèng* 正) within the organizational structure of a 'township' (*lǐ* 里).[657]

8-169+8-233+8-407+8-416+8-1185 IV: 以律令成齎. 來復傳. 敢言之. ☑[658]

Use the statutes and ordinances to complete assistance. Come and send a reply. (He) dares to say. ☑

Robin Yates, Anthony Barbieri-Low, Mǎ Yí and others have argued that the introduction of universal military conscription in the Qín dynasty contributed to a certain degree of literacy among the male population. This demographic, by possessing the ability to read official documents aloud, played a significant role in making these texts more accessible to illiterate segments of society.[659]

8–173 I: 卅一年六月壬午[19]朔庚戌[47], 庫武敢言之: 廷書曰令史操律令詣廷讎[660]

In the 31st year (of King Zhèng) [216 BC], month 6, new moon on a *rénwǔ*, *gēngxū* [day 29], Wǔ from the Warehouse dares to say: The Court document states that the Prefectural Secretary understands the statutes and ordinances and pays a visit to the (County) Court to collate them.

'*Kù* 庫' is an abbreviation for 'Overseer of the Warehouse' (*kùsèfū* 庫嗇夫) in this and similar contexts.[661] According to this text, county-level statutes and ordinances were issued by the '(County) Courts' (*tíng* 廷) and distributed to officials in subordinate institutions for proofreading and collation with existing legal codes. The entire process – illustrated in the text by the Prefectural Secretary bringing documents to the County Court for revision – took one month and two days to complete in this particular case (or from day 29 in month 6 to day 27 in month 7). This process was designed to ensure that all applicable statutes and ordinances were up-to-date and aligned with the versions published and approved by the

656 Chén Wěi 2012: 94.
657 Chén Wěi 2012: 95.
658 Chén Wěi 2012: 102.
659 Barbieri-Low 2007: 63–66; Mǎ Yí 2006; Yates 2009: 40. For insights into the spread of scribal literacy during the Western Hàn dynasty, see Foster 2017.
660 Chén Wěi 2012: 104.
661 Chén Wěi 2012: 90.

County Courts. As indicated by Chén Zhōnglóng, officials at prefectural institutions in the Early Hàn dynasty were required to collaborate with the County Courts to regularly collate statutes and ordinances at both fixed and variable intervals.[662]

'*Lǜ* 律' and '*lǜlìng* 律令' seem to have different usages depending on the context and administrative stage in which they were applied. For example, '*lǜlìng* 律令' was used when high-ranking officials, such as Prefects or Deputy Prefects, issued instructions to lower-ranking officials or institutions under their jurisdiction:

8-63 V: 廿七年十月庚子[37], 遷陵守丞敬告司空主: 以律令從事言. [. . .][663]

In the 27th year (of King Zhèng) [220 BC], month 10 on a *gēngzǐ* [day 12], Jìng, Deputy Prefect of Qiānlíng, reports to the Commissioner of Public Works: Follow up on the matter in accordance with the statues and ordinances and communicate orally. [. . .]

8-140 IV: 臨沮丞禿敢告遷陵丞主、令史, 可以律令從事. 敢告主. [. . .][664]

The Prefect of Línjǔ dares to report to the Prefect of Qiānlíng, ordering the Prefectural Secretary to approve and to follow up on the matter in accordance with the statutes and ordinances. (He) dares to report to the Prefect. [. . .]

Comments: According to Chén Wěi, Línjǔ was a county during the Qín and Hàn dynasties.[665] His transcription of the text implies that the Prefect of Línjǔ reported to both the Prefect and the Prefectural Secretary of Qiānlíng.[666] Chén Jiàn argues that '*lìng* 令' should be read as part of a verb-object collocation in the subordinate clause, meaning 'to order'.[667] Since it is unlikely that the Prefect of Línjǔ would report to the Prefectural Secretary, I agree with Chén Jiàn's interpretation.

8-155 I–II: 四月丙午[43]朔癸丑[50], 遷陵守丞色下: 少内謹案致之. 書到言, 署金布發, 它如┘律令. [. . .][668]

In month 4, new moon on a *bǐngwǔ, guǐchǒu* [day 8], Sè, Deputy Prefect of Qiānlíng, (sends a message to) the lower authorities: The Lesser Treasury should investigate and establish firmly. The document needs to be communicated orally upon delivery, and then forwarded by all Finance Departments. Everything else (should be handled) in accordance with the┘ statutes and ordinances. [. . .]

662 Chén Zhōnglóng 2016.
663 Chén Wěi 2012: 48.
664 Chén Wěi 2012: 80.
665 Chén Wěi 2012: 81.
666 Chén Wěi 2012: 80.
667 Chén Jiàn 2011a.
668 Chén Wěi 2012: 94.

Following Chén Wěi, ‘*ān zhì* 案致’ forms a verb collocation meaning ‘to investigate and establish firmly’.[669]

8-1510 rI: 三月辛亥[48], 遷陵守丞敦狐告司空主, 以律令從事. [. . .][670]

In month 3 on a *xīnhài*, Dūnhú, Deputy Prefect of Qiānlíng, reports to the Commissioner of Public Works. Follow up on the matter in accordance with the statutes and ordinances. [. . .]

8-1525 III–IV: [. . .] • 七月甲子[1]朔乙亥[12], 遷陵守丞配告倉┛ 主: 下券, 以律令從事. [. . .][671]

[. . .] • In month 7, new moon on a *jiǎzǐ*, *yǐhài* [day 12], Yí, Deputy Prefect of Qiānlíng, reports to the Granary Incumbent:┛ Send the tally [to the lower authority] and follow up on the matter in accordance with the statutes and ordinances. [. . .]

Alternatively, the syntax of passages featuring ‘*lǜ* 律’ appears more flexible and less dependent on hierarchical relations within the Qín system of bureaucracy.[672] The term is mentioned only once in a syntactically parallel phrase on manuscripts 8-110+8-669 and 8-1517, stating that “food (provisions) correspond to the statutes” (*shí rú lǜ* 食如律). In other instances, ‘*lǜ* 律’ seems to be quoted from statutes in pre-existing ‘documents’ (*shū* 書) or as part of document titles, as illustrated in the following example:

8-1198 I: ☒守起書言: “傳律”曰☒[673]

☒ *ad Interim* X [pronunciation unknown] takes up the document and communicates orally: The “Statutes Concerning the Transmission of Documents” state ☒

There was a wide array of punishments for officials who violated the law, whether intentionally or due to an error of judgment. The *Lǐyē* texts from layers 5, 6 and 8 do not explicitly mention severe physical punishments for capital offenses, like branding, mutilation or execution. Nonetheless, this absence does not necessarily rule out that such practices were in use. Manuscript 8-148+8-489 gives an overview of penal-

669 Chén Wěi 2012: 94. For more information on this manuscript text, see *Húnán shěng wénwù kǎogǔ yánjiūsuǒ* 2012b: 19.
670 Chén Wěi 2012: 341.
671 Chén Wěi 2012: 349.
672 One interpretation of ‘statutes and ordinances’ (*lǜlìng* 律令) is that they form a legal system in which ‘penal codes’ (*lǜ* 律) address the punishment of criminals, and perhaps also civil codes for non-criminal matters, while ‘administrative codes’ (*lìng* 令) govern the functioning of the administration (Zhāng Zhāoyáng 2008).
673 Chén Wěi 2012: 290.

ties imposed on county officials and conscript soldiers alike, with no clear hierarchical, categorical or numerical order in relation to official titles or the severity of the penalties.

8-149+8-489 AI–VI, BI–VII, CI–VII, DI–VII: A ......【司】空佐敬二甲.┛【司】空守警三甲.┛ 司空守敀三甲.┛ 司空佐沈二甲. 以.┛ □□□一盾. 入.┛ 庫武二甲.┛ B 庫佐駕二甲.┛ 田官佐賀二甲.┛ 髳長忌再□署.┛ 校長予言資二甲.┛ 發弩□二甲.┛ 倉佐平七【盾】.┛ 田佐□一甲.┛ C 令佐圂一盾.┛ 令佐冣七甲.┛ 令佐逌二甲. 已利.┛ □廿錢.┛ 更戍畫二甲.┛ 更戍【五】二甲.┛ 更戍【登】二甲.┛ D 更戍嬰二甲.┛ 更戍□二甲.┛ 更戍裚贖耐.┛ 更戍得贖耐.┛ 更戍堂贖耐.┛ 更戍齒贖耐.┛ 更戍暴贖耐.[674]

A ...... Jìng, Assistant to the【Commissioner of】Public Works, (is fined) two sets of armor.┛ Áo, Assistant to the【Commissioner of】Public Works, (is fined) three sets of armor.┛ Yí, Assistant to the Commissioner of Public Works, (is fined) three sets of armor.┛ Shěn, Assistant to the Commissioner of Public Works, (is fined) two sets of armor. Act.┛ □□□ one shield. Enters.┛ Wǔ from the Warehouse (is fined) two sets of armor.┛ B Jià, Warehouse Assistant, (is fined) two sets of armor.┛ Hè, Assistant to the Office of Agricultural Fields, (is fined) two sets of armor.┛ Jì, the Commanding Official of a Subsection, is again sent into exile at the border area.┛ Yǔyǎn, Section Commander, is supplied with two sets of armor.┛ The Crossbow Archer □ (is fined) two sets of armor.┛ Píng, the Granary Assistant, (is fined) seven【shields】.┛ □ the Assistant to the Office of Agricultural Fields (is fined) one shield.┛ C Prefectural Assistant Hùn (is fined) one shield.┛ Prefectural Assistant Qǔ (is fined) seven shields.┛ Prefectural Assistant Yóu (is fined) two shields. Favorable.┛ □ 20 coins.┛ Boarder Guard Zhòu (is fined) two sets of armor.┛ Border Guard【Wǔ】(is fined) two sets of armor.┛ Border Guard【Dēng】(is fined) two sets of armor.┛ D Border Guard Yīng (is fined) two sets of armor.┛ Border Guard □ (is fined) two sets of armor.┛ Border Guard Jì (is punished by paying the fee for) redemption from being shaved.┛ Border Guard Dé (is punished by paying the fee for) redemption from being shaved.┛ Border Guard Táng (is punished by paying the fee for) redemption from being shaved.┛ Border Guard Chǐ (is punished by paying the fee for) redemption from being shaved.┛ Border Guard Bào (is punished by paying the fee for) redemption from being shaved.

Comments: Official titles such as 'Section Commander' (*xiàozhǎng* 校長), 'Commanding Official of a Subsection' (*máozhǎng* 髳長) and 'Crossbow Archer' (*fānǔ* 發弩) are seen in documents from the "Èrnián lǜlìng" and the "Zòuyànshū".[675] Gāo Mǐn claims that '*xiàozhǎng* 校長' is an ancient title comparable to 'Commanding Official of a Post' (*tíngzhǎng* 亭長) during the Qín dynasty.[676] He also argues that the Qín titles 'Overseer of a Post' (*tíng sèfū* 亭嗇夫) and 'Commanding Official of a Post and Section' (*tíngxiàozhǎng* 亭校長) were consolidated into the single title of 'Section Commander' (*xiàozhǎng* 校長). In the Hàn dynasty, holders of this title were additionally in charge of funerary services.

674 Chén Wěi 2012: 89–90.
675 Applies to strips 471–472 from "Statutes Concerning Salaries" (*Zhì lǜ* 秩律) in the "Èrnián lǜlìng", and strip 16 of the "Zòuyànshū" (Hú Píngshēng 2012a). For a discussion on the title 'Commanding Official of a Subsection' (*máozhǎng* 髳長), see Liào Bóyuán 2003.
676 Gāo Mǐn: 1971.

The character '*qiān* 覀' (*tsha[r,n]) is a variant of '*qiān* 遷' (phonological reconstruction unknown), meaning 'banishment', 'to banish' or 'to send someone into exile to a border area'.[677] Strip 151 from the "Statutes Concerning the Commissioner of Public Works" (*Sīkōng lǜ* 司空律) in the *Shuìhǔdì* texts specifies the regulations related to this form of punishment:

> *Shuìhǔdì*, "Qín lǜ shíbā zhǒng", "Statutes Concerning the Commissioner of Public Works" (*Sīkōng lǜ* 司空律), strip 151: 或遷, 欲入錢者, 日八錢.[678]
>
> If someone is sent into exile, he is someone who wants for financial income, eight coins per day.

The value of one 'suit of armor' (*jiǎ* 甲) was equal to two 'ounces' (*liǎng* 兩) and one 'mace' (*chuí* 錘) of gold, or 1,344 coins. The value of one 'shield' (*dùn* 盾) was equal to two maces of gold, or 384 coins.[679] '*Gèng shù* 更戍' literally means 'to perform military service as punishment'.[680]

The Qín legal system was evidently structured to include a comprehensive set of regulations through which individuals could seek mitigation or redress for unjust legal treatment. As numerous excerpts of 'statutes' (*lǜ* 律) in the *Lǐyē* corpus attest, judicial institutions were accountable to the public and had to publicly justify their use of coercion or re-evaluate legal conflicts upon request. Even after a verdict was delivered, 'lawsuits' (*yù* 獄) could be reopened for review, potentially leading to changes in the sentence, whether through mitigation, reversal or further enforcement. Each transgression or criminal offense was individually examined and carefully assessed in accordance with applicable statutes to avoid arbitrariness and ensure fair rulings.

If a sentence was deemed to be in violation of the statutes and ordinances, 'Judiciary Secretaries' (*yùshǐ* 獄史) had the authority to revoke such rulings and relaunch an investigation. For this reason, Qiānlíng County had at least two 'Judiciary Bureaus' (*yùcáo* 獄曹): the 'Judiciary Bureau of the East' (*yùdōngcáo* 獄東曹) and the 'Judiciary Bureau of the South' (*yùnáncáo* 獄南曹).[681] On manuscript 8-136+8-144, the criminal case of a Bondservant is reviewed by the County Court at the request of a Granary Incumbent:

---

677 Chén Wěi 2012: 90; Hé Yǒuzǔ 2015a; Lau and Staack 2016: 27, 106, 170. This phonological reconstruction of '*qiān* 覀' (*tsha[r,n]) was suggested by Prof. Dr. Wolfgang Behr.

678 Dài Shìjūn 2019.

679 Chén Wěi 2014a.

680 Chén Wěi 2012: 83–84; Lau and Staack 2016: 319.

681 'Judiciary Bureau of the East' (*yùdōngcáo* 獄東曹) appears on manuscripts 5-22, 8-273+8-520, 8-959+8-1291, 8-996 and 8-1155. 'Judiciary Bureau of the South' (*yùnáncáo* 獄南曹) appears on manuscripts 8-728+8-1474, 8-1760 and 8-1886. No text in the *Lǐyē* corpus analyzed to date explicitly identifies the Emperor as the sole originator or representative of Qín law.

8-136+8-144 I–III, rI: ⊠□月己亥[36]朔辛丑[38], 倉守敬敢言之: 令下覆獄遝遷陵隸臣鄧┘ ⊠□□名吏(事)、它坐、遣言. • 問之有名吏(事), 定, 故旬陽隸臣, 以約爲┘ ⊠□□史, 有遝耐辠以上, 毄(繫)遷陵未央(決), 毋遣殹. 謁報覆獄治所, 敢言 └⊠□□刻刻下六, 小史夷吾以來. ノ朝半. 尚手.[682]

⊠□ month, new moon on a *jǐhài*, *xīnchǒu* [day 3], Granary Incumbent Jìng dares to say: Order the lower authorities to review the lawsuit involving (the male) Bondservant Dèng from Qiānlíng (County).┘ ⊠□□ a matter related to (his) personal name, other prosecutions for criminal offenses and dispatched reports. • Clarify whether there is a matter related to (his) personal name and if (he) is recorded (in the register of convicts). Formerly, (he was) a Bondservant in Xúnyáng (County), by agreement becomes┘ ⊠□□ Secretary. There are the shaving punishments or harsher penalties, (but) in Qiānlíng it has not yet been decided if he should be placed under detention. Do not dispatch. Request a response (containing a decision taken by the higher authorities) concerning the reviewing of the lawsuit by the government seat. (He) dares to say └⊠□□ marks, below mark number six, (the document) is brought by minor Secretary Yíwú. / Opened by Cháo. By the hand of Shàng.

Comments: Based on Xǔ Míngqiāng's calendrical reconstructions, the new moon day *jǐhài* could only have occurred during one of the following periods covered by the *Lǐyē* texts: month 5 of the 28th year of King Zhèng (219 BC), month 8 of the 33rd year of King Zhèng (214 BC), or month 9 of the first year of the Second Emperor (209 BC).[683] The term '*bàn* 半' (*pans) was apparently replaced by '*fā* 發' (*pat) after month 9 of the 30th year of King Zhèng (217 BC), which limits the year of this text entry to 219 BC.[684] Manuscripts 8-170 and 8-742 serve as cross-references, with the passages "28th year (of King Zhèng) [219 BC], month 5, new moon on a *jǐhài*, *jiǎyín* [day 16]" (*niàn bā nián wǔ yuè jǐhài shuò jiǎyín* 廿八年五月己亥朔甲寅) and the "28th【year】(of King Zhèng) [219 BC], month 5, new moon on a *jǐhài*, *jǐwèi* [day 21]" (【*niàn*】*bā nián wǔ yuè jǐhài shuò jǐwèi* 【廿】八年五月己亥朔己未) in their opening lines.[685] '*Xīnchǒu* 辛丑' corresponds to 'day 3'.

Lau and Staack interpret '*tà* 遝' (phonological reconstruction unknown) as a term frequently used in Qín and Western Hàn dynasty manuscripts, representing an older variant of '*dài* 逮' (*lə̂s).[686] It has multiple interpretations: 1) as a verb, meaning 'to arrive', 'to reach (to)', 'to come to', 'to include' or 'to cover', 2) as a preposition in '*dài yè* 逮夜', '*dài rì* 逮日' and '*dài àn* 逮闇', or 3) as a conjunction derived from the verbal usage, meaning 'when/then it came to the time of'. As indicated by Baxter and Sagart, '*dài* 逮' (*m-rˤəp-s) is a 'departing tone' (*qùshēng* 去聲) derivation of '*tà* 遝' (*m-rˤəp).[687] Wolfgang Behr notes that '*tà* 遝' often occurs as the conjunction 'and (then)' in Eastern Zhōu bronze inscriptions, and both terms, '*dài* 逮' and '*tà* 遝', are related to '*jí* 及' (*[m-k-]rəp), meaning 'to reach to'.

682 Chén Wěi 2012: 76.
683 Xǔ Míngqiāng 2013a, 2013b; Zhāng Péiyú 1987: 224–225.
684 Chén Wěi 2012; Lóng Jīngshā 2001. Phonological reconstructions are taken from Schuessler 2009: 237, 261.
685 Chén Wěi 2012: 103, 214.
686 Lau and Staack 2016: 133, 182. Phonological reconstruction is taken from Schuessler 2009: 307.
687 Phonological reconstructions are taken from Baxter and Sagart 2014: 19, 107.

I understand '*dìng* 定' as 'to be recorded (in the register of convicts)', following a similar usage on manuscript 8-1515, where '*dìng* 定' precedes the 'register of convicts' (*bù* 簿).[688] '*Yè bào* 謁報' is a fixed compound noun and means 'to request a response (containing a decision taken by the higher authorities)'.[689]

The sovereignty of the Qín state was inherently vulnerable due to its reliance on a legal framework that was in need of careful interpretation. Any misapplication of the law, even by a high-ranking official, could not be enforced. Manuscript 8-754+8-1007 provides a rare example of this vulnerability. It recounts an incident where the Prefect of Qiānlíng County unlawfully imposed a financial penalty on what appears to be a commoner and a district official, for failing to meet him on his way to Èrchūn 貳春 District. After an interrogation ordered by Judiciary Secretary Kān 堪, the unjust sentence was retroactively nullified:

8-754+8-1007 I–II, rI: 卅年□月丙申[33], 遷陵丞昌, 獄史堪【訊】. 昌辤(辭)曰: 上造, 居平□, 侍廷, 爲遷陵丞. □當詣貳春鄉, 鄉【渠、史獲誤詣它鄉, □失】┘ 道百六十七里. 即與史義論貲渠、獲各三甲, 不智(知)劾云貲三甲不應律令. 故皆毋它坐. 它如官書. └☑堪手.[690]

In the 30th year (of King Zhèng) [217 BC], month □ on a *bǐngshēn*, Prefect Chāng of Qiānlíng and Judiciary Secretary Kān【conduct an interrogation (of a legal case)】. Chāng states: (I am a) Holder of the Second Rank □, live in Píng□, serve at the County Court and have become Governor of Qiānlíng. □ When I was heading to Èrchūn District, (the person) Qú from the district and Official【Huò mistakenly headed to another district, □ lost】┘ the way by 167 miles. Then, together with Official Yì, I sentenced Qú and Huò with a fine of three sets of armor each. I did not know that fining this offense with three sets of armor does not comply with the statutes and ordinances. Therefore, they are all no longer prosecuted under it [the fine]. Everything else (should be handled) in accordance with the documents of the offices. └ ☑ By the hand of Kān.

Comment: 167 'miles' (*lǐ* 里) was approximately equivalent 70 kilometers during this period.[691]

Manuscripts 8-755, 8-756, 8-757, 8-758 and 8-759 are rare, dialogue-format documents exchanged between the Governor of Dòngtíng and the Prefect of Qiānlíng. They are concerned with the management of 'agricultural fields' (*tián* 田) (Fig. 5.2).[692] By knowing both the dispatch and reception dates, we can calculate that deliveries took, on

688 For a transcription and translation of manuscript 8-1515, refer to pages 206–207.

689 Lau and Staack 2016: 76.

690 Chén Wěi 2012: 211.

691 Ten 'miles' (*lǐ* 里) corresponded to ca. 4.2 kilometers during the Qín dynasty (with one '*lǐ* 里' being roughly 420 meters) (Lau and Staack 2016: 153).

692 For a detailed discussion of these manuscripts and their potential link to manuscripts 8-1534 and 8-1564, see Gōng Zháijié 2018.

average, around 18 days between these two government seats.[693] The content of these documents addresses a 'Commissioner of Public Works' (*sīkōng* 司空) who was charged with "failing to order the working of the agricultural fields" (*fú lìng tián* 弗令田) due to improper delegation of tasks to his 'convict servants' (*túlì* 徒隸).[694] The case is significant as it unveils how local officials were assessed against performance criteria, which if not met and as already observed in the *Yuèlù* documents, could result in even higher-ranking officials being sentenced to *corvée* labor.[695]

> 8-755 I–II, rI, 8-756 I–II, 8-757 I–II, 8-758 I–II, 8-759 I–II: 8-755 卅四年六月甲午[31]朔乙卯[52], 洞庭守禮謂遷陵丞:┛ 丞言徒隸不田, 奏曰: 司空厭等當坐, 皆有它罪, 8-756 耐爲司寇. 有書, 書壬手. 令曰: 吏僕、養、走、工、組┛ 織、守府門、削匠及它急事不可令田, 六人予田徒 8-757 四人. 徒少及毋徒, 薄(簿)移治虜御史, 御史以均予. 今遷陵┛ 廿五年爲縣, 廿九年田廿六年盡廿八年當田, 司空厭等 8-758 失弗令田. 弗令田即有徒而弗令田且徒少不傳于┛ 奏. 及蒼梧爲郡九歲乃往歲田. 厭失, 當坐論, 即 8-759 如前書律令. ノ七月甲子[1]朔癸酉[10], 洞庭叚(假)守┛ 繹追遷陵. ノ歇手.・以沅陽印行事. └ 8-755 歇手.[696]

> 8-755 In the 34th year (of King Zhèng) [213 BC], month 6, new moon on a *jiǎwǔ*, *yǐmǎo* [day 22], Lǐ, Governor of Dòngtíng, addresses the Prefect of Qiānlíng:┛ The Prefect has said that the convict servants do not work the agricultural fields and submitted the matter in writing, stating: Yàn, Commissioner of Public Works, and others should be prosecuted. They all are liable to punishment. 8-756 By shaving off his beard, he becomes an Overseer of Bandits. There is a document, and this document (was written) by the hand of Rén. The ordinance says: Servants of Officials, Provisioners, Runners, Artisans, Weavers,┛ Gate Keepers, Craftsmen and others are busy with matters and cannot be ordered to work the agricultural fields. If six people are given (to work) the agricultural fields, 8-757 four (should be) convicts. If convicts are few and there are no convicts (at all), the register of convicts (should be) transmitted to the Prosecutor managing the captives, so that the Prosecutor can thereby compensate (the number of convicts) and allocate (more). (What is known) today (as) Qiānlíng┛ became a county in the 25th year (of King Zhèng) [222 BC]. In the 29th year (of King Zhèng) [218 BC], the agricultural fields (were worked), but from the 26th to the 28th year (of King Zhèng) [221–219 BC], the agricultural fields should have been worked also. Yàn, Commissioner of Public Works, and others 8-758 committed the legal offense of failing to order the working of agricultural fields. (The legal offense) of failing to order the working of agricultural fields means that there are (enough) convicts, but they were not ordered to work the agricultural fields. In addition, the low number of convicts was not transmitted in┛ writing (to the higher authorities). Cāngwú has been a commandery for nine years and, in the past, its agricultural fields have (always) been tilled. Yàn committed a legal offense. He should be prosecuted and receive his judgment. 8-759 This is in accordance with the statutes and ordinances in previous documents. / In month 7, new moon on a *jiǎzǐ*, *guǐyǒu* [day

693 The document was sent on day 22 of month 6 and received on day 10 of month 7.

694 For an analysis of the taxation of public fields during the Qín dynasty, see Lǐ Héngquán 2018.

695 Lau and Staack 2016: 133.

696 Chén Wěi 2012: 217.

10], Yì, Governor ad Interim of Dòngtíng,┘ pursues (the suspect) in Qiānlíng. / By the hand of Xiē. • Use the seal of Yuányáng to follow up on the matter. 8-755 └ By the hand of Xiē.

Comments: The term '*zòu* 奏' means 'to submit (a matter) in writing to the ruler'.[697] '*Yǒu zuì* 有罪' is a fixed expression in Qín and Hàn legal vocabulary and means 'to be liable to punishment':

> 8-533 AI–V, BI–II: A 戌有罪爲鬼薪.┘ 齰城旦.┘ 赢城旦.┘ 欬城旦.┘ 瘳城旦.┘ B 滕司寇. ☒┘ 憎司寇.☒ [698]

> A Xū is liable to punishment and becomes a Firewood Gatherer.┘ Cuò (becomes) a Wall Pounder.┘ Yíng (becomes) a Wall Pounder.┘ Kài (becomes) a Wall Pounder.┘ Chōu (becomes) a Wall Pounder.┘ B Téng (becomes) an Overseer of Bandits. ☒┘ Wài (becomes) an Overseer of Bandits. ☒

Rén 壬, referenced on manuscript 8-756, may have served as the Assistant to Jìng 敬, the Granary Incumbent cited on manuscripts 8-764 and 8-1566.[699] The personal name Rén 壬 is recorded in the *Lǐyē* texts between the 31st and 34th year of King Zhèng (216–213 BC) (see Tab. 2.1). Both the Qín and Hàn imperial administrations permitted officials to employ Servants, often drawn from convicts or "soldiers punished to perform military service" (*shùzú* 戍卒).[700] Two manuscript excerpts from the *Lǐyē* corpus provide the following information:

> 8-130+8-190+8-193 I–II: [. . .] 諸徒隸當爲┘ 吏僕養者皆屬倉 [. . .][701]

> [. . .] All convicts who are meant to become Servants or Provisioners of officials belong to the Granary. [. . .]

> 8-106: ☒遷陵戍卒多爲吏僕, 吏僕☒ [702]

> ☒ border soldiers from Qiānlíng County often become Servants for officials. Servants for officials ☒

According to Chén Wěi, '*zǔshí* 組識' were weavers.[703] '*Lùn* 論' was used when an official 'passed judgment (on an alleged criminal)'.[704] If Cāngwú 蒼梧 Commandery was already nine years old by the 34th year of King Zhèng (213 BC), it must have been established three years prior to the proclamation of the empire in the 25th year of King Zhèng (220 BC).[705]

---

697 Lau and Staack 2016: 322.
698 Chén Wěi 2012: 175.
699 For a transcription of manuscripts 8-764 and 8-1566, see Chén Wěi 2012: 219, 362. Manuscript 8-1566 is transcribed and translated on page 164.
700 Lau and Staack 2016: 222–223, 319.
701 Chén Wěi 2012: 68.
702 Chén Wěi 2012: 63.
703 Chén Wěi 2012: 217. For a study on textiles in the Hàn dynasty, see Langford 2009.
704 Lau and Staack 2016: 317, 322.
705 For more information on commanderies and counties during the Qín dynasty, see Hé Jièjūn 2005 or refer to chapter four.

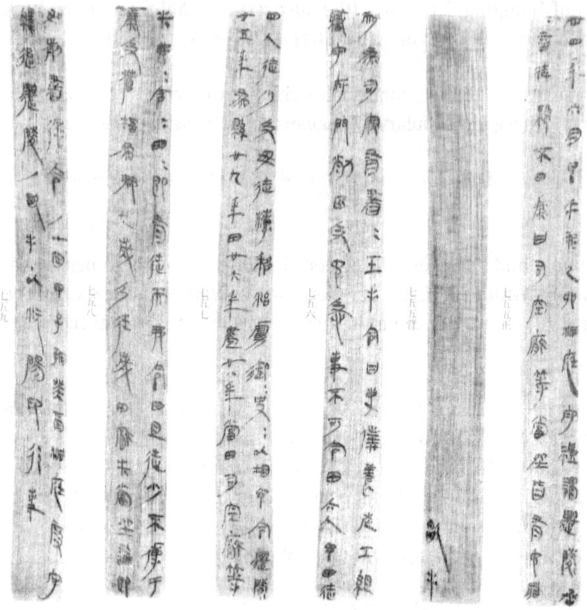

Fig. 5.2: Monochrome photographs of *Lǐyē* manuscripts 8-755, 8-756, 8-757, 8-758 and 8-759. Ink on wood. Dimensions unknown.[706]

Another example can be found on manuscript 8-137. It recounts the failure of a 'Servant' (*pú* 僕) from the Office of Domestic Animals to forward evaluation reports to higher authorities, ultimately exposing the wrongdoings of both his superior and the Official in Charge of the Documents.

> 8-137 I–III, rI–II: ☒□朔戊午[55], 遷陵丞遷告畜官僕足, 令┘ ☒□毋書史, 畜官課有未上, 書到亟日 ┘ ☒□守府事已, 復視官事如故, 而子弗 └☒事, 以其故不上, 且致劾論子, 它承┘ ☒就手.[707]

> ☒□ new moon on a *wùwǔ*, Governor Qiān of Qiānlíng reports to Servant Zú from the Office of Domestic Animals, orders┘ ☒□ not the Official in Charge of the Documents, the evaluations of the Office of Domestic Animals exists, (but have) not yet been submitted to the higher authorities. The documents are urgently due┘ ☒□ the matters of the Governor have ended. (He) repeatedly inspected the matters of the office as he did in the past, but he did not └☒ matters, for this reason did not declare to a higher authority, and moreover exposed the crimes of his master and brought about his judgment, it received┘ ☒ By the hand of Jiù.

706 *Húnán shěng wénwù kǎogǔ yánjiūsuǒ* 2012a: 109, 110.
707 Chén Wěi 2012: 77.

Curtailing criminal activity was achieved not only through surveillance, punishment, coercion and other disciplinary measures[708] but also by giving financial remuneration and rewards for actions deemed beneficial to the empire. This balanced approach to rewards and punishments had deep roots in early Chinese governance, as reflected in the material from Lǐyē. Moreover, imperial administrative units were committed to principles of due process, ensuring that legal frameworks were in place to resolve differences and conflicts among individuals of the same or differing social standings. These measures formed part of a perpetual effort to advance and maintain social order by promoting some level of fairness and a structured system of incentives and deterrents.

708 For an examination of the order of punishments in Qín and Hàn legal documents, refer to Lau and Staack 2016: 27–28.

6 Resource Management

Developments in agriculture were driven by the efforts of hardworking farmers, the labor of forced workers, and policies adopted by the Emperor that prioritized agricultural output. Transmitted texts concerning the Qín dynasty, such as the *Book of Lord Shāng* (*Shāngjūn shū* 商君書), categorize agricultural production as a 'fundamental occupation' (*běn yè* 本業), ranking it above other forms of manual labor or commercial activity.[709] The first section of a stone stela inscription at Mount Lángyá 琅邪, attributed to the First Emperor, declares the following:[710]

> Inscription on Mount Lángyá 琅邪: 維廿六年, 皇帝作始, 端平法度, 萬物之紀, 以明人事, 合同父子, 聖智仁義, 顯白道理, 東撫東土, 以省卒士, 事已大畢, 乃臨于海, 勤勞本事, 上農除末, 黔首是富, 普天之下, 摶心揖志, 器械一量, [. . .][711]
>
> Now, in his 26th year, the Emperor created a beginning: He rectified and balanced the rules and measures as guidelines for the ten thousand beings. In order to illuminate human affairs, he united and led to concord father and son. Sage, wise, humane and right: He made manifest and clear the way and its inner pattern! Eastwards, he tours the eastern land, to inspect the common soldiers and officers. His achievements already largely completed, he now looks down on the (lands by the eastern) sea. The Emperor, he has merit: Diligently, he labors on principal tasks. He exalts agriculture, eliminates peripheral (occupations), the black-haired people, these he enriches! Everywhere under vast heaven he unifies minds and integrates wills. Vessels and implements have their identical measures, [. . .][712]

---

709 See Pines 2017b.

710 The First Emperor conducted five imperial 'inspection tours' (*xúnshǒu* 巡守), during which he performed sacrifices and left behind inscriptions of rhymed poetry, primarily credited to Lǐ Sī 李斯 (ca. 280–208 BC; served 246–208 BC) and other scholars under his supervision. These inscriptions were engraved on stone stelae erected for this purpose. Six of the seven inscription texts are transmitted in the *Shǐjì*. One originates from a Táng dynasty rubbing and is preserved on a Sòng dynasty stone stela exhibited in the Stele Museum (*Bēilín* 碑林) in modern-day Xī'ān. Additionally, small fragments of the Mount Lángyá 琅邪 inscription (219 BC) are kept in the National Museum of China (*Zhōngguó guójiā bówùguǎn* 中國國家博物館). In preparation for the Emperor's visit, thousands of 'households' (*hù* 戶) were relocated to Lángyá to ensure that his presence was widely recognized, partly contributing to the widespread awareness of his rule. Manuscripts recovered from Shuìhǔdì 睡虎地 corroborate certain accounts found in transmitted sources and document two notable events during the Emperor's passage through a nearby region in 219 BC: 1) the erection of an imperial stone stela, and 2) the punishment of spirits upon his return journey to the Qín capital of Xiányáng 咸陽, performed at Mount Xiāng 湘 near Dòngtíng Lake (*Dòngtíng hú* 洞庭湖) in present-day Húnán Province (Kern 2000; Sanft 2008).

711 Kern 2000: 25–29.

712 Translation by Kern 2000: 25–29, with minor modifications.

https://doi.org/10.1515/9783110791853-006

Particular attention should be paid to agriculture, which arguably was the most important sector in and around Qiānlíng County. Manuscript 8-355 reads:

8-355: ☒【黔】首習俗好本事不好末作, 其習俗槎田歲更, 以異中 縣.[713]

☒ The social habits of the【black】-headed people involve agricultural production and a disliking for manufacturing. Their social habits include hewing the arable fields with the passing of the years. By this they differ from the counties in the center (of the empire).

Comments: Chén Wěi understands *'běn shì* 本事' as 'agriculture' or 'agricultural production' in this particular context.[714] Elements of culture from the peripheral regions of the early Chinese empires, as well as from Central and South Asia, seem to have had a strong impact on the identity, trade and policy formation of the Qín and Hàn dynasties.[715]

The administrative oversight of business and trade in Qiānlíng County was likely underdeveloped. Therefore, one of the government's primary objectives may have been to implement measures aimed at protecting and enhancing agricultural production, such as reclaiming unused fields, institutionalizing compulsory labor and assigning households to fixed units of land. The Qín administration placed significant emphasis on increasing agricultural productivity in the region. It also prioritized grain loans for farmers, redistributing people and livestock and maintaining control of surplus food.[716]

'*Tú(lì)* 徒(隸)' were 'convict (servants)' who were stripped of their rights and forced to perform manual labor for the imperial institutions. These individuals faced harsh treatment and were traded in markets as if they were inanimate commodities.[717] Manuscript 8-154 indicates that, on the 1st day of each month, the Prefect of Qiānlíng was required to report the number of convicts purchased during the previous month to his superiors:

8-154 I–II, rI: 卅三年二月壬寅[39]朔朔日, 遷陵守丞都敢言之: 令曰恒以┛ 朔日上所買徒隸數.·問之, 毋當令者, 敢言之. └ 二月壬寅[39]水十一刻刻下二, 郵人得行. 圂手.[718]

713 Chén Wěi 2012: 136.
714 Chén Wěi 2012: 136–137.
715 See, for example, Demandt 2020; Huo 2008. For more information on the sense of unity in early Chinese empires, see Loewe 1994.
716 Hulsewé 1985: 41–42. The *Lǐyē* corpus contains numerous texts addressing the redistribution of food rations and money among the general population. See, for example, manuscripts 8-50+8-422, 8-143, 8-217 or 8-1002+8-1091.
717 Cáo Lǚnìng 2015; Yates 1995: 345–346. For more information on *corvée* labor, see chapter five.
718 Chén Wěi 2012: 93.

> In the 33rd year (of King Zhèng) [214 BC], month 2, new moon on a *rényín*, new moon day [day 1], Dū, Deputy Prefect of Qiānlíng, dares to say: The ordinance states to make regular┘ on the new moon day (of each month) to submit to the higher authorities the number of convict servants that have been bought. • Clarify whether this does not correspond to the ordinance. (He) dares to say. └In month 2 on the new moon day [day 1], at water below mark 2 of 11 marks, Postman Dé delivers (the document). By the hand of Hùn.

According to this report and similar ones, convicts were reassigned to government units most in need of manual labor.

> 8-199+8-688 AI–III, BI–III, rI–II: A 卅年十二月乙卯[52], 畜□□□作徒薄(簿).☒┘ 受司空居貲一人.☒┘ 受倉隸妾三人.☒┘ B ☒□☒┘ 【凡】☒┘ 【一人】☒ └十二月乙卯[52], 畜官守丙敢言之: 上. 敢言☒┘ 十二月乙卯[52]水十一刻刻下一, 佐貳以來.☒[719]

> A In the 30th year (of King Zhèng) [217 BC], month 12 on a *yǐmǎo* [day 26], domestic animals □□□ creates the register of convicts. ☒┘ Receive one person who is redeeming the fines from the Commissioner of Public Works. ☒┘ Receive (female) Bondservants from the Granary, three people. ☒┘ B ☒□☒┘ 【In total】 ☒┘ 【One person】 ☒ └In month 12 on a *yǐmǎo* [day 26], Bǐng, Incumbent of the Office of Domestic Animals, dares to say: Send (the document) to the higher authorities. (He) dares to say ☒┘ In month 12 on a *yǐmǎo* [day 26], at water below mark 1 of 11 marks, (the document) is brought by Assistant Èr. ☒

> Comments: '*Yǐmǎo* 乙卯' in month 12 of the 30th year of King Zhèng corresponds to 'day 26'.[720]

One agricultural season comprised long months of cultivation and comparatively short harvest periods, leaving crops vulnerable to floods, droughts, fires and pests. To mitigate these risks, local and regional governments were advised to adopt countercyclical policies to store food and other supplies in facilities like 'Granaries' (*cāng* 倉) and 'Warehouses' (*kù* 庫). This strategy helped stabilize potentially dangerous fluctuations in annual prices and yields.

No Granary or Warehouse-like structures have been discovered in and around Lǐyē. However, later archeological findings from the Early Hàn dynasty suggest that such buildings were considered logistical hubs of great strategic importance.[721] For example, the Huá Granary (*Huá cāng* 華倉) at Huáyīn 華阴, present-day Shǎnxī Province (dated 141–87 BC), and the Wǔ Warehouse (*Wǔ kù* 武庫) built within the Hàn dynasty capital of Cháng'ān 長安 (dated 198 BC), were not constructed on simple earth cores or low foundations, as was common with most structures of the time.[722] Instead, the perimeter of these buildings was surrounded by deep, massive rectangular wall footings. One section of the Wǔ Warehouse mea-

719 Chén Wěi 2012: 111.
720 Compare with Xǔ Míngqiāng 2013b.
721 *Hàn Huá cāng wājué duì* 1982.
722 *Hàn chéng wājué duì* 1978.

sured 197 meters east to west and 24 meters north to south, with outer walls reaching 4 meters into the ground and measuring up to 4.5 meters in width. Within the outer walls, the building was subdivided into two chambers, each ranging between 90 by 16 meters. The internal structures supported by free-standing columns spaced 4 and 5 meters apart, each between 70 and 90 centimeters in diameter (Figs. 6.1, 6.2).

Fig. 6.1: View of footings from the east of the Huá Granary (*Huá cāng* 華倉), constructed between 141 and 87 BC at Huáyīn 華阴, present-day Shǎnxī Province. Monochrome photograph taken by the excavation team in 1982.[723]

The exact factors behind the extraordinary design of Granaries and Warehouses are still subject to ongoing analysis. Influential factors may include the climatic conditions of the Shǎnxī region, the general tendency towards monumental architecture in and around Cháng'ān, and the necessity for protection against intruders, both human and animal. Major battles fought over food supplies stored in Granaries, as documented in transmitted sources from the civil uprisings toward the end of the Qín empire, may have also played a role in determining the location and construction of critical storage facilities.[724]

There is limited evidence prior to the Hàn dynasty regarding local or regional food shortages or relief measures for starving populations. Fortunately, a variety

723 Thorp 1986: 370.
724 See Cài Wànjìn 1996.

Fig. 6.2: View of footings from the southeast of the Wǔ Warehouse (*Wǔ kù* 武庫), constructed in 198 BC in the Hàn dynasty capital of Cháng'ān 長安. Monochrome photograph taken by the excavation team in 1978.[725]

of county Granary and 'Grain Store' (*kuài* 廥) documents dealing with food storage and reallocation have survived through the millennia in the *Lǐyē* corpus. In Qiānlíng County, Granary officials autonomously compiled both estimates and emergency plans, which were then forwarded to the Prefects of other counties as well as to the Military Commandant and Governor of Dòngtíng. This decentralized process ensured the rapid and efficient delivery of aid to affected areas, without direct intervention from central agencies.[726]

Manuscripts 8-1517 and 8-169+8-233+8-407+8-416+8-1185, both written in the 35th year of King Zhèng (212 BC), provide insights into the intricate workings of the Qín dynasty's food alert system. A report on manuscript 8-169+8-233+8-407+8-416+8-1185, submitted by a Granary official on day 19 of month 2 (212 BC), requests that Granary officials from other districts and counties report their food inventories, signaling the initiation of a county-wide attempt to redistribute food supplies:

725 Thorp 1986: 369.

726 According to the *Shuìhǔdì* legal texts, the state of crops, including any damages, had to be reported fast and thoroughly by local officials (Hulsewé 1985: 21).

8-169+8-233+8-407+8-416+8-1185 I–IV, rI: 卅五年二月庚申[57]朔戊寅[15], 倉□擇敢言之: 隸□饋爲獄行辟┘書彭陽, 食盡二月, 謁告過所縣鄉以次牘(續)食. 節(即)不┘ 能投宿齎. 遷陵田能自食. 未入關縣鄉, 當成齍,┘ 以律令成齍. 來復傳. 敢言之. ☑ └☑擇手.[727]

In the 35th year (of King Zhèng) [212 BC], month 2, new moon on a *gēngshēn*, *wùyín* [day 19], Granary □ Zé dares to say: Bondservant □ X [pronunciation unknown] filed a lawsuit and sent the documentation of legal cases┘ to Péngyáng County. The food (supplies) will be exhausted in two months. Request a report to pass on key points of transportation in the county and commandery following a proper order to supply food. If he cannot┘ find accommodation, offer it. The arable lands of Qiānlíng are able to provide food for its own (population). (The document) has not yet entered the frontier passes of the counties and districts, but it should be verified.┘ Verify (the document) by the statutes and ordinances. Come and send a reply. (He) dares to say. ☑ └☑ By the hand of Zé.

Comments: '*Bì shū* 辟書' was a document containing criminal hearings and investigations.[728] The term also appears on manuscript 8-680. Strip 276 of the "Èrnián lǜlìng" mentions a "document concerning all lawsuits and investigations" (*zhū yù bì shū* 諸獄辟書). Péng Hào suggests that '*bì shū* 辟書' was part of a reference document containing complex legal cases resolved at the county level and possibly forwarded to a senior commandery official or a related department of the central government.[729]

Due to the structural similarities between this manuscript and manuscripts 8-50+8-422 and 8-1517, I recommend placing a comma between '*xiāng* 鄉' and '*yǐ* 以' in the sentence "*yè gào guò suǒ xiàn xiāng yǐ cì dú (dú) shí* 謁告過所縣鄉以次牘(續)食."

Manuscript 8-1517, submitted one month later on day 22 of month 3 in 212 BC, refers to what appears to be a previous 'document' (*shū* 書) concerning the past, ongoing and planned assignments of officials and convicts. Considering the state of the food stores and the number of personnel in the region at that time, the Granary's food rations were projected to run out within three months. The report explicitly urges the Military Commandant of Dòngtíng to review and balance food supplies across all district and county Granaries:

8-1517 I–III: 卅五年三月庚寅[27]朔辛亥[48], 倉銜敢言之: 疏書吏、徒上事尉府┘ 者牘北(背), 食皆盡三月, 遷陵田能自食. 謁告過所縣, 以縣鄉次續┘ 食如律. 雨留不能投宿齎. 當騰騰. 來復傳. 敢言之.[730]

In the 35th year (of King Zhèng) [212 BC], month 3, new moon on a *gēngyín*, *xīnhài* [day 22], Granary Official Xián dares to say: The document containing (a list of) tasks for officials and convicts has been sent to the Military Commandant (of Dòngtíng Commandery)┘ on the back side of the tablet. The food (supplies) will be exhausted in three months. The arable

727 Chén Wěi 2012: 102.
728 Chén Wěi 2012: 102; *Lǐyē Qín jiǎndú jiàoshì xiǎozǔ* 2014.
729 Pēng Hào 2002.
730 Chén Wěi 2012: 344–345.

> lands of Qiānlíng are able to provide food for its own (population). Request a report from the counties in which (this message) goes through. By hierarchic order of counties and districts, increase food (supplies)┘ according to the statutes. As the rain continues, we cannot provide accommodation and assistance. Spread (the message) via the imperial postal system. Come and send a reply. (He) dares to say.

In some cases, proactive requests for food redistribution were coordinated with adjacent counties, as illustrated on manuscript 8-50+8-422, where the Prefect of Qiānlíng directly contacts Yǒuyáng 酉陽 and Línyuán 臨沅, both of which are counties within Dòngtíng Commandery:

> 8-50+8-422 I–IV: ☒□倉□建□□□畜官適□☒┘ ☒□謁告過所縣 鄉, 以次續食. 雨☒┘ ☒騰騰. 遷陵田能自食. 敢言之. ☒┘ ☒□□□丞遷移酉陽、臨沅. ノ得☒[731]

> ☒□ Granary □ Jiàn □□□ Office of Domestic Animals Shì □☒┘ ☒□ Request a report from the counties and districts in which (this message) goes through. By hierarchic order increase food (supplies). The rain ☒┘ ☒ via the imperial postal system. The arable lands of Qiānlíng are able to provide food for its own (population). (He) dares to say. ☒┘ ☒□□□ Prefect from Qiān[líng] to Yǒuyáng and Línyuán. / Dé ☒

Previous analyses of the *Shuìhǔdì* texts indicate that animals raised by Granary employees were likely intended for the use and consumption by high-ranking officials.[732] There is no evidence in these texts of meat being distributed to low-ranking officials, conscripts or convicts. Both the "Èrnián lǜlìng" and strips 16–18 of the "Qín lǜ shíbā zhǒng" further contain legislation explicitly prohibiting the trade of meat.[733] Yet, the remaining parts of animals, such as horns, bones, fur and skin, were permitted for separate sale and purchase.

Documents concerning the veneration of an 'Agricultural Ancestor' (*Xiān Nóng* 先農) were excavated from layers 14 and 15. These documents offer further insights into the exaltation of agriculture and the imperial Qín dynasty's ritual system inherited from earlier periods.[734] Granaries appear to have prepared sacrificial meat,

731 Chén Wěi 2012: 41.

732 Hulsewé 1985: 45.

733 Chén Wěi 2012: 168.

734 The term '*Xiān Nóng* 先農' also appears on bamboo manuscripts discovered in a Qín tomb at Zhōujiātái 周家台 (Péng Hào 2008). Both ancient and modern scholars have often, though unconvincingly, associated the 'Agricultural Ancestor' (*Xiān Nóng* 先農) with the 'Divine Farmer' (*Shén Nóng* 神農) or 'Blazing Thearch' (*Yándì* 炎帝). This association is based on a passage from the *Ancient Rules of the (Former) Hàn* (*Hàn jiùyí* 漢舊儀), a commentary fragment on the *Book of the Later Hàn* (*Hòu Hànshū* 後漢書) believed to have been written by Wèi Hóng 魏宏 in the 1st century BC (Sanft 2014b: 331–332).

food and alcoholic beverages, with the leftovers sold to the wider population as a means of generating additional profits.[735] A few examples of this practice are provided below.[736]

14-649+14-679: 卅二年三月丁丑[14]朔丙申[33], 倉是佐狗出祠先農餘徹豚肉一斗半斗賣于城旦赫所, 取钱四. 令史尚視平.[737]

In the 32nd year (of King Zhèng) [215 BC], month 3, new moon on a *dīngchǒu*, *bǐngshēn* [day 20], Shì from the Granary and his Assistant Gǒu give out leftover pork meat offered (as sacrifice) to the Agricultural Ancestor. 1.5 bushels are sold to Wall Pounder Hèsuǒ, receiving four coins. Prefectural Secretary Shàng oversees the weighing (of the food).

14-656+15-434: 卅二年三月丁丑[14]朔丙申[33], 倉是佐狗出黍米四斗以祠先農.[738]

In the 32nd year (of King Zhèng) [215 BC], month 3, new moon on a *dīngchǒu*, *bǐngshēn* [day 20], Shì from the Granary and his Assistant Gǒu give out four bushels of glutinous millet and hulled rice offered (as sacrifice) to the Agricultural Ancestor.

14-698: 卅二年三月丁丑[14]朔丙申[33], 倉是佐狗出祀先農餘徹酒一斗半斗賣于城[. . .][739]

In the 32nd year (of King Zhèng) [215 BC], month 3, new moon on a *dīngchǒu*, *bǐngshēn* [day 20], Shì from the Granary and his Assistant Gǒu give out 1.5 bushels of leftover alcoholic beverages offered (as sacrifice) to the Agricultural Ancestor. They sell them to Wall (Pounder) [. . .]

15-451: 鹽四分升一以祠先農[740]

0.4 pints of salt offered [as sacrifice] to the Agricultural Ancestor

15-490: 卅二年三月丁丑[14]朔丙申[33], 倉是佐狗出祠先農餘徹肉二斗賣于大.[741]

In the 32nd year (of King Zhèng) [215 BC], month 3, new moon on a *dīngchǒu*, *bǐngshēn* [day 20], Shì from the Granary and his Assistant Gǒu give out leftover pork meat offered (as sacrifice) to the Agricultural Ancestor. Two bushels are sold (to a person called) Dà.

All well-preserved manuscripts from layers 14 and 15 unanimously record the activities of Granary Official Shì 是 and his Assistant Gǒu 狗 preparing salt, rice,

735 For a discussion of *Lǐyē* texts concerning 'sacrificial offerings' (*cí* 祠) to the 'Agricultural Ancestor' (Xiān Nóng 先農), see Péng Hào 2008; Shǐ Zhìlóng 2009.
736 A more comprehensive list of relevant sources would also include manuscripts 14-66, 14-300 +14-764, 14-639+14-762, 14-650+14-652, 14-654, 14-656+15-434, 14-675, 14-685, 14-719, 15-480, 15-490 and 15-511.
737 Shǐ Zhìlóng 2009.
738 Shǐ Zhìlóng 2009.
739 Shǐ Zhìlóng 2009.
740 Péng Hào 2008.
741 Shǐ Zhìlóng 2009.

millet and other offerings for sacrifice to the Agricultural Ancestor. On the 20th day (*bǐngshēn* 丙申) of month 3 in the 32nd year of King Zhèng (215 BC), they are also documented selling leftover items, in particular alcoholic beverages, mutton, oxen and pork. The recipients are either disclosed by name or referred to collectively as 'Wall Pounders' (*chéngdàn* 城旦).

Warehouses, in addition to their other responsibilities, also provided ritual food for ancestral worship. Numerous texts from the 12th day (*jǐsì* 己巳) of month 6 in the 35th year of King Zhèng (212 BC) record the sale of leftover ritual meat and alcohol to commoners and convicts. This practice too seemingly functioned as an additional source of state income.

8-845: 卅五年六月戊午[55]朔己巳[6], 庫建、佐般出賣【祠】☒衛(率)之, 斗二錢.☒[742]

In the 35th year (of King Zhèng) [212 BC], month 6, new moon on a *wùwǔ*, *jǐsì* [day 12], Jiàn from the Warehouse and Assistant Bān sell【offered】(as sacrifice to) ☒ calculate it, two coins per bushel. ☒

Comments: This manuscript has a notch depicting the number 'six' (*liù* 六) carved into its left side.

8-847 I–II: ☒己巳[6], 庫建、佐般出賣祠☒┘ ☒令史歡☒ [743]

☒ on a *jǐsì*, Jiàn from the Warehouse and Assistant Bān sell ☒ offered (as sacrifice) to┘ ☒ by Prefectural Secretary X [pronunciation unknown] ☒

8-907+8-923+8-1422 I–II: 卅五年六月戊午[55]朔己巳[6], 庫建、佐般出賣祠⿱穴言餘徹酒二斗八升于□☒┘ 衛(率)之, 斗二錢. 令史歡監.☒ [744]

In the 35th year (of King Zhèng) [212 BC], month 6, new moon on a *wùwǔ*, *jǐsì* [day 12], Jiàn from the Warehouse and Assistant Bān sell the leftover alcoholic beverages offered (as sacrifice) to X [pronunciation unknown], two bushels and eight pints go to □☒┘ calculate it, two coins per bushel. Supervised by Prefectural Secretary X [pronunciation unknown]. ☒

'*Cí* 祠' is best understood as 'to offer sacrifices to gods and ancestors', while '⿱穴言' (pronunciation unknown) could be the name of an agricultural deity or a spirit residing in an 'underground storehouse' (*yìn* 窨), a 'shore bank' (*àn* 岸), a place or a temple compound.[745] The character '⿱穴言' does not seem to appear in any transmitted texts.[746] This manuscript also has a notch depicting the number 'six' (*liù* 六) carved into its left side.

742 Chén Wěi 2012: 236.
743 Chén Wěi 2012: 236–237.
744 Chén Wěi 2012: 246.
745 Chén Wěi 2012: 246–247; Péng 2008: 199; Sanft 2014b: 345–346; Zhāng Chūnlóng 2007.
746 Sanft 2014b: 344–346.

8-993 I–II: 卅五年六月戊午[55]朔己巳[6], 庫建、佐般出賣祠窨□☒┘ 令史歊監.☒[747]

In the 35th year (of King Zhèng) [212 BC], month 6, new moon on a *wùwǔ*, *jǐsì* [day 12], Jiàn from the Warehouse and Assistant Bān sell □☒ offered (as sacrifice) to X [pronunciation unknown]┘ Supervised by Prefectural Secretary X [pronunciation unknown]. ☒

This manuscript has a notch depicting the number 'one' (*yī* 一) carved into its left side.

8-1002+8-1091 I–II: 卅五年六月戊午[55]朔己巳[6], 庫建、佐般出賣祠窨□□□一朐于隸臣徐, 所取錢一.┘ 令史歊監. 般手.[748]

In the 35th year (of King Zhèng) [212 BC], month 6, new moon on a *wùwǔ*, *jǐsì* [day 12], Jiàn from the Warehouse and Assistant Bān sell □□□ offered (as sacrifice) to X [pronunciation unknown]. One strip goes to Bondservant Xú, and what they receive is one coin.┘ Supervised by Prefectural Secretary X [pronunciation unknown]. By the hand of Bān.

The character '*qú* 朐' is a measure word used for a long strip of dried meat.[749] This manuscript has a notch depicting the number 'ten' (*shí* 十) carved into its left side.

8-1055+8-1579 I–II: 卅五年六月戊午[55]朔己巳[6], 庫建、佐般出賣祠窨餘徹脯一朐于□□□, 所取錢一.┘ 令史歊監. 般手.[750]

In the 35th year (of King Zhèng) [212 BC], month 6, new moon on a *wùwǔ*, *jǐsì* [day 12], Jiàn from the Warehouse and Assistant Bān sell the leftover dried meat offered (as sacrifice) to X [pronunciation unknown]. One strip goes to □□□, and what they receive is one coin.┘ Supervised by Prefectural Secretary X [pronunciation unknown]. By the hand of Bān.

This manuscript also has a notch depicting the number 'one' (*yī* 一) carved into its left side.

8-1162: ☒般出賣祠窨餘徹食☒[751]

☒ Bān sell the leftover food offered (as sacrifice) to X [pronunciation unknown] ☒

A few of these wooden documents or 'tallies' (*quàn* 券), concerning the redistribution or reselling of food, convey additional administrative information through the so-called 'notches' (*kèchǐ* 刻齒).[752] Some of the first scholars to systematically decode these markings were Zhāng, Okhawa and Momiyama in 2015, who contend that different shapes of notches represent varying amounts of food, cash or other

747 Chén Wěi 2012: 258.
748 Chén Wěi 2012: 259.
749 Chén Wěi 2012: 260; Hé Yǒuzǔ 2014.
750 Chén Wěi 2012: 269.
751 Chén Wěi 2012: 286.
752 Ma Tsang Wing 2020a. The term 'notch' (*kèchǐ* 刻齒) is used by Ma Tsang Wing. Chén Wěi uses the term '*chǐ* 齒', which carries the same meaning.

commodities mentioned in the texts. Notches were also used to verify the transfer of documents by marking pairs of tallies as matching counterparts (Fig. 6.3).[753]

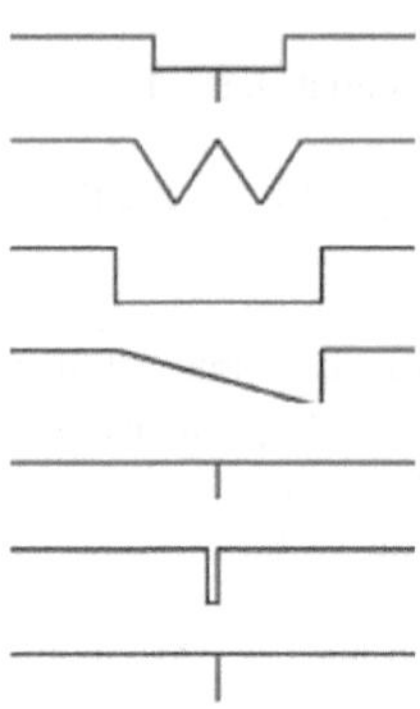

Fig. 6.3: The meaning of different 'notches' (*chǐ* 齒) according to shape, as seen on some wooden documents in the *Lǐyē* corpus.[754]

The practice of offering sacrifices to agricultural ancestors was already well-established in the Western Zhōu dynasty, as documented in written sources like the *Book of Rites* (*Lǐ jì* 禮記) from the Springs and Autumns period, or the *Book of Hàn* (*Hànshū* 漢書) and the *Book of Later Hàn* (*Hòu Hànshū* 後漢書) from the Eastern Hàn dynasty. Before the discovery of the *Lǐyē* and *Zhōujiātái* 周家台 manuscripts, no detailed records of institutionalized worship of the Agricultural Ancestor had been attributed to the Qín dynasty. Similarly, there are no references to such practices in transmitted Qín or pre-Qín texts, such as the *Shǐjì* or the *Book of Hàn.*[755] Some of the earliest reliable sources referencing the veneration of an Agricultural Ancestor date back to the 1st century BC. Because of the First Emperor's alleged cultural autocracy, further overshadowed by the infamous "burning of books and killing of scholars" (*fén shū kēng rú* 焚書坑儒), many modern scholars believe that the vast majority of institutionalized rituals were abolished under imperial rule. The Qín administration, adhering to "legalist" principles, harshly punished those who violated its laws. It was not until the Early Hàn dynasty that sacrificial rituals and ancestor worshop were allegedly restored.[756] In fact, expressions like 'burn' (*shāo* 烧), 'public executions (in the marketplace)'

753 Zhāng Chūnlóng, Ohkawa und Momiyama 2015: 53–69.

754 This graphic is taken from Zhāng Chūnlóng, Ohkawa und Momiyama 2015: 54.

755 Sanft 2014: 332.

756 For reasons why the term 'Legalism' is questionable in the study of Early China, see, for example, Creel 1970: 92–120; Goldin 2011. For different views on the alleged, large-scale biblioclasm that happened under imperial Qín rulership, see Kern 2000: 183–196; Petersen 1995.

(*qìshì* 弃市) and the 'hundred schools of thought' (*bǎijiā* 百家) are absent from the *Lǐyē* texts. Neither these terms nor any other in the corpus remotely allude to a large-scale biblioclasm or systematic persecution of scholars representative of certain schools of thought. Nevertheless, the absence of such references in the *Lǐyē* material does not necessarily disprove the existence of these practices.

In examining the continuation of rituals from the Eastern Zhōu to the Qín dynasty, manuscript 8-138+8-174+8-522+8-523 and line BIII of manuscript 8-681 warrant special attention.[757] Both manuscripts underscore the importance of ritualized ancestral performances and the institutionalized obligations they entailed. These obligations included "building temples" (*zuò miào* 作廟), "inspecting temples" (*xíng miào* 行廟) and repairing temples.[758] Although not directly related, line BII of manuscript 8-461 draws parallels between sacrifices made to the Emperor and those dedicated to the 'Celestial Thearch' (*Tiāndì* 天帝). Not only does this analogy indicate the deification of the Emperor, it also suggests that the Qín dynasty may have derived its rule from religious authority or divine right.[759]

> 8-138+8-174+8-522+8-523 I–II, rAI–IV, rBI–IV, rCI–IV, rDI–IV: 廿六年六月壬子[49], 遷陵□,【丞】敦狐爲令史更行廟詔: 令史行⊠┘ 失期. 行廟者必謹視中□各自署廟所質日. 行先道旁曹始, 以坐次相屬. └A 十一月己未[56], 令史慶行廟.┘ 十一月己巳[6], 令史廱行廟.┘ 十二月戊辰[5], 令史陽行廟.┘ 十二月己丑[26], 令史夫行廟.┘ B □□□□令史韋行.┘ 端月丁未[44], 令史廱行廟.┘ □□□□, 令史慶行廟.┘ □月癸酉[10], 令史犯行廟.┘ C 二月壬午[19], 令史行行廟.┘ 二月壬辰[29], 令史莫

757 Column B, line III of manuscript 8-681 states: "three people inspect the temple" (*sān rén xíng miào* 三人行廟) (Chén Wěi 2012: 202). This is controversial, given that the text appears to record this task being assigned to criminals. Gāo Yīzhì hence claims that "*xíng miào* 行廟" means "*zuò miào* 作廟" (Gāo Yīzhì 2013: 239–244). He bases this assumption on texts in the *Yüèlù* collection which only record high-level officials inspecting temples. Evidence in the *Lǐyē* corpus is too scarce to offer any conclusions about the usage of '*xíng* 行' and '*zuò* 作' in the context of sacrificial rituals and ancestral worship.

758 For a discussion of the phrase "inspecting the temple" (*xíng miào* 行廟), see Chén Wěi 2012: 79; Fàn Yúnfēi 2016; Gāo Yīzhì 2013: 239–244; Lǔ 2014; Zhōu Zhènhè 2010: 30–52. "Building the temple" (*zuò miào* 作廟) appears twice on the manuscrips from layer 8, namely on manuscripts 8-145 (line EVII) and 8-162 (line BIII). Both address the work assignments of convicts (Chén Wěi 2012: 85, 98). According to the *Yuèlù* documents, officials started to "inspect temples" (*xíng miào* 行廟) in the second year of the Second Emperor (209 BC) (Chén Sōngcháng 2015: 226). *Lǐyē* manuscript 8-138+8-174+8-522+8-523 already verifies the existence of this institutionalized practice in 221 BC. For a study on local ritual practices as reflected in the *Shuìhǔdì* documents, see Sterckx 2009, 2010. For a comparison of reforms made to imperial cults and calendars during the formative years of the Qín, Hàn and Roman empires, see Robinson 2016.

759 See also Kern 2000; Puett 2002; Yóu Yìfēi 2013. For a transcription and translation of manuscript 8-461, see chapter 4.1.

邪行廟.┘ 二月壬寅[39], 令史釦行廟.┘ 四月丙申[33], 史戎夫行廟. ┘D 五月丙午[43], 史釦行廟.┘ 五月丙辰[53], 令史上行廟.┘ 五月乙丑[2], 令史□□□┘ 六月癸巳[30], 令史除行廟.[760]

In the 26th year (of King Zhèng) [221 BC], month 6 on a *rénzǐ*, Dūnhú, Deputy【Prefect】of Qiānlíng, orders Prefectural Secretary Gèng to inspect the temple and instructs: (I) order the Prefectural Secretaries to inspect ☒┘ miss the (right) time.┘ The one who inspects the temple must carefully watch the middle □ each proceeds from the day verified in all temples. Inspection starts first in the adjacent Bureaus of the marches. Prosecute as a criminal offense if the order is mixed up. └A In month 11 on a *jǐwèi*, Prefectural Secretary Qìng inspects the temple.┘ In month 11 on a *jǐsì*, Prefectural Secretary Kàng inspects the temple.┘ In month 12 on a *wùchén*, Prefectural Secretary Yáng inspects the temple.┘ In month 12 on a *jǐchǒu*, Prefectural Secretary Fū inspects the temple.┘ B □□□□ Prefectural Secretary Wéi inspects.┘ In the 1st month on a *dīngwèi*, Prefectural Secretary Kàng inspects the temple.┘ □□□□, Prefectural Secretary Qìng inspects the temple.┘ In month □ on a *guǐyǒu*, Prefectural Secretary Fàn inspects the temple.┘ C In month 2 on a *rénwǔ*, Prefectural Secretary Xíng inspects the temple.┘ In month 2 on a *rénchén*, Prefectural Secretary Mòyé inspects the temple.┘ In month 2 on a *rényín*, Prefectural Secretary Kòu inspects the temple.┘ In month 4 on a *bǐngshēn*, Official Róngfū inspects the temple.┘ D In month 5 on a *bǐngwǔ*, Official Kòu inspects the temple.┘ In month 5 on a *bǐngchén*, Prefectural Secretary Shàng inspects the temple.┘ In month 5 on a *yǐchǒu*, Prefectural Secretary □□□┘ In month 6 on a *guǐsì*, Prefectural Secretary Chú inspects the temple.

Comments: Chén Wěi's transcription of line I reads "*Qiānlíng* □、【*chéng*】遷陵□、【丞】", which need not be separated in my view.[761] The missing character could plausibly be '*shǒu* 守', forming the term 'Deputy Prefect' (*shǒuchéng* 守丞).

The term '*zhào* 詔', traditionally believed to exclusively designate the Emperor's edicts from the Qín dynasty onwards, is presented here with an alternative, broader meaning and a more universal applicability. In this context, '*zhào* 詔' means 'to instruct', 'to order', 'instruction' or 'edict'.[762]

Chén Wěi understands '*shǔ* 署' as 'to record' or 'to make notes', and '*zhì rì* 質日' as the so-called 'event calendars' kept by local authorities during the Qín and Hàn dynasties.[763] On this manuscript, '*zuò cì* 坐次' is commonly translated as the 'sitting order' of the Prefectural Secretaries.[764]

The term '*zuò* 坐', explicitly used in legal contexts in other *Lǐyē* manuscripts means 'to be prosecuted (for a criminal offense)'. I would argue for a similar interpretation in this instance.

Tab. 6.1 highlights the days of the 'sexagenary cycle' (*gānzhī* 干支) when Qiānlíng County officials, predominantly 'Prefectural Secretaries' (*lìngshǐ* 令史), conducted temple inspections. Despite three dates containing illegible characters (lines rBI,

760 Chén Wěi 2012: 78.

761 Chén Wěi 2012: 78.

762 Lau and Staack 2016: 286.

763 Chén Wěi 2012: 79; Chén Wěi 2018.

764 See, for example, Hé Yǒuzǔ 2015a.

rBIII–IV), there does not seem to be a regular pattern or standardized inspection protocol applicable to all 13 inspections performed in the 26th year of King Zhèng (221 BC). A closer look at the number of temple inspections across the sexagenary cycle reveals no significant correlations (Fig. 6.4). Based on the dates provided on manuscript 8-138+8-174+8-522+8-523, most of the inspections in 221 BC occurred during months 2 and 5.

Tab. 6.1: Days according to the 'sexagenary cycle' (*gānzhī* 干支) in 221 BC when Qiānlíng County officials "inspected the temples" (*xíng miào* 行廟), as cited on manuscript 8-138+8-174+8-522+8-523. Month 11 is highlighted in blue, month 12 in orange, month 1 in brown, month 2 in purple, month 4 in green, month 5 in yellow, and month 6 in red.

1	*jiǎzǐ* 甲子	2	*yǐchǒu* 乙丑	3	*bǐngyín* 丙寅	4	*dīngmǎo* 丁卯	5	*wùchén* 戊辰
6	*jǐsì* 己巳	7	*gēngwǔ* 庚午	8	*xīnwèi* 辛未	9	*rénshēn* 壬申	10	*guǐyǒu* 癸酉
11	*jiǎxū* 甲戌	12	*yǐhài* 乙亥	13	*bǐngzǐ* 丙子	14	*dīngchǒu* 丁丑	15	*wùyín* 戊寅
16	*jǐmǎo* 己卯	17	*gēngchén* 庚辰	18	*xīnsì* 辛巳	19	*rénwǔ* 壬午	20	*guǐwèi* 癸未
21	*jiǎshēn* 甲申	22	*yǐyǒu* 乙酉	23	*bǐngxū* 丙戌	24	*dīnghài* 丁亥	25	*wùzǐ* 戊子
26	*jǐchǒu* 己丑	27	*gēngyín* 庚寅	28	*xīnmǎo* 辛卯	29	*rénchén* 壬辰	30	*guǐsì* 癸巳
31	*jiǎwǔ* 甲午	32	*yǐwèi* 乙未	33	*bǐngshēn* 丙申	34	*dīngyǒu* 丁酉	35	*wùxū* 戊戌
36	*jǐhài* 己亥	37	*gēngzǐ* 庚子	38	*xīnchǒu* 辛丑	39	*rényín* 壬寅	40	*guǐmǎo* 癸卯
41	*jiǎchén* 甲辰	42	*yǐsì* 乙巳	43	*bǐngwǔ* 丙午	44	*dīngwèi* 丁未	45	*wùshēn* 戊申
46	*jǐyǒu* 己酉	47	*gēngxū* 庚戌	48	*xīnhài* 辛亥	49	*rénzǐ* 壬子	50	*guǐchǒu* 癸丑
51	*jiǎyín* 甲寅	52	*yǐmǎo* 乙卯	53	*bǐngchén* 丙辰	54	*dīngsì* 丁巳	55	*wùwǔ* 戊午
56	*jǐwèi* 己未	57	*gēngshēn* 庚申	58	*xīnyǒu* 辛酉	59	*rénxū* 壬戌	60	*guǐhài* 癸亥

It is striking that no inspection days overlap within the sexagenary cycle (Tab. 6.1, Fig. 6.4–6.5). Three inspections were scheduled for months 2 and 5, marking the highest frequency of inspections per month in 221 BC. These inspections were spaced exactly ten days apart, except for the one on '*yǐchǒu* 乙丑' in position 2. The majority of inspections were concentrated at the beginning of the lunar year, with additional peaks observed in spring and summer. During month 3 or between months 7 and 10, there were no inspections at all. The illegible sections in lines rBI, rBIII and rBIV prevent precise dating. However, they likely indicate one inspection between months 12 and 1, and two more between months 1 and 2, providing further support to the observed pattern.

Aside from their role in preparing and selling food and alcohol for rituals, Warehouses also stored weaponry and reported parts of their statistics to the 'Finance Departments' (*jīnbù* 金布) – frequently referred to by scholars as 'Finance

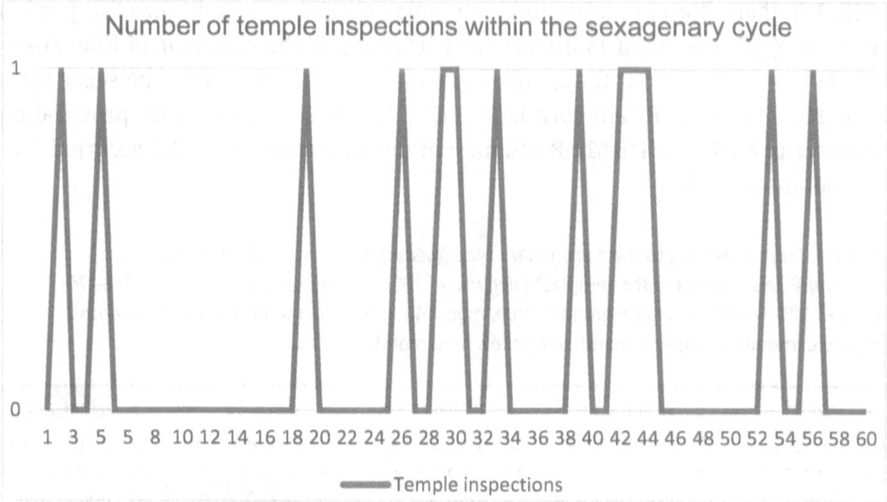

Fig. 6.4: Diagram of Qiānlíng County temple inspections in 221 BC, according to the 'sexagenary cycle' (*gānzhī* 干支) and as seen on manuscript 8-138+8-174+8-522+8-523.

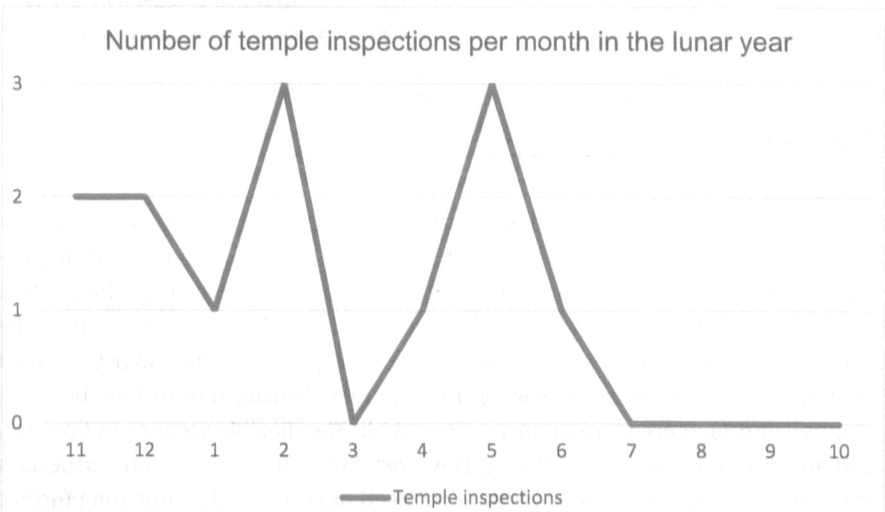

Fig. 6.5: Diagram of temple inspections per month in 221 BC, as seen on manuscript 8-138+8-174+8-522+8-523.

Bureaus.'[765] The texts from layers 5, 6 and 8 lack actual Warehouse evaluations or statistics, although manuscripts 8-493 and 8-458 help reconstruct information regarding "Warehouse weaponries" (*kù bīng* 庫兵) and related inventories (Fig. 6.6). Manuscript 8-493 states:

8-493 AI–III, BI–III, CI–II: A 金布計錄:┘ 庫兵計,┘ 車計,┘ B 工用計,┘ 工用器計,┘ 少內器計,┘ C【金】錢計.┘ 凡六計.[766]

A Statistical registers of the Finance Department:┘ Statistics of the Warehouse weaponry,┘ statistics of carriages,┘ B statistics of tools for public use,┘ statistics of vessels for public use,┘ statistics of vessels by the Lesser Treasury,┘ C statistics of【metal】and coins.┘ In total, there are six statistics.

Comment: The *Lǐyē* texts specifically distinguish between 'carriages (for military use)' (*chē* 車) and 'carriages for agricultural use' (*tián chē* 田車), as indicated on the following manuscript:

8-410 I: 廿八年, 遷陵田車計付鴈(雁)門泰守府☑[767]

In the 28th year (of King Zhèng) [219 BC], Qiānlíng's statistics of carriages for agricultural use are handed over to the Governor of Yànmén (Commandery) ☑

The statistics on Warehouse weaponry was one of six types of records either compiled by the Finance Departments or submitted to them. Another text from layer 8 contains an inventory list from a Warehouse in Qiānlíng County and provides an overview of the quantities and types of weaponry that may have been stored in other county Warehouses as well:

8-458 AI–IV, BI–IV, CI–II: A 遷陵庫真□┘ 甲三百卌九.┘ 甲宛廿一.┘ 鞮瞀卅九.┘ B 冑廿八.┘ 弩二百五十一.┘ 臂九十七.┘ 弦千八百一.┘ C 矢四萬九百□.┘ 軗(戟)二百五十一.[768]

A Zhēn □ of the Qiānlíng Warehouse ┘ body armors, 349.┘ Armor of the type X [pronunciation unknown], 21.┘ Armored battle helmets, 39.┘ B *Zhòu*-type helmets, 28.┘ Bows, 251.┘ Crossbows, 97.┘ Strings, 1801.┘ C Arrows, 40,9 □.┘ Halberds with crescent blades, 251.

It remains uncertain how imperial institutions stored information from the Warehouses, or whether inventory lists were systematically exchanged and compared between the Finance Departments of different counties and commanderies. The inflow and outflow of weaponry was painstakingly documented, as was the logistical reallocation of goods, recorded on manuscript 8-151:

---

765 For more information on 'Finance Departments' (*jīnbù* 金布), refer to Sūn Wénbó 2014b.
766 Chén Wěi 2012: 169.
767 Chén Wěi 2012: 144.
768 Chén Wěi 2012: 154.

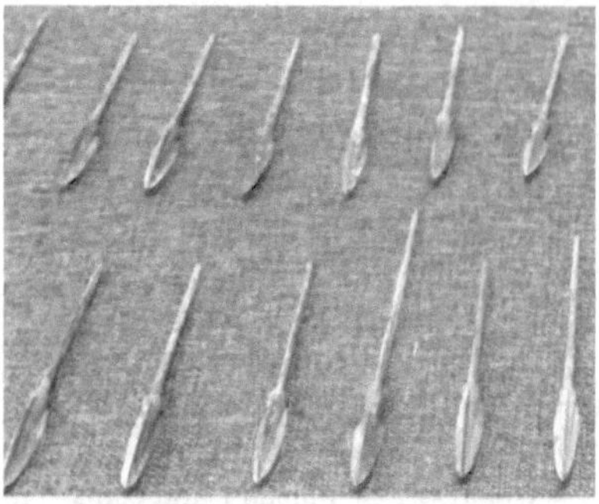

Fig. 6.6: Photograph of arrowheads made of bronze. Dimensions unknown. On display at the Lǐyē Museum of Qín Slips (*Lǐyē Qín jiǎn bówùguǎn* 里耶秦簡博物館).[769]

8-151 I–VI: 遷陵已計: 卌四年餘見弩臂百六十九.┘ · 凡百六十九.┘ 出弩臂四輸益陽.┘ 出弩臂三輸臨沅.┘ · 凡出七.┘ 今九月見弩臂百六十二.[770]

Qiānlíng has already (received) the statistics: In the 34th year (of King Zhèng) [213 BC], a surplus is seen of 169 bows.┘ · In total, there are 169 bows.┘ (Of the) outgoing bows, four are transported to Yìyáng County.┘ (Of the) outgoing bows, three are transported to Línyuán County.┘ · In total, seven go out.┘ Now, in month 9, 162 bows are seen.

Comments: Yìyáng 益陽 County also appears in the "Event Calendar of the 34th Year" in the *Yuèlù* documents.[771] As observed by Zhōng Wěi, Zhuāng Xiǎoxiá and Yàn Chāngguì, Yìyáng was under the administrative jurisdiction of Cāngwú 蒼梧 Commandery.[772] *Lǐyē* manuscripts 16-5 and 16-6 confirm that Yìyáng was not part of Dòngtíng Commandery.[773]

Relative to our previous understanding of the Qín dynasty, this meticulous attention to detail in the writing, usage and archiving of imperial documents revealed by these texts is unparalleled. This level of precision is also evident in other *Lǐyē* texts, such as in manuscripts 8-216+8-351, 8-217, 8-1002+8-1091 and 8-143, which even record minuscule movements of food and money, and the treatment provided for a single sick horse:

8-216+8-351 I–II: ☐□司空守茲、佐得出以食舂、小城旦卻等五十二人, 積五十二日, 日四升六分升一.┘ ☐令史尚視平. 得手.[774]

☐□ Zī, Commissioner of Public Works, and Assistant Dé give out food (rations) to (female) Grain Pounders, (male) Minor Wall Pounder Què and others. (In total, there are) 52 people

769 My photograph, taken at the Lǐyē Museum of Qín Slips (*Lǐyē Qín jiǎn bówùguǎn* 里耶秦簡博物館) on May 14, 2019.

770 Chén Wěi 2012: 91–92.

771 Zhū Hànmín and Chén Sōngzhǎng 2010: 70, 74.

772 Yàn Chāngguì 2014; Zhōng Wěi 2005; Zhōng Wěi and Yàn Chāngguì 2008.

773 Yóu Yìfēi 2015.

774 Chén Wěi 2012: 116.

for an accumulated period of 52 days. Per day, (this amounts to) four and one-sixth pints.┘ ☒ Prefectural Secretary Shàng oversees the weighing (of the food). By the hand of Dé.

Comments: The daily food rations of "*sì shēng liù fēn shēng yī* 四升六分升一" for Grain Pounders, (a male) Minor Wall Pounder and possibly other convicts are calculated to be equivalent to four and one-sixth 'pints' (*shēng* 升), or twenty-five-sixths of a pint. This amounts to a total of 216 and two-thirds pints, or 216.67 pints, for a period of 52 people over 52 days (52 days x 25/6 pints).[775]

8-217 I–II: 稻四斗八升少半半升. 卅一年八月壬寅[39], 倉是、史感、稟人堂出稟隸臣嬰自〈兒〉槐庫.┘ 令史悍平. 六月食. 感手.[776]

(The amount of) paddy is four bushels plus eight and five-sixth pints. In the 31st year, month 8 on a *rényín* [day 22], Shì, Secretary Gǎn and Intermediate Táng from the Granary give out (food rations) to the toddler of (female) Bondservant Huáituí.┘ Balanced by Prefectural Secretary Hàn. Food supplies for month 6. By the hand of Gǎn.

The expression "*shǎo bàn bàn shēng* 少半半升" means one-third of a pint (or two-sixths of a pint) plus one half a pint (or three-sixths of a pint). This is equivalent to five-sixths of a pint. The new moon day in month 8 of the 31st year of King Zhèng was a '*xīnsì* 辛巳', which makes '*rényín* 壬寅' correspond to 'day 22'.[777]

The term '*píng* 平' (*breŋ) is commonly understood as 'flat', 'even', 'level', 'equal', 'impartial' or 'peaceful'.[778] In this particular context, I understand the character as a graphic variant of '*chèng* 秤' (*tʰjəŋ) and translate it as 'to make even' or 'to make equal', referring to the balanced weighing of goods.[779]

8-1002+8-1091 I–II: 卅五年六月戊午[55]朔己巳[6], 庫建、佐般出賣祠窨□□□一朐于隸臣徐, 所取錢一.┘ 令史歃監. 般手.[780]

In the 35th year (of King Zhèng) [212 BC], month 6, new moon on a *wùwǔ*, *jǐsì* [day 12], Jiàn from the Warehouse and Assistant Bān sell □□□ offered to X [pronunciation unknown]. One strip goes to Bondservant Xú, and what they receive is one coin.┘ Supervised by Prefectural Secretary X [pronunciation unknown]. By the hand of Bān.

8-143 I–IV, rI–II: 卅四年九月癸亥[60]朔乙酉[22], 畜□☒┘ 蓋侍食羸病馬無小, 謁令官遣☒┘ 病者無小, 令止行書徒更戍城父柘□☒┘ 之. ノ卅五年十一月辛卯[28]朔朔日, 遷陵□☒┘ 如律令. ノ履手. ノ十一月【壬】☒┘ 十一月辛卯[28]旦, 史穫以來. ノ☒[781]

775 See also Chén Wěi 2012: 116.
776 Chén Wěi 2012: 116.
777 Compare with Xǔ Míngqiāng 2013b.
778 Schüssler 2009: 415; Lau and Staack 2016: 225–226.
779 See also Sanft 2014: 338.
780 Chén Wěi 2012: 259.
781 Chén Wěi 2012: 83.

In the 34th year (of King Zhèng) [213 BC], month 9, new moon on a *guǐhài, yǐyǒu* [day 23], the domestic animals □▨┘ cogon grass is fed to the exhausted, sick horse named Nothing Small About It. Request an ordinance from the office to dispatch ▨┘ sick Nothing Small About It, now prevents the convict delivering the documents from performing military service at the border in the township Zhè of Chéngfù County □▨┘ Walks. / In the 35th year (of King Zhèng) [212 BC], month 11, new moon on a *xīnmǎo*, new moon day [day 1], Qiānlíng □▨ └ According to the statutes and ordinances. / By the hand of Lǚ. / In month 11, 【*rén*】 ▨┘ In month 11, in the morning of the new moon day [day 1], (the documents) are brought by the Assistant Secretary. / ▨

The bottom part of this manuscript is damaged. Nothing Small About It (Wúxiǎo 無小) is most likely a horse's name.[782] More examples of horse names are found in the Hàn dynasty "Documents of Registered Horse Names" (*Chuán mǎ míng jí* 傳馬名籍) from the *Suspended Spring Postal Station* (*Xuán quán zhì* 懸泉置) texts discovered in Dūnhuáng.[783] Other examples include Quánjiù 全厩, Huángquè 黃雀 and Wánxìng 完幸, taken from manuscripts 12, 14 and 18, respectively.

'*Gèng shù* 更戍' literally translates to 'to perform military service as punishment'.[784] As mentioned in the "Treatise on Geography" (*Dìlǐ zhì* 地理志) from the *Book of Hàn* (*Hànshū* 漢書), Zhè 柘 was a township located in Chéngfù 城父 County.[785] Due to the illness of Nothing Small About It, some of the documents could not be delivered to Zhè. It ultimately took over a year – specifically, one year, one month and seven days – to complete this administrative process.

Taking care of horses and cattle was a costly affair for local officials. A variety of texts unearthed at Shuìhǔdì 睡虎地 related to the theft of domestic animals bear witness to their value as a source of food and as essential assets for warfare and communication.[786] The "Clay Seals of Qín" (*Qín fēngní* 秦封泥) distinguish 21 officials responsible for raising horses and managing the 'Stables' (*jiù* 厩).[787] These officials were often granted a fixed number of cattle, horses and carriages, which were maintained on public grounds by lower-ranking officials, local farmers and convicts.[788]

Although horses were most certainly raised and kept in Stables, they are not classified as 'domestic' (*chù* 畜) in the *Lǐyē* corpus.[789] From the texts excavated in layers 5, 6 and 8, manuscript 8-163 is unique for emphasizing the importance of

782 Chén Wěi 2012: 83; Hé Yǒuzǔ 2015b.

783 Hú and Zhāng 2001: 82–84.

784 Chén Wěi 2012: 83-84; Lau and Staack 2016: 319.

785 Chén Wěi 2012: 84.

786 Hulsewé 1985: 122–133, 189.

787 Zhōu Xiǎolù and Lù Dōngzhī 2000: 183–98.

788 Hulsewé 1985: 63.

789 The 'Large Stable' (*dà jiù* 大廄), 'Middle Stable' (*zhōng jiù* 中廄) and 'Palace Stable' (*gōng jiù* 宮廄) were different kinds of 'Stables (*jiù* 廄) in the Qín dynasty. In the Hàn dynasty, the 'Large Stable' (*dà jiù* 大廄) became one of the six Stables of the Emperor, while the carriages and horses of the Empress were kept in the 'Middle Stable' (*zhōng jiù* 中廄) (Chén Wěi 2012: 99).

Stables in the Qiānlíng area. The fact that the Assistant to the Commissioner of Public Works was assigned to oversee a Stable at daily intervals indicates the elevated status horses held in Qín society. It may also imply that the Stables and the officials responsible for managing them were given special attention and priority.

8-163 I–III, rI: 廿六年八月庚戌[47]朔壬戌[59], 厩守慶敢言之: 令曰┘ 司空佐貳今爲厩佐言視事日.・今以戊申[45]┘ 視事. 敢言之. └貳手[790]

In the 26th year (of King Zhèng) [221 BC], month 8, new moon on a *gēngxū*, *rénxū* [day 13], Stable Incumbent Qìng dares to say: The ordinance states┘ that Èr, Assistant to the Commissioner of Public Works, now becomes Assistant of the Stable in charge of communicating and overseeing the matter on a daily basis. • Now, by day *wùshēn*, ┘(he) oversees the matter. (He) dares to say. └By the hand of Èr.

Comments: If the new moon day in month 8 of the 26th year of King Zhèng was a '*gēngxū* 庚戌' in position 47, then '*wùshēn* 戊申' would correspond to position 45. This discrepancy means that the sentence in question might either be unrelated to the first part of the manuscript (i.e., a quotation from another source) or it could result from a clerical error.

The 'statistics of oxen' (*niú jì* 牛計) mentioned in line BIV of manuscript 8-481 was just one of many responsibilities assigned to the 'Office of Domestic Animals' (*chùguān* 畜官).[791] This governmental body also compiled statistics of animals that died under the care of convicts, those used for butchering and selling meat, and tracking other fluctuations in the number of horses, oxen and sheep across the region. A more complete sample of evaluation records conducted by the Office of Domestic Animals is provided on manuscript 8-490+8-501:

8-490+8-501 AI–IV, BI–V: A 畜官課志:┘ 徒隸牧畜死負、剝賣課,┘ 徒隸牧畜畜死不請課,┘ 馬產子課,┘ B 畜牛死亡課,┘ 畜牛產子課,┘ 畜羊死亡課,┘ 畜羊產子課.┘ ・凡八課.[792]

A Evaluation records of the Office of Domestic Animals:┘ Evaluation of the compensation payments for domestic animals that died under the husbandry of convict servants, as well as their butchering and selling,┘ evaluation of untold (cases of) domestic animals that died under the husbandry of convict servants,┘ B evaluation of horses that gave birth,┘ evaluation of domestic oxen that died,┘ evaluation of domestic oxen that gave birth,┘ evaluation of domestic sheep that died,┘ evaluation of domestic sheep that gave birth.┘ • In total, there are eight types of evaluations.

790 Chén Wěi 2012: 99.
791 For a transcription and translation of manuscript 8-481, see chapter 4.4.
792 Chén Wěi 2012: 168.

Comments: Chén Wěi translates 'qǐng 請' as 'to tell' in this context.[793] Even though the text appears complete, the summary line records eight types of evaluations instead of the seven actually listed.

The Qín regime had already developed efficient agricultural techniques prior to the imperial unification under the First Emperor. However, with the newly conquered territories under its control, the empire was compelled to introduce additional wide-scale monetary and fiscal reforms, along with the registration of households. These initiatives standardized units of metal and currency, allowing bulky or perishable goods to be converted into wealth that could be efficiently stored, transported or used as currency. Such reforms made vital contributions to labor, military services, commercial networks and even criminal proceedings, where wages became integral.

The "Statutes Concerning the Finance Department" (*Jīnbù lǜ* 金布律) from the *Shuìhǔdì* documents deal with the imperial administration of currency, finances, and the assignment of convicts, livestock and public goods to "Minor Offices" (*xiǎo guān* 小官).[794] In the *Lǐyē* documents, 'Finance Departments' (*jīnbù* 金布) are involved in receiving and compiling various statistics. These statistics relate to carriages, precious metals, money, Warehouse weaponries, vessels and tools administered by the 'Lesser Treasuries' (*shǎonèi* 少内) or other vessels designated for public use.

Although Finance Departments are nowadays often equated with the Lesser Treasuries or with other units that share similar responsibilities, this categorization requires revision.[795] Lesser Treasuries appear to have functioned as administrative units tasked with overseeing the Finance Departments. They received and redistributed funds across the empire's territories and implemented fiscal policies.[796] Finance Departments, by contrast, primarily coordinated resources and investments with the '(County) Courts' (*tíng* 廷). Contrary to common assumptions, Finance Departments were not part of the County Courts. The frequent mistranslation of '*tíng jīnbù* 廷金布' on certain *Lǐyē* manuscripts as the 'Finance Department of the (County) Court' (*tíng jīnbù* 廷金布) has led to this misconception.[797] Instead, such phrases should be considered inherent to the imperial Qín postal system, and the correct translation should be "from the (County) Court to the Finance Department."[798]

793 Chén Wěi 2012: 168.

794 Hulsewé 1985: 46–56. Hulsewé translates '*Jīnbù lǜ* 金布律' as "Statutes on Currency".

795 See, for example, Sūn Wénbó 2014b.

796 For a discussion of the 'Lesser Treasury' (*shǎonèi* 少内), based on transmitted text passages, see, for example, Hulsewé 1985: 195–200.

797 Sūn Wénbó 2018: 86.

798 For more on this topic, refer to chapter 2.3.

These Finance Departments regularly supplied the County Courts and other higher administrative institutions with statistical registers and evaluations. The information they made available covered a wide range of topics, including currency, weaponry, precious metals, water ponds, gardens, convicts, fines, tools and vessels for public use, lacquer and bamboo cultivation, losses due to fire or floods, and the collection of raw materials for iron production. All of these factors significantly influenced the financial outlook and bookkeeping of the imperial accounts.

In the manuscript that follows, copies of evaluation records compiled by other Bureaus were submitted to the Finance Department. Use of the term '*shàng* 上', meaning 'to submit', implies the "handing over (of something) to a higher authority," suggesting that the Finance Department held a superior administrative position to the Bureaus in Qiānlíng County:

8-454 AI–V, BI–V, CI–IV: A 課上金布副.┘ 桼課.┘ 作務.┘ 疇竹.┘ 池課.┘ B 園栗.┘ 采鐵.┘ 市課.┘ 作務徒死亡.┘ 所不能自給而求輸.┘ C 縣官有買用錢.ノ 鑄段(鍛).┘ 竹箭.┘ 水火所敗亡.ノ園課. 采金.┘ 貲、贖、責(債)毋不收課.[799]

A Copies of evaluations submitted to the Finance Department.┘ Evaluation of varnish trees.┘ (Evaluation of) revenue generated (by convicts).┘ (Evaluation of) bamboo cultivation.┘ Evaluation of water ponds.┘ B (Evaluation of) gardens and chestnuts.┘ (Evaluation of) collecting (raw materials for the production of) iron.┘ Evaluation of marketplaces.┘ (Evaluation of) convicts doing manual labor who died or fled.┘ (Evaluation of) requested transportation of those who cannot support themselves.┘ C (Evaluation of) County Offices that have coins to buy (goods). / (Evaluation of) metal casting.┘ (Evaluation of) arrows made of bamboo.┘ (Evaluation of things that) were destroyed and lost by water or fire. / Evaluation of gardens. (Evaluation of) collecting metals.┘ Evaluation of fines, redemption fees and debts that were received.

Comments: Resin from 'varnish trees' (*qī* 桼) (*tshit) was used to produce 'lacquer' (*qī* 漆) (*tshit), two terms which were synonymous and interchangeable in early texts.[800] '*Zuò wù* 作務' in line three could be an abbreviation of "*zuò wù chǎn qián kè* 作務產錢課" found in the "evaluation records of the Granary" (*cāng kè zhì* 倉課志) on manuscript 8-495.[801]

The "evaluation of things that were destroyed or lost by water or fire" (*shuǐ huǒ bài wáng zhě kè* 水火敗亡者課) also appears on manuscript 8-1511.[802]

'*Jīnbù* 金布' was an administrative unit positioned hierarchically between the County Court and other county Bureaus. For this reason, I specifically refer to the Finance Department as a Department and not as a Bureau or Office.

799 Chén Wěi 2012: 152–153.

800 Phonological reconstructions are taken from Schuessler 2009: 302.

801 See chapter 4.4.

802 For a transcription of manuscript 8-1511, see Chén Wěi 2012: 341–342.

Records show that the Deputy Prefect of Qiānlíng instructed the Finance Departments to confirm and communicate an issue that had already been investigated and resolved by the Lesser Treasury. Once the matter was finalized, the document was forwarded from the Governor's administrative unit to the Lesser Treasury. Like the Finance Department, the Lesser Treasury acted as an intermediary between the county Bureaus, Offices and the County Court. Manuscript 8-164+8-1475 further corroborates the hierarchical position of the Lesser Treasury within the institutional order of Qiānlíng County. It also records an instance in which an official from the Lesser Treasury reported an unfinished matter to the County Court:

8-155 I–II: 四月丙午[43]朔癸丑[50], 遷陵守丞色下: 少内謹案致之. 書到言, 署金布發, 它如┛ 律令.ノ欣手.ノ四月癸丑[50]水十一刻刻下五, 守府快行少內.[803]

In month 4, new moon on a *bǐngwǔ*, *guǐchǒu* [day 8], Sè, Deputy Prefect of Qiānlíng, sends (the document) to the lower authorities: The Lesser Treasury has carefully investigated (the issue) and established (the outcome) firmly. Communicate orally upon delivery. To be signed by all Finance Departments and sent (back). Everything else (should be handled) in accordance with the ┛ statutes and ordinances. / By the hand of Xīn. / In month 4 on a *guǐchǒu* [day 8], at water below mark 5 of 11 marks, Governor Kuài dispatches (the document) to the Lesser Treasury.

8-164+8-1475 I–III, rI: □□年後九月辛酉[58]朔丁亥[24], 少內武敢言之: 上計☒┛ □□而後論者獄校廿一牒, 謁告遷陵將計丞☒┛ 上校. 敢言之.☒┛ ☒九月丁亥[24]水十一刻刻下三, 佐欣行廷. 欣手. ☒[804]

□□ year, intercalary month 9, new moon on a *xīnyǒu*, *dīnghài* [day 27], Wǔ from the Lesser Treasury dares to say: Submit the statistics to the higher authorities ☒┛ □□ and pass a second judgment based on reviewing the criminal cases on the 21 tablets. Request to report to the Prefect of Qiānlíng in lead of the statistics ☒┛ submit the crosschecking to the higher authorities. (He) dares to say. ☒ ┗☒ Month 9, *dīnghài* [day 27], at water below mark 3 of 11 marks, Assistant Xīn dispatches (the document) to the (County) Court. By the hand of Xīn. ☒

Comment: The "Statutes Concerning Crosschecking" (*Xiao lü* 校律) in the *Shuìhǔdì* documents outline procedures for compiling inventory lists and mechanisms for holding officials accountable for missing objects.[805]

The limited interaction between central and regional government units, including the Lesser Treasuries, Finance Departments, Granaries and Warehouses, points to a considerable degree of decentralization and structural independence within

803 Chén Wěi 2012: 94.
804 Chén Wěi 2012: 100.
805 Hulsewé 1985: 78–82, 93–101.

the imperial Qín dynasty. Instead of transferring resources between distant counties and the capitals institutions, resources were largely managed locally, thereby saving time and reducing expenses. A fully centralized administrative approach would have been impractical due to the vast executive reach required to sustain the empire without relying on semi-autonomous regional authorities.[806] At the same time, the rigid hierarchies among local institutions, where the chain of command was theoretically final, reflect an idealized notion of governance. In practice, however, and depending on the issue at hand, both commanderies and counties occasionally overstepped hierarchical boundaries and bypassed administrative units that were, strictly speaking, formally superior.

One of the most lucrative means of generating resources, aside from the Qín administration's detailed records of assets, was through reforms affecting the registration of 'households' (*hù* 戶), the organization of labor and the taxation of laborers. The head of a household was typically a male who would pass on his position and parcels of land to his eldest son, or in the absence of a male heir, to his widow or daughter.[807] Manuscript 8-19 lists female heads of households who presumably inherited land and official titles from their husbands or fathers:

> 8-19 AI–VI, BI–V: A ▨□二戶.┛ 大夫一戶.┛ 大夫寡三戶.┛ 不更一戶.┛ 小上造三戶.┛ 小公士一戶.┛ B 士五(伍)七戶.▨┛ 司寇一【戶】.┛ 小男子□▨┛ 大女子□▨┛・凡廿五▨[808]
>
> A ▨□ two households.┛ Holders of the Fifth Rank, one household.┛ Widows of Holders of the Fifth Rank, three households.┛ Holders of the Fourth Rank, one household.┛ Minor Holders of the Second Rank, three households.┛ Minor Holders of the First Rank, one household.┛ B Commoners without Rank, seven households. ▨┛ Overseers of Bandits, one 【household】.┛ Minor males □▨┛ Minor females □▨┛・In total, there are 25 ▨
>
> Comments: The 'Fifth Rank' (*dàifu* 大夫), 'Fourth Rank' (*búgèng* 不更), 'Second Rank' (*shàngzào* 上造) and 'First Rank' (*gōngshì* 公士) belonged to the 20 'ranks' (*jué* 爵) used for officials during the Qín and Hàn dynasties.[809] The First Rank was the lowest of these ranks. 'Minor' (*xiǎo* 小) officials and commoners were further distinguished from 'adult' (*dà* 大) officials and commoners.[810]

In the last two months of the lunar year, 'Household Bureaus' (*hùcáo* 戶曹) counted and arranged the population into nuclear 'Units of Five People' (*shìwǔ* 士

806 Hulsewé 1985: 62.

807 For more on household records, see Chén Jié 2009; I-tien Hsing 2013: 155–86.

808 Chén Wěi 2012: 32–33.

809 For an overview of the system of 20 'ranks' (*jué* 爵) during the Qín and Hàn imperial era, see Lau and Lüdke 2012; Nishijima 1961.

810 For the distinction between 'minor' (*xiǎo* 小) and 'adult' (*dà* 大) officials and commoners, see Chén Pán 1975; Liú Mǐn 2004; 2009.

五 or *wŭ* 伍) (or households), following the military-inspired model of interchangeable, modular groupings. Unlike other excavated Qín manuscripts, the *Lĭyē* corpus contains groupings divided into 'Units of Ten People' (*shí* 什) (or households).[811] The extent to which the Qín authorities merged or divided households remains unknown and cannot be fully discerned from the *Lĭyē* materials currently available.

We learn from manuscript 8-488 – the only manuscript discovered in layers 5, 6 and 8 with a table of statistical registers forwarded to, or conducted by, the Household Bureau – that Household Bureaus kept statistics on individuals, taxes, lacquer production and criminal investigations. 'Being registered' (*fù* 傅) to a 'household' (*hù* 戶) was likely mandatory for those belonging to a Unit of Five People who were at least 1.5 meters tall or aged 16 to 17.[812] Once registered, individuals became subject to imperial authority and were allocated specific partitions of land, assigned agricultural labor, engaged in legal proceedings, mobilized for warfare and taxed accordingly.[813]

Michael Loewe identifies five types of registers forming the basis for annual poll and land tax calculations, which were forwarded to the central government. These registers included:

1) The details of the dwelling,
2) The extent of adjoining arable lands,
3) The ages of the residents,
4) The sum total of land taxes due,
5) The quality of the land and its suitability for certain purposes.[814]

Manuscript 8-488 lists seven types of statistical registers that served various aims, in addition to being essential tools for calculating the annual mandatory grain tax and other taxes:

> 8-488 AI–V, BI–IV: A 戶曹計錄:┘ 鄉戶計,┘ 繇(徭)計,┘ 器計,┘ 租質計,┘ B 田提封計,┘ 髤計,┘ 鞫計.┘ · 凡七計.[815]

---

811 Robin Yates, in his analysis of the *Yúnmèng Lónggăng* 雲夢壟崗 and other Qín bamboo manuscripts, observes that the term 'Unit of Ten People' (*shí* 什) is notably absent (Yates 1987: 223). For a transcription and translation of *Lĭyē* manuscript 8-439+8-519+8-537, an arrest warrant containing the official title 'Commander of a Unit of Ten People' (*shízhăng* 什長), refer to chapter five.

812 Yates 1987: 231.

813 For a discussion on the state dominance of local communities in the Qín dynasty, based mainly on transmitted and recovered sources available in the 1960s, see Perelomov 1961: 66–84.

814 See Loewe 2006: 139.

815 Chén Wĕi 2012: 167.

A Statistical registers of the Household Bureau:┘ Statistics of the district households,┘ statistics of work assignments,┘ statistics of vessels,┘ statistics of grain taxes and loans,┘ B statistics of promotional letters for agricultural land,┘ statistics of lacquer,┘ statistics of criminal investigations. ┘ · In total, there are seven (types of) statistics.

Comment: In criminal investigations, the term '*jū* 鞫' means 'finding of fact' as a noun or 'to find the facts' as a verb.[816] This step was important for determining the legal consequences for individuals accused of crimes, as it formed the basis upon which guilt or innocence was assessed, and the appropriate legal measures were applied.

The 'poll taxes' (*fù* 賦) levied by the imperial government, payable either in cash or grain, were, by traditional explanation, so high – reaching up to 66% of total peasant output at times – that households were repeatedly pushed to the brink of starvation.[817] The severity of penalties inflicted on legal transgressors reportedly exacerbated the situation, exposing peasant households to the risks of convict labor and military service, all of which allegedly contributed to the peasant uprisings and ultimate downfall of the Qín regime.[818] This view is rooted in two texts compiled at the beginning of the first century CE, namely the section "On Military Strategy" (*Bīnglüè xùn* 兵略訓) from the *(Writings of) the Huáinán Masters* (*Huáinánzǐ* 淮南子) (ca. 139 BC), and the "Treatise on Food and Money" (*Shí huò zhì* 食貨志) from the *Book of Hàn*. As suggested by contemporary Hàn-critical scholars, however, this narrative that the Qín's demise ultimately resulted from incompatible tax rates and defamatory allegations against the authorities should be rethought.[819] As we have no reliable evidence to determine actual poll tax rates or to establish whether these taxes were uniformly imposed across all counties and commanderies, the discussion leaves much room for speculation.

With regard to the grain tax, *Lǐyē* manuscript 8-1519 allows for the calculation of precise grain tax rates in Qiānlíng County during the 35th year of King Zhèng (212 BC):

8-1519 I–III, AI–IV, rB: 遷陵卅五年豤(墾)田輿五十二頃九十五畝, 稅田四頃□□┘ 戶百五十二, 租六百七十七石. 衛(率)之, 畝一石五; ┘ 戶嬰四石四斗五升, 奇不衛(率)六斗.┘ A 啓田九頃十畝, 租九十七石六斗.┘ 都田十七頃五十一畝, 租二百卌一石.┘ 貳田廿六頃卅四畝, 租三百卅九石三.┘ 凡田七十頃卌二畝.· 租凡九百一十. └ B 六百七十七石.[820]

816 Lau and Staack 2016: 43, 108, 125, 143, 316.
817 See Fù Zhèngyuán 1993: 111; Hui 2001: 383.
818 For more information on the peasant uprisings that ultimately led to the collapse of the Qín dynasty, see Dull 1983.
819 Dull 1983: 292–294; Lín Jiànmíng 1981: 204–205. Both argue that, to some extent, economic well-being and legal protection for the majority of the population needed to be safeguarded to ensure obedience and prevent rebellion caused by suppression.
820 Chén Wěi 2012: 345–346.

Qiānlíng, in the 35th year (of King Zhèng) [212 BC], the newly acquired agricultural fields are equal to 52 hectares and 95 acres. The taxable fields are four hectares □□┘ A 152 households, the grain tax is 677 piculs. Calculated, (that is) 1.5 piculs per acre; ┘ The households should (give) four piculs, four bushels and five pints. The remaining not calculated (amount) is six bushels.┘ The agricultural fields of Qǐ are nine hectares and ten acres, the grain tax is 97 piculs and six bushels.┘ The agricultural fields of Dū are 17 hectares and 51 acres, the grain tax is 241 piculs.┘ The agricultural fields of Èr are 26 hectares and 34 acres, the grain tax is 339.3 piculs. ┘In sum (including the existing and the newly acquired agricultural fields), the agricultural fields are 70 hectares and 42 acres. • The overall grain tax (for the existing and the newly acquired agricultural fields) is 910. └ B (The grain tax is) 677 piculs.

Comments: In this text, decimals are expressed in two ways: 1) by providing the subordinate units of measurement, as in "four piculs, four bushels and five pints" (*sì dàn sì dǒu wǔ shēng* 四石四斗五升), which equates to 4.45 piculs, or 2) by omitting the subordinate units, as in "339.3 piculs" (*sānbǎi shìjiǔ dàn sān* 三百卌九石三).

The aggregated size of newly acquired agricultural fields in Qǐ, Dū and Èr amounted to 52 hectares and 95 acres, or 5,295 acres (910 acres + 1,751 acres + 2,634 acres). When compared to the 1,747 acres of existing agricultural fields (7,042 – 5,295 acres), this is a significant increase, in fact 303.1% or more than three times the size of the existing agricultural territory.

The calculations on manuscript 8-1519 are extremely accurate and enable us to draw conclusions about grain taxes for both existing and newly acquired agricultural land as distributed across a given number of households. A step-by-step reconstruction of these calculations is provided below:

1) The total size of existing and newly acquired agricultural fields in the districts Qǐ(líng) 啓(陵), Dū 都 and Èr(chūn) 貳(春) is 70 hectares and 42 acres, or 7,042 acres (1 hectare = 100 acres).[821]
2) The newly acquired agricultural fields in Qǐ are 9 hectares and 10 acres, or 910 acres. The aggregated grain tax for these fields is 97 piculs and 6 bushels, or 97.6 piculs.
3) The newly acquired agricultural fields in Dū are 17 hectares and 51 acres, or 1,751 acres. The aggregated grain tax for these fields is 241 piculs.
4) The newly acquired agricultural fields in Èr are 26 hectares and 34 acres, or 2,634 acres. The aggregated grain tax for these fields is 339.3 piculs.
5) The total size of newly acquired agricultural fields (in Qǐ, Dū and Èr) is 52 hectares and 95 acres, or 5,295 acres (910 + 1,751 + 2,634 acres).
6) The total size of existing agricultural fields is 17 hectares and 47 acres, or 1,747 acres (7,042 – 5,295 acres).

821 One 'hectare' (*qǐng* 頃) was equivalent to approximately 45,710 square meters or 11.3 acres today (Wilkinson 2013: 557).

7) 152 households pay a total of 677 piculs of grain tax for the newly acquired agricultural fields. Strictly speaking, the aggregated grain tax for the newly acquired fields is 677.9 piculs (97.6 + 241 + 339.3 piculs). However, this figure is rounded down to 677.
8) The grain tax for the newly acquired agricultural fields is 1.5 piculs per acre, meaning, for this region, each household pays an average of 4 piculs, 4 bushels and 5 pints, or 4.45 piculs (677 piculs / 152 households). Strictly speaking, the average household grain tax for the newly acquired fields is slightly higher. However, this figure is rounded from 4.45395 to 4.45, and the difference to 676.4 piculs, aggregated for the total number of households, is given in line III: "the remaining not calculated (amount) is 6 bushels" (*qí bú lǜ liù dǒu* 奇不率六斗).
9) If the entire size of newly acquired agricultural fields were taxed, then, for this region, the 152 households would pay a total of 7,942.5 piculs (5,295 acres x 1.5 piculs). Instead, they pay 677 piculs, meaning only 8.52% of newly acquired fields (677 / 7,942.5 piculs = 8.5238%), equivalent to 451.33 acres (8.5238% of 5,295 acres), are considered 'taxable fields' (*shuì tián* 稅田). The missing two characters at the end of line I are most likely '5' (*wǔ* 五) and '1' (*yī* 一), resulting in '51 acres'. As such, the total taxable area of newly acquired fields in Qǐ, Dū and Èr is 4 hectares and 51 acres, or 451 acres.
10) The aggregated grain tax for the existing agricultural fields is 233 piculs (910 – 677 piculs), meaning, for this region, each household pays an average of 1 picul, 5 bushels and 3 pints, or 1.53 piculs (233 piculs / 152 households).
11) If the entire size of existing agricultural fields were taxed, then, for this region, the 152 households would pay a total of 2,620.5 piculs (1,747 acres x 1.5 piculs), assuming a tax rate of 1.5 piculs per acre. Instead, they pay 233 piculs, meaning only 8.89% of existing fields (233 / 2,620.5 piculs = 8.8914%), equivalent to 155.33 acres (8.8914% of 1,747 acres), are considered 'taxable fields' (*shuì tián* 稅田).
12) 152 households pay a total of 910 piculs of grain tax for the existing and newly acquired agricultural fields. Strictly speaking, the aggregated grain tax for the existing and newly acquired fields is 908.96 piculs (4.45 + 1.53 = 5.98 piculs, multiplied by 152 households). However, this figure is rounded up from 908.96 to 910.
13) If the entire size of existing and newly acquired agricultural fields were taxed, then, for this region, the 152 households would pay a total of 10,563 piculs (7,042 acres x 1.5 piculs), assuming a tax rate of 1.5 piculs per acre. Instead, they pay 910 piculs, meaning only 8.62% of new and existing fields (910 / 10,563 piculs = 8.6150%), equivalent to 607 acres (8.6150% / 7,042 acres), are considered 'taxable fields' (*shuì tián* 稅田). This amounts to 451.33 acres of newly acquired fields (see step 9), and 155.33 acres of existing fields (see step 11).

From these calculations, we are able to draw the following conclusions:

a) If the total size of existing and newly acquired agricultural fields in Qǐ, Dū and Èr was 70 hectares and 42 acres (or 7,042 acres), then one household was an average of 46.33 acres – or almost half a hectare – in size.[822]
b) The overall grain tax (on taxable land) for the existing agricultural fields, imposed on a total of 152 households in Qǐ, Dū and Èr, was 8.89% as of 212 BC.
c) The overall grain tax (on taxable land) for the newly acquired agricultural fields, imposed on a total of 152 households in Qǐ, Dū and Èr, was 8.52% as of 212 BC. This is slightly lower than the grain tax for the existing agricultural fields, which could have created moderate incentives for acquiring land and for an overall reduction in grain tax for arable areas (both new and existing).
d) The overall grain tax (on taxable land) for the existing and newly acquired agricultural fields, imposed on a total of 152 households in Qǐ, Dū and Èr, was 8.62% as of 212 BC.
e) Because the respective number of households in Qǐ, Dū and Èr is not given, we are unable to calculate the grain tax rates for these areas individually. Hence, we do not know whether taxable land differed from region to region within the county.
f) The respective size of existing agricultural fields in Qǐ, Dū and Èr is not specified. There is also no information on population density, agricultural productivity or the ratio between public and private land ownership in these areas.
g) It also remains unknown whether grain tax rates differed at the household level. From mere calculation, it would seem as if taxes in a given area were equally distributed across all households. If this is true, as of 212 BC, each household in Qǐ, Dū and Èr paid 1.53 piculs for the existing agricultural fields and 4.45 piculs for the newly acquired agricultural fields, or 5.98 piculs for both. This would mean that those households assigned to either larger, more fertile or more accessible plots, and having the necessary means to cultivate them (i.e., tools, convicts, farm animals), were also better off tax-wise. This may have been intentional, encouraging a sense of collective responsibility within communities, as well as mutual support and competition among different households and areas. It may have also incentivized social mobility, potentially deterring criminal behavior.

822 According to Brian Lander's calculations, this figure is just over one-third of a hectare, or roughly 4.2 acres, for each household (Lander 2015: 320). However, he only takes the newly acquired agricultural fields into consideration. One 'picul' (*dàn* 石) was a measure of both weight and volume. As a unit of weight, it is equal to 120 'catties' (*jīn* 斤). As a unit of volume, it is equal to 100 'pints' (*shēng* 升) (Lander 2015: 319–320).

Additionally, these calculations suggest that the Qín dynasty adjusted its grain tax rates annually, taking into account factors such as the amount of existing and newly acquired land, the official categorization of plots as cultivated or uncultivated fields, the overall state of harvests, and external determinants, such as fire, floods, pestilence, meteorological conditions, and the surplus food available for redistribution in the imperial Granaries.[823]

Direct comparisons between these specific tax rates and tax rates in the later periods of the Hàn dynasty are extremely challenging, provided that "taxation from produce and earnings from land" (*zū* 租), one of the many taxes levied in the Hàn dynasty, and a term perhaps closely linked to the 'grain tax' (*zū* 租) in the *Lǐyē* texts, was often paid in cash as opposed to kind.[824] Dǒng Zhòngshū 懂仲舒 (ca. 179–104 BC), a Western Hàn dynasty philosopher and politician, criticized the agrarian reforms of the Qín as the main cause of deprivation, over-taxation (a twentyfold rise since ancient times), increase in *corvée* labor and the mandatory conscription of farmers.[825] In opposition, he advocated an ideal land tax ratio of 10%.[826] Land taxes varied from period to period in the Early Hàn dynasty, fluctuating between 1/15–1/10 (or 6.67–10%) under Emperor Gāozǔ 高祖 (r. 202–195 BC) and Emperor Huìdì 惠帝 (r. 195–188 BC), to 1/30 (or 3.33%) under Emperor Wéndì 文帝 (r. ca. 180–157 BC).[827] Jiǎ Yì 賈誼 (201–169 BC), an influential Western Hàn poet and advisor to Emperor Wéndì, further proposed to abolish all taxation from produce and land earnings, a measure that was possibly never implemented.[828] The assessment of produce and land taxes during the Hàn dynasty was influenced by surface calculation methods, measurement standards, inconsistencies in reporting, land ownership, and overall grain productivity (determined by factors such as location, land quality and available labor). These factors introduced dynamic variables and had a significant impact on tax return estimations.[829]

Aside from references on the previous manuscript, the term 'grain tax' (*zū* 租) and the more general term 'tax' (*shuì* 稅) are sparsely used in the *Lǐyē* corpus.

---

823 For more information on annually adjusted tax rates based on other excavated manuscripts, see Hulsewé 1985: 21, 42, 163; Sūn Zhānyǔ 2013; Yates 1985–1987: 247.

824 Sterckx 2020: 468, footnote 11; Wáng Yànhuī 2016; Zāng Zhīfēi 2017. Other kinds of tax burdens imposed on Early Hàn dynasty households included, for example, 'poll taxes' (*suàn fù* 算賦), 'hay and straw taxes' (*chú gǎo* 芻稾), 'land taxes' (*tián zū* 田租) and specific forms of labor (von Glahn 2017: 109).

825 Loewe 2011: 101–104; Sterckx 2020: 479–482.

826 Sterckx 2020: 480, footnote 51.

827 Yú Kūnqí 2012: 92–100.

828 Hsu Cho-yun 1980: 158–160.

829 Chén Yīng 2010: 46–47, 65; Yú Kūnqí 2012: 130–135, 147–148.

Another instance of the term 'taxable fields' (*shuì tián* 稅田) occurs on manuscript 8-1535, which is heavily damaged. The texts from layers 5, 6 and 8 do not contain the terms 'field tax' (*tián shuì* 田稅) or 'field levy' (*tián fù* 田賦). Instead, they mention four distinct types of taxes:

1) 'Household levy' (*hù fù* 戶賦); collected from households in the form of silkworm cocoons,
2) 'Hay and straw tax' (*chú gǎo* 芻稾); calculated on the size of each farm and the amount of land given to each household, often paid in cash,[830]
3) 'Cash levy' (*qián fù* 錢賦); possibly based on the amount of cash owned by households,
4) 'Feather levy' (*yǔ fù* 羽賦).[831]

These four taxes each appear once, but additional information allowing for a reconstruction of exact rates is unfortunately missing.

> 8-518 I–II: 卅四年, 啓陵鄉見戶、當出戶賦者志: ☑ ⌟ 見戶廿八戶, 當出繭十斤八兩.☑[832]
>
> In the 34th year (of King Zhèng) [213 BC], the district of Qǐlíng sees the households. It should give out the records of those who are subject to the household levy: ☑ ⌟ Seeing the households (results in) 28 households. Ten catties and eight ounces of cocoons should be given out. ☑
>
> Comments: There is a notable semantic difference between 'counting the households' (*jī hù* 積戶) and 'seeing the households' (*jiàn hù* 見戶) in the *Lǐyē* corpus.[833] 'Counting the households' (*jī hù* 積戶) appears on manuscripts 8-552 and 8-1716, where it seems to apply to large numbers of households counted over the course of the entire year. 'Seeing the households' (*jiàn hù* 見戶) appears on manuscript 8-518 and 8-1236+8-1791; on the latter as a complete administrative process, ostensibly involving small numbers of households (fewer than 200) as well as 'settlements' (*yì* 邑) or 'townships' (*lǐ* 里), to be conducted once per year:
>
>> 8-1236+8-1791: 今見一邑二里: 大夫七戶, 大夫寡二戶, 大夫子三戶, 不更五戶, □□四戶, 上造十二戶, 公士二戶, 從廿六戶.☑[834]
>>
>> Today, (the households of) one settlement and two townships were seen: Seven households of Holders of the Fifth Rank, two households of Widows of the Fifth Rank, three households of Sons of the Fifth Rank, five households of Holders of the Fourth Rank,

830 See manuscripts 8-559 and 8-1165 for listings of 'hay taxes' (*chú* 芻) imposed on households, payable in cash.
831 Compare with Hulsewé 1985: 23. For the trade, taxation and use of feathers in the Qín dynasty, mainly based on readings of the *Lǐyē* documents, see Wáng Zǐjīn 2016.
832 Chén Wěi 2012: 172.
833 For an analysis of the terms 'seeing the households' (*jiàn hù* 見戶) and 'counting the households' (*jī hù* 積戶), see Wáng Wěi and Sūn Zhàohuá 2014.
834 Chén Wěi 2012: 297.

four households □□, 12 households of Holders of the Second Rank, two households of Holders of the First Rank, from 26 households. ▨

Comments: According to Chén Wěi, the meaning of the character '*cóng* 從' is unclear and requires further research.[835] It unlikely to designate one of the 20 official ranks. Given that the last section of the manuscript is missing, it could be part of a calculation comparing the numbers in the previous seeing of households with those in the current one. Alternatively, '*cóng* 從' could stand for '*cóngrén* 從人', a term possibly designating 'followers', mostly high-ranking adherents of conquered states and their associates who sought to evade subjugation by the Qín regime.[836] Once defeated and taken into captivity, 'followers' may have been referred to as 'captives' (*lǔ* 虜) and – depending on the circumstances – could have been assigned a slave-like or, in cases of legal transgressions, a convict-like status.[837]

8-1165: 戶芻錢六十四. 卅五年. ▨[838]

Hay tax for households is 64 coins. 35th year. ▨

8-1483 I, rI: ▨一見芻稾數言 └▨ 詘手[839]

▨ one, see the number of the hay and straw tax └▨ By the hand of Qū.

8-1199: ▨【首】當出義賦者令皆□▨ [840]

▨【headed】(people) should give out to the ones who are subjected to the cash levy. Order everyone □▨

The term '*yì* 義' is read as '*qián* 錢' in this context.[841]

8-1735: 廿七年羽賦二千五【百】▨ [842]

In the 27th year (of King Zhèng) [220 BC], the feather levy is two thousand five【hundred】▨

'*Zū* 租' was collected in relation to the size of cultivated areas and occurs nine times in the *Lǐyē* corpus with very limited context.[843] References to "grain tax

835 Chén Wěi 2012: 297.
836 Lǐ Hóngcái 2016; Yates 2022: 7.
837 Yates 2022: 8–9; 15.
838 Chén Wěi 2012: 286.
839 Chén Wěi 2012: 337.
840 Chén Wěi 2012: 290.
841 Chén Wěi 2012: 290.
842 Chén Wěi 2012: 384.
843 Specifically, it occurs on manuscripts 8-488, 8-985, 8-1180, 8-1246, 8-1302 (twice), 8-1499, 8-1535 and 8-2226+8-2227.

money" (*zū qián* 租錢) on manuscript 8-1180, "statistics of grain taxes and loans" (*zū zhì jì* 租質計) on manuscript 8-488, and "incoming money from the grain tax and loans" (*zū zhì rù qián* 租質入錢) on manuscript 8-2226+8-2227, however, indicate that taxes could also be settled in cash instead of kind, and that both grain taxes and loans were considered similar in terms of government revenue.

Grain taxes paid in kind most likely consisted of local crops. The southern region of Qiānlíng County mainly produced rice.[844] Yet, 'millet' (*sù* 粟) and its glutinous variety '*shú* 秫' also appear to have been readily available. 'Millet' (*sù* 粟) is listed alongside 'hulled rice' (*mǐ* 米) in three out of six *Lǐyē* texts, two of which pertain to the supervised distribution of food rations (see Tab. 6.2).[845]

Tab. 6.2: Appearances of the terms 'hulled rice' (*mǐ* 米) and 'millet' (*sù* 粟) on the *Lǐyē* manuscripts from layers 5, 6 and 8.

No.	Term	Occurrence: Manuscript(s) number
1	'□ hulled rice' (□*mǐ* 米) The character preceding '*mǐ* 米' is illegible	**1:** 5-33
2	'Hulled rice' (*mǐ* 米)	**2:** 8-439+8-519+8-537, 8-2205
3	'Hulled rice and millet' (*mǐ sù* 米粟)	**1:** 8-1374
4	'Millet and hulled rice' (*sù mǐ* 粟米)	**2:** 8-1540, 8-1576
5	'Glutinous millet' (*shú* 秫)	**2:** 8-200+8-296 (twice)

Glutinous millet was frequently used to make alcoholic beverages and sugar. In this context, it was linked to the intake of alcohol and its role in establishing 'peace' (*ān* 安) among consumers, contributing to social cohesion in the region.

> 8-200+8-296 I, rI: 秫巵秫求請得以巵求 └ 聿聿建安.[846]
>
> (With) a wine cup of glutinous millet and (with) glutinous millet, seek to invite and obtain. Seek by means of the wine cup └ to establish peace.

Another type of 'glutinous millet' (*shǔ* 黍) appears on manuscript 14-656+15-434.[847] In this text, glutinous millet, in its natural and unrefined form, is used as a sacrificial offering to the 'Agricultural Ancestor' (*Xiān Nóng* 先農):

844 Bray 1984.
845 Manuscript 8-439+8-519+8-537 is an arrest warrant. For a transcription and translation, see chapter five. The other three manuscripts are damaged and provide little context.
846 Chén Wěi 2012: 111.
847 Péng Hào 2008.

14-656+15-434: 卅二年三月丁丑[14]朔丙申[33], 倉是佐狗出黍米四斗以祠先農.[848]

In the 33rd year (of King Zhèng) [214 BC], month 3, new moon on a *dīngchǒu*, *bǐngshēn* [day 20], Shì from the Warehouse and his Assistant Gǒu give out four bushels of glutinous millet and hulled rice (to be offered as) sacrifice to the Agricultural Ancestor.

Whether 'millet' (*sù* 粟) and 'glutinous millet' (*shú* 秫, *shǔ* 黍) were locally cultivated or transported to Qiānlíng County via trade routes connecting the north and the south is not specified. Presuming taxes paid in natural produce were local, at least two factors have not been taken into account: 1) the weight differences between equal volumes of rice, millet and other grains, and 2) inconsistencies in the quality or yield potential of different fields. The *Lǐyē* documents provide no further information regarding the weight differences between taxes paid in rice, millet or other grains. However, according to documents unearthed from layer 9 – which are displayed at the Lǐyē Museum of Qín Slips (*Lǐyē Qín jiǎn bówùguǎn* 里耶秦簡博物館) – there were disparities in the quality or yield potential of cultivated areas.[849] Some of these documents also detail methods of land distribution and provide records of individual landholdings.[850]

On manuscript 9-2350, a widow petitions to re-register her 'mulberry fields' (*sāng tián* 桑田), which had previously been catalogued as 'cleared grassy fields' (*kěn cǎo tián* 豤草田). Her reasons for doing so are unknown. Perhaps, cleared grassy fields were allocated solely to grazing rather than crop cultivation and thus entitled her to employ convicts or a specific class of convicts. Another reason may have been that mulberry trees, being immobile and essential for silk production, traditionally involved female labor in fieldwork and fabric production.[851] This gives us an extraordinary overview of land ownership and the precision employed in recordkeeping, even for small plots within Qín-controlled territories.

Distinct species of trees were also mapped and recorded by the Qiānlíng administration. For example, manuscripts 8-455 and 8-1527 specify the size, location and condition of rare 'oriental raisin trees' (*zhījǔ* 枝枸) in Èrchūn District:[852]

848 Péng Hào 2008.
849 See Lander 2015: 318–320.
850 Lander 2015: 320.
851 Lander 2015: 318. For a study on financial administration in the Táng dynasty, see Twitchett 1963.
852 For more information on the cultivation of 'oriental raisin trees' (*zhījǔ* 枝枸) in Early China, see Hú Píngshēng 2009.

8-455 AI–III, BI–III: A 貳春鄉枝(枳)枸志.┘ 枝(枳)枸三木.┘ ☒下廣一畝,┘ B 格廣半畝, 高丈二尺.┘ 去鄉七里.┘ 卅四年不實.[853]

A Record of the oriental raisin trees in Èrchūn District.┘ There are three oriental raisin trees.┘ ☒ at the bottom cover one acre,┘ Their branches cover half an acre, and their height is (one) yard and two feet.┘ They are seven miles away from the district.┘ For 34 years they have not borne any fruit.

8-1527 I–II, rI: 卅四年八月癸巳[30]朔丙申[33], 貳春鄉守平敢言之:┘ 貳春鄉樹枝(枳)枸卅四年不實. 敢言之. └ 平手.[854]

In the 34th year (of King Zhèng) [213 BC], month 8, new moon on a *guǐsì*, *bǐngshēn* [day 4], Píng, Incumbent of Èrchūn District, dares to say:┘ (Considering the) trees in Èrchūn District, the oriental raisin trees have not borne fruit for 34 years. (He) dares to say. └ By the hand of Píng.

Both manuscripts could refer to the same oriental raisin trees, but only the handwriting on manuscript 8-1527 can be attributed to an official named Píng 平 (Fig. 6.7). If

Fig. 6.7: Monochrome photographs of manuscript 8-455 (l) as well as the *verso* (r) and *recto* (m) of manuscript 8-1527. Ink on wood. Dimensions unknown.[855]

853 Chén Wěi 2012: 153.
854 Chén Wěi 2012: 350.
855 *Húnán shěng wénwù kǎogǔ yánjiūsuǒ* 2012a: 67, 196.

these records are correct, then the oriental raisin trees in Èrchūn District had not borne fruit since the inception of the Qín empire. Whether this should be understood as an implicit critique of the First Emperor is not specified and remains open to debate.

Detailed records concerning resource surpluses or shortages helped to review and discipline officials, laborers and convicts over extended periods.[856] As demonstrated in the previous chapter, officials who failed to record information accurately or submit documents on time to higher authorities faced, in lenient cases, 'fines' (*zī* 訾) paid in 'armors' (*jiǎ* 甲) and 'shields' (*dùn* 盾). If these fines were not paid, offenders were subjected to harsher consequences, occasionally resulting in conscript labor until the debt was worked off.

In essence, many efforts in the Qiānlíng region were geared towards improving agricultural productivity. The cultivation of land, redistribution of people, livestock and goods, and the control over surplus food were all key priorities. Especially at the local and regional levels, the Qín administration paid keen attention to individuals and households contributing goods and services to the empire, providing them a somewhat flexible framework within which they could improve their social standing.

856 See Xú Fùchāng 1992: 482–499.

7 Conclusion

This book has explored overall governance at the southern fringes of the Qín dynasty as reflected in excavated manuscripts from Lǐyē well J1. Through an analysis of new textual evidence, the study aims to offer a revised understanding of the Qín empire, suggesting that the centralization of authority in imperial times was more moderate than previously assumed and largely counterbalanced by relatively autonomous regional administrations. In a not so dissimilar vein, a key objective has been to demonstrate the own sense of internal diversification, intellectual openness, historical continuity and the Qín's desire for unification through the integration of standards set by earlier rulers.

The second chapter summarized the Lǐyē archeological reports and examined the physical properties of the documents, including their quantitative data and the way they were handled by Qín officials. Compared to other Qín writings, the *Lǐyē* documents are unique in terms of material, shape, dimensions, calligraphy and content. They do not represent an idealization of how the system *should* work but reveal, in fact, how officials and administrative relations *did* work. There is no sign of the popular appropriation of elite beliefs by low-ranking officials or any indication that the *Lǐyē* corpus was actively integrated into ritual practice.

The material of the documents conveys additional information through factors such as length, width, shape, cut-off corrections, empty spaces between characters, blackened regions, side notches, and more, which, being rarely documented, could help us significantly extend our knowledge of early textual culture, particularly how messages were transmitted and received. Invaluable to the reconstruction efforts is the abundance of records containing the exact date, rank and names of officials involved in the administration of government. At the end of certain entries, the signatory clause 'by the hand of' (*shǒu* 手), preceded by a slash mark, is followed by the signature of the official in charge of producing or handling the document (or parts of it). In total, the *Lǐyē* manuscripts excavated from layers 5, 6 and 8 identify at least 85 individuals across 208 documents who were responsible for the production or handling of texts in and around Dòngtíng Commandery. While most of the personnel identified are 'Secretaries' (*shǐ* 史) or 'Assistants' (*zuǒ* 佐), their skills and assignments were not limited to these titles.

Strictly speaking, the term 'scribe' is too narrow to apply to clerical personnel in the texts, since it does not capture the full range of responsibilities taken on by these individuals. Moreover, the term '*shǐ* 史', which at times means 'scribe', especially in received literature from the Eastern Zhōu and later periods, is best trans-

https://doi.org/10.1515/9783110791853-007

lated as 'Secretary' in the *Lǐyē* corpus. Scribal literacy or administrative literacy – categories denoting a group of specially trained workers able to read and write characters in short, formulaic sentences, and handle the materiality of manuscripts – may more aptly describe the responsibilities of Lǐyē officials, as these terms encompass a diverse range of skills and ranks.

In sharp contrast to views advocating the amalgamation of terms such as 'Bureau' and '(County) Court' (*tíng* 廷) to denote designations like 'Court Household Bureau' (*tínghùcáo* 廷戶曹) or 'Court Personnel Bureau' (*tínglìcáo* 廷吏曹), I argue that these designations were originally intended to represent the dispatch of documents or goods from the '(County) Court' (*tíng* 廷) to other administrative units. Phrases such as "*tínghùcáo* 廷戶曹" or "*tínglìcáo* 廷吏曹" should be translated as "from the (County) Court to the Household Bureau" and "from the (County) Court to the Bureau for Officials," respectively. In all these instances, the resulting expression should be understood as referring to two separate entities, namely the (County) Court and the respective 'Bureau' under the Court's authority.

The third chapter dealt with calendar and timekeeping systems that were used for various activities, including the cultivation of land, the coordination of administrative duties, the redistribution of resources, and the collection, archiving and retrieval of information. An understanding of imperial calendar and timekeeping systems in the Qín dynasty has also allowed us to date damaged manuscripts and manuscript fragments or reconstruct geographical distances between locations.

In general, the *Lǐyē* corpus does not support the time formats put forward to this date. Double-hours are used, yet these are neither assigned to chrononyms nor divided into 'quarters' of 100. The division of double-hours into quarters, which are further partitioned into tenths, is also missing in the texts. Instead, water clocks seem to have had eleven one-hour 'marks' (*kè* 刻), dividing up the time between 5 am and 5 pm, or the 12 hours from sunrise onward. The commencement of a new water clock cycle was labeled 'beginning' (*qǐ* 起), whereas its ending was referred to as 'exhaustion' (*jìn* 盡). Marks '0' and '12' never appear in the *Lǐyē* manuscripts and would have been redundant on the clepsydra scale.

A closer look at the time frame of the texts has also provided interesting insights into the status of certain terms and expressions. With a reasonable degree of probability, homophones of the First Emperor's personal name were not strictly banned but remained part of the official vocabulary.

The fourth chapter focused on the imperial government's consolidation of authority at various levels. Earlier institutions and terminologies were not simply discarded but underwent a process of adaptation and renewal, and conventional titles and expressions used in earlier periods became associated with a new era of governance.

In order to incorporate newly conquered territories, the Qín administration sought to replace the feudal system of government with a system based on the parceling of land into 'commanderies' (*jùn* 郡), 'counties' (*xiàn* 縣), 'marches' (*dào* 道) and 'districts' (*xiāng* 鄉), which were further subdivided into 'settlements' (*yì* 邑), 'townships' (*lǐ* 里), 'sections' (*xiào* 校), 'subsections' (*máo* 髳), 'posts' (*tíng* 亭) and 'postal units' (*yóu* 郵). Officials at the Qiānlíng County level supervised at least the three districts Èrchūn 貳春, Qǐlíng 啓陵 and Dū 都. This kind of administrative division had already been introduced in the Chǔ 楚 and Qín kingdoms. The texts from layers 5, 6, 8, 12, 14 and 15 mention, in total, 47 counties within 15 commanderies. Qiānlíng is not mentioned as part of the Qín dynasty in any other transmitted or excavated historiography and appears for the first time as a county in a *Lǐyē* text entry made one year before the proclamation of the empire.

High-level commandery officials worked at long distances from the empire's center and were sometimes hundreds of kilometers away from each other. This situation did not allow for immediate consultations in cases of emergency, and certain procedures took months or years to be completed. Officials were often forced to wait for approval on matters by the higher authorities, risking long delays, or act independently and potentially face legal charges. In some cases, the chain of command was explicitly circumvented to delegate more autonomy to subordinate, regional government units.

The limited interaction between central and regional government units demonstrate a high degree of decentralization and structural independence within the Qín dynasty. Rather than moving resources back and forth between remote counties and the capital institutions, time and expenses were saved by providing services locally. In this respect, a centralized administration of resources or finances also seems unrealistic, given the vast executive reach that would have been required to sustain the empire without the help of semi-autonomous regional authorities.

'Bureaus' (*cáo* 曹) and 'Offices' (*guān* 官) seem to have differed primarily in terms of purpose and function. Even though Bureaus relied on the logistical efforts of the Offices – and vice versa – they also appear to have been direct auxiliaries to the County Courts and assigned to higher-level duties. Like Bureaus, Offices were located within Districts, however, they appear to have operated outside of direct Court authority. They were also predominantly headed by 'Overseers' (*sèfū* 嗇夫) and more numerous and geographically dispersed. Some Bureaus, like the 'Bureau of the Commissioner of Public Works' (*sīkōngcáo* 司空曹), were discontinued in later periods.

The fifth chapter offered an analysis of the lowest strata of Qín society. The *Lǐyē* corpus groups convict servants into at least five different categories based on social status, official rank, skills, age, sex and the amount of compensation received for their sentences. Some documents not only refer to explicit numbers of convicts but also include their assigned tasks and personal names. Just like officials, convicts had multiple responsibilities and were deployed flexibly according to their abilities and the requirements of the administration. When forced labor necessitated specific skills, those skills were prioritized over factors such as age or level of punishment.

Punishment, surveillance, coercion and other disciplinary measures were not the only means of the Qín empire for curtailing criminal behavior. Approaches incorporating a balance between rewards and punishments had a long-standing tradition in Early China, which is likewise reflected in the material from Lǐyē. Convicts were given financial remuneration and rewards for actions deemed beneficial and, on the whole, this formed part of a perpetual effort to advance and maintain social order.

As numerous excerpts of statutory provisions in the *Lǐyē* corpus demonstrate, judicial institutions were answerable to the general populace, meaning they had to publicly justify coercion or re-evaluate conflicts in relation to the law upon request. If a sentence was considered in violation of the statutes and ordinances, 'Judiciary Secretaries' (*yùshǐ* 獄史) had the authority to revoke the ruling and reopen an investigation. Qín state sovereignty was vulnerable insofar as it relied upon a legal framework that upheld principles of due process. In effect, any incorrect application of the law could not be enforced, even when backed by a high-ranking official.

The last chapter covered the generation, storage and redistribution of resources, including preventive reallocations of food and detailed grain tax calculations for district households. In Qiānlíng County, estimations as well as emergency plans were compiled by local officials and forwarded to the authorities of other counties and commanderies. What set the Qín dynasty apart from previous dynasties was not merely the creation of administrative records, but also the mastery of information archiving, retrieval and exchange across different levels of government. The obsession with statistics, the intense focus on reporting lines and regulations, and the attention to detail in handling imperial documents exceeds our previous understanding of the empire. Indeed, the Qín administration seems to have invested heavily in the training and education of its subjects, thereby facilitating the secure, accurate, rapid and semi-independent transmission of information.

By far, the most lucrative means of resource generation, aside from the meticulous records of assets kept by the administration, were reforms affecting the registration of households, the organization of labor and the recordkeeping of laborers including the taxes imposed on them. The head of a household was typically a male who passed on his position and certain parcels of his land to his eldest son or – in cases he had no male successor – to his widow or daughter. In the last two months of the lunar year, 'Household Bureaus' (*hùcáo* 戶曹) counted and arranged the population into nuclear 'Units of Five People' (*shìwǔ* 士五 or *wǔ* 伍) following the military model of interchangeable, modular groupings. Unlike other excavated Qín manuscripts, the *Lǐyē* corpus also contains groupings divided into 'Units of Ten People' (*shí* 什).

Regarding the continuation of rituals from the Eastern Zhōu to the Qín dynasty, several manuscripts highlight the importance of ancestral temples and the institutionalized obligations accompanying them. These included, "building the temples" (*zuò miào* 作廟), "inspecting the temples" (*xíng miào* 行廟) and repairing the temples. Additionally, manuscript 8-461 draws analogies between the sacrifices offered to the Emperor and those offered to the 'Celestial Thearch' (*Tiāndì* 天帝). This not only indicates the deification of the Emperor but also suggests that the Qín dynasty may have been governed based on religious authority or divine right. These findings offer new perspectives on the Qín "legalist" tradition and draw attention to customs adopted from earlier periods that were partially institutionalized under the rule of the First Emperor and maintained throughout succeeding periods.

The *Lǐyē* documents examined in this book lack any reference to the so-called "burning of books and killing of scholars" (*fén shū kēng rú* 焚書坑儒) from the "Annals of Qín Shǐhuáng" (*Qín Shǐhuáng běnjì* 秦始皇本纪) in the *Shǐjì*. In fact, expressions like 'burn' (*shāo* 烧), 'public executions (in the marketplace)' (*qìshì* 弃市) or the 'hundred schools of thought' (*bǎijiā* 百家) do not appear in the texts, either separately or in combination with one another, nor is there anything else in the corpus that would remotely allude to a large-scale biblioclasm or systematic persecution of scholars. Admittedly, the texts provide no indication that the *Book of Odes* (*Shījīng* 詩經) or the *Book of Document* (*Shūjīng* 書經) – works allegedly banned during imperial times – were read or taught publicly. However, this could be, in part, due to the almost non-existent correspondence between *Lǐyē* and the capital and the administrative nature of the manuscripts, which focus on the organization of government in a peripheral region.

On the subject of the 'grain tax' (*zū* 租), the *Lǐyē* corpus offers remarkable insights into precise grain tax rates in Qiānlíng County. The overall grain tax for the existing and newly acquired agricultural fields, imposed on a total of 152 households in Qǐ(líng) 啓(陵), Dū 都 and Èr(chūn) 貳(春) Districts, was an average of 8.62% in the 35th year of King Zhèng (212 BC). The calculations suggest that the Qín dynasty adjusted its grain tax rates annually based on the amount of existing and newly acquired land, the official categorization of land plots as fields with or without crops, the overall state of harvests and external determinants, including fire, floods, pestilence, meteorological conditions and the surplus food available for redistribution in the imperial Granaries.

Research on the *Lǐyē* cache is currently underway to resolve historical gaps and inaccuracies and provide additional insights into the very rare imperial Qín materials currently available. Although findings may, at times, clash or contradict more conventional understandings of the empire, they should not be viewed simply as refutations of transmitted sources. As ongoing archeological work will undoubtedly enable new discoveries in the years and decades to come, both excavated and transmitted sources should be considered complementary rather than mutually exclusive in reshaping our knowledge of Early Chinese empires.

The material presented in this book is intended as a contribution to a revised understanding of the Qín dynasty. While the discussion has been selective of a relatively small archive of administrative texts local to a peripheral region, it has offered a glimpse into some of the microstructures and complexities of the imperial regime in Qiānlíng and other counties within Dòngtíng Commandery. The *Lǐyē* documents analyzed to date, along with those expected to be excavated from wells J2 and J3, will continue on as a foundational reference for future research in the field. On that note, I look forward to peer reviews of the arguments presented in this book and to future discoveries that may either confirm existing sources or further enrich our understanding of the Qín dynasty at large.

Appendix

Table A1: Conversion of time formats that appear in the *Lǐyē* corpus.

Years of entry	Conversion
25th year of King Zhèng	222 BC
26th year of King Zhèng	221 BC
27th year of King Zhèng	220 BC
28th year of King Zhèng	219 BC
29th year of King Zhèng	218 BC
30th year of King Zhèng	217 BC
31st year of King Zhèng	216 BC
32nd year of King Zhèng	215 BC
33rd year of King Zhèng	214 BC
34th year of King Zhèng	213 BC
35th year of King Zhèng	212 BC
36th year of King Zhèng	211 BC
37th year of King Zhèng	210 BC
1st year of the Second Emperor	209 BC
2nd year of the Second Emperor	208 BC

Table A2: Positions and names of the 60 days in the 'sexagenary cycle' (*gānzhī* 干支).

1	*jiǎzǐ* 甲子	2	*yǐchǒu* 乙丑	3	*bǐngyín* 丙寅	4	*dīngmǎo* 丁卯	5	*wùchén* 戊辰
6	*jǐsì* 己巳	7	*gēngwǔ* 庚午	8	*xīnwèi* 辛未	9	*rénshēn* 壬申	10	*guǐyǒu* 癸酉
11	*jiǎxū* 甲戌	12	*yǐhài* 乙亥	13	*bǐngzǐ* 丙子	14	*dīngchǒu* 丁丑	15	*wùyín* 戊寅
16	*jǐmǎo* 己卯	17	*gēngchén* 庚辰	18	*xīnsì* 辛巳	19	*rénwǔ* 壬午	20	*guǐwèi* 癸未
21	*jiǎshēn* 甲申	22	*yǐyǒu* 乙酉	23	*bǐngxū* 丙戌	24	*dīnghài* 丁亥	25	*wùzǐ* 戊子
26	*jǐchǒu* 己丑	27	*gēngyín* 庚寅	28	*xīnmǎo* 辛卯	29	*rénchén* 壬辰	30	*guǐsì* 癸巳
31	*jiǎwǔ* 甲午	32	*yǐwèi* 乙未	33	*bǐngshēn* 丙申	34	*dīngyǒu* 丁酉	35	*wùxū* 戊戌
36	*jǐhài* 己亥	37	*gēngzǐ* 庚子	38	*xīnchǒu* 辛丑	39	*rényín* 壬寅	40	*guǐmǎo* 癸卯
41	*jiǎchén* 甲辰	42	*yǐsì* 乙巳	43	*bǐngwǔ* 丙午	44	*dīngwèi* 丁未	45	*wùshēn* 戊申
46	*jǐyǒu* 己酉	47	*gēngxū* 庚戌	48	*xīnhài* 辛亥	49	*rénzǐ* 壬子	50	*guǐchǒu* 癸丑
51	*jiǎyín* 甲寅	52	*yǐmǎo* 乙卯	53	*bǐngchén* 丙辰	54	*dīngsì* 丁巳	55	*wùwǔ* 戊午
56	*jǐwèi* 己未	57	*gēngshēn* 庚申	58	*xīnyǒu* 辛酉	59	*rénxū* 壬戌	60	*guǐhài* 癸亥

https://doi.org/10.1515/9783110791853-008

List of Figures

Chapter 1

Chapter 2

https://doi.org/10.1515/9783110791853-009

Chapter 3

Chapter 4

Chapter 5

Chapter 6

List of Tables

Chapter 2

Chapter 3

Chapter 4

Chapter 5

https://doi.org/10.1515/9783110791853-010

Chapter 6

Appendix

Bibliography

Allan, Sarah. 1984. “The Myth of the Xià Dynasty”. *The Journal of the Royal Asiatic Society of Great Britain and Ireland*. Cambridge: Cambridge University Press, CXVI, 2, 242–256.

Ān, Zuòzhāng 安作璋, and Xióng Tiějī 熊鐵基. 2007. *Qín Hàn guānzhì shǐ gǎo* 秦漢官制史稿. 2nd edition. Jǐnán: *Qí Lǔ shūshè* 齊魯書社.

Andreini, Attilio. 2015. “Alcune Considerazioni sulla Natura del Pensiero Giuridico Cinese Antico”. *Sulla via del Catai*, VIII, 12, 9–27.

Bān, Gù 班固, and Yán Shīgǔ 顏師古. 1962. *Hànshū* 漢書. 12 vols. Běijīng: *Zhōnghuá shūjú* 中華書局.

Barbieri-Low, Anthony J. 2007. *Artisans in Early Imperial China*. Seattle: University of Washington Press.

Barbieri-Low, Anthony J. 2019. “Coerced Migration and Resettlement in the Qín Imperial Expansion”. *Journal of Chinese History*. Cambridge: Cambridge University Press, V, 2, 181–202.

Barbieri-Low, Anthony J., and Robin D.S. Yates. 2015. *Law, State, and Society in Early Imperial China (2 vols): A Study with Critical Edition and Translation of the Legal Texts from Zhāngjiāshān Tomb no. 247.* Leiden: Brill.

Beck, Mansvelt B.J. 1987. “The First Emperor’s Taboo Character and the Three Day Reign of Xiàowén: Two Moot Points Raised by the Qín Chronicle Unearthed in Shuìhǔdì in 1975”. *T’oung Pao*. Leiden: Brill, LXXIII, 1/3, 68–85.

Behr, Wolfgang. 2004. “Language Change in Premodern China: Notes on its Perception and Impact on the Idea of a ‘Constant Way’”. Vol. I. *Historical Truth, Historical Criticism, and Ideology: Chinese Historiography and Historical Culture from a New Comparative Perspective*, by Axel Schneider and Susanne Weigelin-Schwiedrzik. Leiden and Boston: Brill, 13–51.

Behr, Wolfgang. 2007. “Placed into the Right Position – Etymological Notes on *tú* 圖 and Congeners”. *Graphics and Text in the Production of Technical Knowledge in China: The Warp and the Weft*, by Francesca Bray, Vera Dorofeeva-Lichtmann and Georges Métailié. Leiden: Brill, 109–134.

Bielenstein, Hans. 1980. *The Bureaucracy of Hàn Times.* Cambridge: Cambridge University Press.

Bitsakis, Yanis, and Cornelius Hartz. 2015. “Ein Antiker Mechanischer Kosmos: Der Mechanismus von Antikythera”. *Antike Welt*, 5, 27–32.

Bodde, Derk. 1938. *China’s First Unifier*. Leiden: Brill.

Bodde, Derk. 1963. “Basic Conceptions of Chinese Law: The Genesis and Evolution of Legal Thought in Traditional China”. *Proceedings of the American Philosophical Society*, CVII, 5, 375–398.

Boltz, William. 1994. *The Origin and Early Development of the Chinese Writing System*. Ann Harbor: American Oriental Society.

Brashier, Ken E. 2011. *Ancestral Memory in Early China*. Cambridge: Harvard University Asia Center for the Harvard-Yenching Institute.

Bray, Francesca. 1984. “Agriculture”. *Science and Civilisation in China*, by Joseph Needham. Cambridge: Cambridge University Press, 491–95.

Brindley, Erica Fox. 2015. *Ancient China and the Yuè: Perceptions and Identities on the Southern Frontier, c. 400 BCE–50 CE*. Cambridge: Cambridge University Press.

Brosseder, Ursula B. 2015. “A Study on the Complexity and Dynamics of Interaction and Exchange in Late Iron Age Eurasia”. *Complexity of Interaction Along the Eurasian Steppe Zone in the First Millennium CE*, by Jan Bemmann and Michael Schmauder. Bonn: Bonn Contributions to Asian Archeology, XII, 199–332.

Brown, Miranda. 2007. *The Politics of Mourning in Early China*. Albany: State University of New York.

https://doi.org/10.1515/9783110791853-011

Brown, Miranda, and Charles Sanft. 2011. “Categories and Legal Reasoning in Early Imperial China: The Meaning of *fǎ* in Recovered Texts”. *Oriens Extremus*. Wiesbaden: Harrassowitz Verlag, 49, 283–306.

Bǔ, Xiànqún 卜憲群. 1997. “Qín Hàn gōngwén wénshū yǔ guānliáo xíngzhèng guǎnlǐ 秦漢公文文書與官僚行政管理”. *Lìshǐ yánjiū* 歷史研究, 6.

Bǔ, Xiànqún 卜憲群, and Liú Yáng 劉楊. 2013. “Qín Hàn rìcháng zhìxù zhōng de shèhuì yǔ xíngzhèng guānxì chūtàn – Guānyú ‘zì yán’ yī cí de jiědú 秦漢日常秩序中的社會與行政關系初探–關於‘自言’一詞的解讀”. *Wénshǐ zhé* 文史哲, 4, 81–92.

Caboara, Marco. 2015. “Yě 也 in Excavated Texts”. *Encyclopedia of Chinese Language and Linguistics*, by Rint Sybesma. Leiden: Brill.

Cài, Wànjìn 蔡萬進. 1996. *Qín guǒ liángshí jīngjì yánjiū* 秦國粮食經濟研究. Hohot: *Nèiměnggǔ rénmín* 內蒙古人民.

Cài, Wànjìn 蔡萬進. 2008. “Qín ‘suǒ qǔ jīng xīndì’ yǔ Cāngwú jùn shèzhì 秦‘所取荊新地’與蒼梧郡設置”. *Zhèngzhōu dàxué xuébào (Zhéxué shèhuì kēxué bǎn)* 鄭州大學學報(哲學社會科學版), 41.5, 103–5.

Cáo, Lǚnìng 曹旅寧. 2002. *Qín lǜ xīn tàn* 秦律新探. Běijīng: *Běijīng Zhōngguó shèhuì kēxué* 中國社會科學.

Cáo, Lǚnìng 曹旅寧. 2015. “Yuèlù Qín jiǎn (sì) suǒ jiàn Qín xíng tú zhōngshēn fúyì de xīn zhèng 岳麓秦簡(四)所見秦刑徒終身服役的新證”. *Jiǎnbó* 簡帛. http://www.bsm.org.cn/?qinjian/6309.html (last acc. May 31, 2024)

Cáo, Wǎnrú 曹婉如. 1983. “Maps 2,000 Years Ago and Ancient Cartographical Rules”. *Ancient China’s Technology and Science*. Běijīng: Foreign Language Press, 251–257.

Chang, Kwang-chih. 1986. *The Archaeology of Ancient China*. New Haven, Connecticut: Yale University Press.

Chén, Jiàn 陳劍. 2011a. *Chūtǔ wénxiàn yǔ gǔ wénzì yánjiū* 出土文獻與古文字研究. Shànghǎi: *Shànghǎi gǔjí chūbǎnshè* 上海古籍出版社.

Chén, Jiàn 陳劍. 2011b. “Dú Qín Hàn jiǎn zhájì sān piān 讀秦漢簡札記三篇”. *Fùdàn dàxué chūtǔ wénxiàn yǔ gǔwénzì yánjiū zhōngxīn* 復旦大學出土文獻與古文字研究中心. Shànghǎi: *Shànghǎi gǔjí chūbǎnshè* 上海古籍出版社, 371–374.

Chén, Jié 陳絜. 2009. “Lǐyē ‘hù jí jiǎn’ yǔ Zhànguó mòqī de jīcéng shèhuì 里耶‘戶籍簡’與戰國末期的基層社會”. *Lìshǐ yánjiū* 歷史研究, 5, 23–40.

Chén, Kǎnlǐ 陳侃理. 2014. “Lǐyē Qín fāng yǔ ‘shū tóng wénzì’ 里耶秦方與‘書同文字’”. *Wénwù* 文物, 79–81.

Chén, Měidōng 陳美東. 1989. “Shìlùn Xī Hàn lòuhú de ruògān wèntí 試論西漢漏壺的若干問題”. *Zhōngguó gǔdài tiānwén wénwù lùn jí* 中國古代天文文物論集, 137–144.

Chén, Mèngjiā 陳夢家. 1980a. “Hàn jiǎn suǒ jiàn jū yán biānsài yǔ fángyù zǔzhī 漢簡所見居延邊塞與防禦組織”. *Hàn jiǎn zhuì shù* 漢簡綴述, by Chén Mèngjiā陳夢家. Běijīng: *Zhōnghuá shūjú* 中華書局, 37–95.

Chén, Mèngjiā 陳夢家. 1980b. “Hàn jiǎn suǒ jiàn tàishǒu, jùnwèi èr fǔ shǔlì 漢簡所見太守、郡尉二府屬吏”. *Hàn jiǎn zhuìshù* 漢簡綴述. Běijīng: *Zhōnghuá shūjú* 中華書局.

Chén, Mèngjiā 陳夢家. 1985. *Shàngshū tōnglùn* 尚書通論. Běijīng: *Zhōnghuá shūjú* 中華書局.

Chén, Pán 陳槃. 1975. “Hàn Jìn yí jiǎn shí xiǎo qī zhǒng 漢晉遺簡識小七種”. *Zhōngyāng yánjiūyuàn* 中央研究院. Shànghǎi: *Shànghǎi gǔjí chūbǎnshè* 上海古籍出版社, 27–31.

Chén, Sōngcháng 陳松長. 2004. “Xiāngxī Lǐyē Qín dài jiǎndú xuǎnshì jiàodú 湘西里耶秦代簡牘選釋校讀”. *Jiǎndú xué yánjiū* 簡牘學研究, 4, 17–21.

Chén, Sōngcháng 陳松長. 2015. *Yuèlù shūyuàn cáng Qín jiǎn* 嶽麓書院藏秦簡. Shànghǎi: *Shànghǎi císhū chūbǎnshè* 上海辭書出版社.

Chén, Shùguó 陈戍国. 1991. *Xiān Qín lǐzhì yánjiū* 先秦禮制研究. Chángshā: *Húnán jiàoyù chūbǎnshè* 湖南教育出版社.

Chén, Wěi 陳偉. 2009. "Guānyú Qín jiǎn dú zònghé zhěnglǐ yǔ yánjiū de jǐ diǎn sīkǎo 關於秦簡牘綜合整理與研究的幾點思考". *Jiǎnbó* 簡帛, 4, 1–10.

Chén, Wěi 陳偉. 2011a. "Qín Cāngwú, Dòngtíng èr jùn chú lùn 秦蒼梧, 洞庭二郡芻論". *Lìshǐ yánjiū* 歷史研究, 5, 168–172.

Chén, Wěi 陳偉. 2011b. "Qín zhì Hàn chū Xiāo xiàn dì wàng bǔshuō 秦至漢初銷縣地望補説". *Jiǎnbó* 簡帛. Wǔhàn: *Wǔhàn dàxué chūbǎnshè* 武漢大學出版社.

Chén, Wěi 陳偉. 2012. *Lǐyē Qín jiǎndú jiàoshì (yī)* 里耶秦簡牘校釋(壹). Vol. I. Wǔhàn: *Wǔhàn dàxué chūbǎnshè* 武漢大學出版社.

Chén, Wěi 陳偉. 2014a. "Lǐyē Qín jiǎn suǒ jiàn Qín dài xíngzhèng yǔ suànshù 里耶秦簡所見秦代行政與算術". *Jiǎnbó* 簡帛. http://www.bsm.org.cn/?qinjian/6164.html (last acc. May 31, 2024)

Chén, Wěi 陳偉. 2014b. "Qín bìhuì 'zhèng' zì wèntí zài kǎochá 秦避諱'正'字問題再考 察". *Jiǎnbó* 簡帛. http://www.bsm.org.cn/?qinjian/6240.html (last acc. May 31, 2024)

Chén, Wěi 陳偉. 2016. "Qín jiǎn dú zhōng de 'bān' yǔ 'zuì' 秦簡牘中的'辠'與'罪'". *Jiǎnbó* 簡帛 http://www.bsm.org.cn/?qinjian/7421.html (last acc. May 31, 2024)

Chén, Wěi 陳偉. 2018. "'Event Calendars' in the Early Imperial Era: A Re-assessment". *Bamboo and Silk*. Leiden: Brill, 446–468.

Chén, Yīng 陈英. 2010. *Hàndài pín fù chàjù yǔ zhèngfǔ kòngzhì yánjiū* 漢代貧富差距與政府控制研. Běijīng: *Zhōngguó shèhuì kēxué chūbǎnshè* 中國社會科學出版社.

Chén, Zhèngxián 陳正賢. 2020. "Qín Shǐhuáng shíshī shūtóng wénzì de nián yuè xīn kǎo – cóng Lǐyē Qín jiǎn suǒ jì niàn nián jíqí yòngzì tèzhēng kǎozhèng 秦始皇實施書同文字的年月新考–從里耶秦簡所記廿六年及其用字特徵考證". *Fùdàn dàxué chūtǔ wěnxiàn yǔ gǔwénzì yánjiū zhōngxīn* 复旦大學出土文獻與古文字研究中心. http://www.gwz.fudan.edu.cn/Web/Show/4556

Chén, Zhōnglóng 陳中龍. 2016. "Cóng Qín dài guānfǔ niándù lǜlìng jiàochóu de zhìdù lùn Hàn chū 'Èrnián lǜlìng' de 'èr nián' 從秦代官府年度律令校讎的制度論漢初 '二年律令'的'二年'". *Jiǎnbó* 簡帛. http://www.bsm.org.cn/?qinjian/6707.html (last acc. May 31, 2024)

Chén, Zhìguó 陳治國. 2008. "Cóng Lǐyē Qīn jiǎn kàn Qín de gōngwén zhìdù 從里耶秦簡看秦的公文制度". *Jiǎnbó* 簡帛. http://www.bsm.org.cn/?qinjian/4988.html (last acc. May 31, 2024)

Ch'en, Chih-Mai. 1966. *Chinese Calligraphers and their Art*. London and New York: Cambridge University Press.

Christian, David. 2000. "Silk Roads or Steppe Roads? The Silk Roads in World History". *Journal of World History*. Honolulu: University of Hawai'i Press, II, 1, 1–26.

Cordier, Henri. 1920. *Histoire Générale de la Chine et de ses Relations avec les Pays Étrangers: Depuis les temps les plus anciens jusqu'á la chute de la dynastie Mandchoue*. Paris: Geuthner.

Creel, Herrlee G. 1960. *Confucius and the Chinese Way*. New York: Harper.

Creel, Herrlee G. 1970. "The Fa-chia: 'Legalists' or 'Administrators'?" *What Is Taoism? and Other Studies in Chinese Cultural History*, by Herrlee G. Creel, 92–120. Chicago and London: University of Chicago Press.

Csikszentmihalyi, Mark, and Michael Nylan. 2003. "Constructing Lineages and Inventing Traditions through Exemplary Figures in Early China". *T'oung Pao*. Leiden: Brill, LXXXIX, 1/3: 59–99.

Cullen, Christopher. 2018. *Heavenly Numbers: Astronomy and Authority in Early Imperial China*. Oxford: Oxford University Press.

Dài, Shìjūn 戴世君. 2010. "Qín sīfǎ wénshū 'dāng téngténg' yòngyǔ shìyì 秦司法文書'當騰騰'用語釋義". *Jiǎnbó* 簡帛. http://www.bsm.org.cn/?qinjian/5429.html (last acc. May 31, 2024)

Dài, Yìchún 戴奕純. 2019. "Qín dài shú qiān jīn'é kǎo 秦代贖遷金額考". *Jiǎnbó* 簡帛. http://www.bsm.org.cn/?qinjian/8025.html (last acc. May 31, 2024)

Dān, Yùchén 單育辰. 2007. "Qín jiǎn 'bī' zì shìyì 秦簡'柀'字釋義". *Jiānghàn kǎogù* 江漢考古, 4, 81–84.

Demandt, Michèle H.S. 2020. "Reaching 'the Southern Wilderness': Expansion and the Expansion of the Lǐngnán Transportation Network during the Qín and Hàn Dynasties". *Journal of the Economic and Social History of the Orient*. Leiden: Brill, 63, 157–194.

Dīng, Yuánzhí 丁原植. 2004. *Guōdiàn chǔ jiǎn: Rújiā yì jí sì zhǒng shì xī* 郭店楚簡: 儒家佚籍四種釋析. Wǔnán: *Wǔnán túshū chūbǎn* 五南圖書出版.

Dawson, Raymond. 2007. *The First Emperor: Selections from the Historical Records*. Oxford: Oxford University Press.

Di Cosmo, Nicola. 2015. "China-Steppe Relations in Historical Perspective". *Complexity of Interaction Along the Eurasian Steppe Zone in the First Millennium CE*, by Jan Bemmann and Michael Schmauder. Bonn: Bonn Contributions to Asian Archeology, XII, 49–72.

Dull, Jack L. 1983. "Anti-Qín Rebels: No Peasant Leaders Here". *Modern China*. Los Angeles: Sage Publications, IX, 2, 285–318.

Fán, Guódòng 凡國棟. 2009. "Lǐyē Qín jiǎn yánjiū huígù yǔ qiánzhān 里耶秦簡研究回顧與前瞻". *Jiǎnbó* 簡帛, 37–57.

Fàn, Yúnfēi 范雲飛. 2016. "Cóng xīn chū Qín jiǎn kàn Qín Hàn de dìfāng miào zhì 從新出秦簡看秦漢的地方廟制". *Jiǎnbó* 簡帛. http://www.bsm.org.cn/?qinjian/6698.html (last acc. May 31, 2024)

Fields, Lanny B. 1989. "The Ch'in Dynasty: Legalism and Confucianism". *Journal of Asian History*. Wiesbaden: Harrassowitz Verlag, XXIII, 1, 1–25.

Fölster, Max Jakob. 2018. "Libraries and Archives in the Former Hàn Dynasty (206 BCE–9 CE): Arguing for a Distinction". *Manuscripts and Archives*. Berlin: De Gruyter, 201–230.

Foster, Christopher J. 2017. *Study of the Cāng Jié piān: Past and Present*. Doctoral Dissertation. Cambridge: Harvard University.

Fù, Jiāyí 傅嘉儀. 2007. *Qín fēngní huìkǎo* 秦封泥彙攷. Shànghǎi: *Shànghǎi shūdiàn* 商海書店.

Fù, Zhèngyuán 傅正元. 1993. *Autrocratic Tradition and Chinese Politics*. Cambridge: Cambridge University Press.

Galambos, Imre. 2004. "The Myth of the Qín Unification of Writing in Hàn Sources". *Acta Orientalia Academiae Scientiarum Hungaricae*. Budapest: Akadémiai Kiadó, LVII, 2, 181–203.

Galambos, Imre. 2006. *Orthography of Early Chinese Writing: Evidence from newly excavated manuscripts*. Budapest: Budapest Monographs in East Asian Studies.

Gāo, Héng 高恒. 1983. "Qín lǜ zhōng de xíng tú jí xíngqī wèntí 秦律中的刑徒及刑期問題". *Fǎxué yánjiū* 法學研究.

Gāo, Mǐn 高敏. 1971. "Lùn Qín Hàn shíqī de 'tíng': dú Yúnmèng Qín jiǎn zhájì 論秦漢時期的'亭': 讀雲夢秦簡札記". *Yúnmèng Qín jiǎn chūtàn* 雲夢秦簡初探, by Gāo Mǐn 高敏. Zhèngzhōu: *Hénán rénmín chūbǎnshè* 河南人民出版社.

Gāo, Mǐn 高敏. 2000. *Shuìhǔdì Qín jiǎn chūtàn* 睡虎地秦簡初探. Táiběi: *Wànjuànlóu* 萬卷樓.

Gāo, Mǐn 高敏. 2004. "Hàn chū fǎlǜ xì quánbù jìchéng qín lǜ shuō: dú Zhāngjiāshān Hàn jiǎn 'Zòuyànshū' zhájì 漢初法律系全部繼承秦律說: 讀張家山漢簡'奏讞書'札記". *Qín Hàn Wèi Jìn Nánběi cháo shǐ lùnkǎo* 秦漢魏晉南北朝史論考, 76–84.

Gāo, Míng 高明. 1982. *Gǔ wénzì lèi biān* 古文字類編. Běijīng: *Zhōnghuá shūjú* 中華書局.

Gāo, Yīzhì 高一致. 2013. *Lǐyé Qín jiǎn (yī) jiàoshì sìzé* 里耶秦簡(壹)校釋四則. Shànghǎi: *Shànghǎi gǔjí chūbǎnshè* 上海古籍出版社.

Gāo, Zhènhuán 高震寰. 2013. "Cóng 'Lǐyē' Qín jiǎn (yī)' 'zuòtúbù' guǎnkuī Qín dài xíng tú zhìdù 從'里耶秦簡(壹)''作徒簿'管窺秦代刑徒制度". *Chūtǔ wénxiàn yánjiū* 出土文獻研究. Běijīng: *Zhōngxī shūjú* 中西書局, 132–143.

Gāo, Chóngwén 高崇文. 2011. "The Evolution of Funerary Ritual from the Pre-Qín to the Hàn Era (2007)". *Chinese Archaeology*. Berlin: De Gruyter, 145–149.

Gassmann, Robert H., and Wolfgang Behr. *Antikchinesisch: Ein Lehrbuch in drei Teilen*. Bern: Lang, 2005.

Giele, Enno. 2005. "Signatures of 'Scribes' in Early Imperial China". *Asiatische Studien/Études Asiatiques*. Berlin: De Gruyter, 353–387.

Giele, Enno. 2009. "Kodai no shikiji nōryoku o ikani hantei suru no ka—Kandai gyōsei monjo no jirei kenkyū 古代の識字能力を如何に判定するのか–漢代行政文書の事例研究". *Kanji bunka sanzennen* 漢字文化三千年, by Takata Tokio 高田時 雄. Kyoto: *Rinsen Shoten* 臨川書店.

Giele, Enno. 2009. "Private Letter Manuscripts from Early Imperial China". *A History of Chinese Letters and Epistolary Culture*, by Antje Richter. Leiden: Brill, 399–474.

Goldin, Paul R. 1972. *Confucianism*. London and New York: Routledge.

Goldin, Paul R. 2000. "Personal Names in Early China". *Journal of the American Oriental Society*. Ann Harbor: American Oriental Society, CXX, 1, 77–81.

Goldin, Paul R. 2011. "Persistent Misconceptions about Chinese 'Legalism'". *Journal of Chinese Philosophy*. Leiden: Brill, 88–104.

Goldin, Paul R. 2017. "Some Shāng Antecedents of Later Chinese Ideology and Culture". *Journal of the American Oriental Society*. Ann Harbor: American Oriental Society, CXXXVII, 1, 121–127.

Gōng, Zháijié 宫宅潔. 2018. "Guānyú Lǐyē Qín jiǎn (8)755-759 jiǎn yǔ (8)1564 de biānlián 關於里耶秦簡(8)755-759簡與(8)1564簡的編聯". *Jiǎnbó* 簡帛. http://m.bsm.org.cn/?qinjian/7728.html (last acc. May 31, 2024)

Gorodetskaya, Olga. 2008. "Qín Shǐhuángdì de chēnghào 秦始皇帝的稱號". *Lìshǐ wénwù* 歷史文物. Táiběi: *Lìshǐ bówùguǎn* 歷史博物館, CXXXIX.

Granet, Marcel. 1934. *La Pensée Chinoise*. Paris: La Renaissance du Livre.

Grant, Hardy. 1999. *Worlds of Bronze and Bamboo: Sīmǎ Qiān's Conquest of History*. New York: Columbia University Press.

Guǎn, Xīhuá 管錫華. 2002. *Zhōngguó gǔdài biāodiǎn fúhào fāzhǎn shǐ* 中國古代標點符號發展史. Chéngdū: *Bā Shǔ shūshè* 巴蜀書社.

Guǎn, Zhòng 管仲, and Allyn W. Rickett. 1985. *Guǎnzǐ: Political, Economic, and Philosophical Essays from Early China. A Study and Translation*. Princeton: Princeton University Press.

Guō, Tāo 郭濤. 2017. *Wénshū xíngzhèng yǔ qín dài Dòngtíng jùn de xiàn jì wǎngluò* 文書行政與秦代洞庭郡的縣際網絡. *Shèhuì kēxué* 社會科學, X.

Guō, Zǐzhí 郭子直. 1986. "Zhànguó Qín fēng zōngyì wǎshū míngwén xīnshì 戰國秦封宗邑瓦書銘文新釋". *Gǔwénzì yánjiū* 古文字研究. Běijīng: *Zhōnghuá shūjú* 中華書局, 14, 177–180.

Guójiā wénwù jú 國家文物局. 2003. *Zhōngguó zhòngyào kǎogǔ fāxiàn* 中國重要考古發現. Běijīng: *Wénwù chūbǎnshè* 文物出版社.

Gutkind Bulling, Annaliese. 1978. "Ancient Chinese Maps". *Expedition Magazine*. Philadelphia: Penn Museum, XX, 2, 16–25.

Hán, Zhàoqí 韩兆琦. 2004. *Shǐjì jiān zhèng* 史记笺证. Nánchāng: *Jiāngxī rénmín chūbǎnshè* 江西人民出版社.

Hàn chéng wājué duì 漢城挖掘隊. 1978. "Hàn Cháng'ān chéng wǔkù yízhǐ fājué de chūbù shōuhuò 漢長安城武庫遺址發覺的初步收穫". *Kǎogǔ* 考古, 4, 261–269.

Hàn Huá cāng wājué duì 漢華倉挖掘隊. 1982. "Hàn Huá cāng yízhǐ yīhào cāng jiànzhù fùyuán tàntǎo 漢華倉遺址一號倉建築復原探討". *Kǎogǔ yǔ wénwù* 考古與文物, 6, 29–32.

Hànyǔ dà cídiǎn biānjí wěiyuánhuì 漢語大詞典編輯委員會. 2001. *Hànyǔ dà cídiǎn* 漢語大詞典. Shànghǎi: *Hànyǔ dà cídiǎn* 漢語大詞典.

Harper, Donald. 1999. "Warring States Natural Philosophy and Occult Thought". *The Cambridge History of Ancient China: From the Origins of Civilization to 221 B.C.*, by Michael Loewe and Edward L. Shaughnessy. Cambridge and New York: Cambridge University Press, 813–884.

Harris, William V. 1991. *Ancient Literacy*. Cambridge: Harvard University Press.

Hé, Jièjūn 何介鈞. 2005. *'Qín sānshíliù jùn' hé Xī Hàn zēng zhì jùnguó kǎozhèng* '秦三十六郡'和西漢增置郡國考證. Běijīng: *Zhōngguó jiàoyù wénhuà chūbǎnshè* 中國教育文化出版社.

Hé, Línyí 何琳儀. 1998. *Zhànguó gǔwén zìdiǎn* 戰國古文字典. Běijīng: *Zhōnghuá shūjú* 中華書局.

Hé, Yǒuzǔ 何有祖. 2014. "Jīngzhōu Gāotái 46 hào Xī Hàn mù mùdú jiàodú jì 荊州高臺46號西漢墓木牘校讀記". *Jiǎnbó* 簡帛. http://www.bsm.org.cn/?hanjian/6278.html (last acc. May 31, 2024)

Hé, Yǒuzǔ 何有祖. 2015a. "Dú Lǐyē Qín jiǎn zhájì 讀里耶秦簡札記". *Jiǎnbó* 簡帛. http://www.bsm.org.cn/?qinjian/6431.html (last acc. May 31, 2024)

Hé, Yǒuzǔ 何有祖. 2015b. "Lǐyē Qín jiǎn dú zhuìhé (jiǔ) 里耶秦簡牘綴合(九)". *Jiǎnbó* 簡帛. http://www.bsm.org.cn/?qinjian/6527.html (last acc. May 31, 2024)

Hé, Yǒuzǔ 何有祖. 2016a. "Lǐyē Qín jiǎn 14-469, 14-638 hào jiǎn bǔ shì 里耶秦簡14-469、14-638號簡補釋". *Jiǎnbó* 簡帛. http://www.bsm.org.cn/?qinjian/6767.html (last acc. May 31, 2024)

Hé, Yǒuzǔ 何有祖. 2016b. "Lǐyē Qín jiǎn 15-259 hào jiǎn bǔshì 里耶秦簡15-259號簡補釋". *Jiǎnbó* 簡帛. http://www.bsm.org.cn/?qinjian/6765.html (last acc. May 31, 2024)

Hé, Yǒuzǔ 何有祖. 2017. "Lǐyē Qín jiǎn suǒ jiàn tōngjī lèi wénshū xīn tàn 里耶秦簡所見通緝類文書新探". *Jiǎnbó* 簡帛. http://www.bsm.org.cn/?qinjian/7463.html (last acc. May 31, 2024)

Henricks, Robert G. 1984. "The Hero Pattern and the Life of Confucius". *Journal of Chinese Studies*. New York: American Association for Chinese Studies, I, 3, 241–260.

Higham, Charles. 1996. *The Bronze Age of Southeast Asia*. Cambridge: Cambridge University Press.

Hinsch, Bret. 2002. *Women in Early Imperial China*. Maryland: Rowman & Littlefield Publishers, Inc.

Hsu, Cho-yun,許倬雲. 1980. *Hàn Agriculture: The Formation of Early Chinese Agrarian Economy (206 B.C.–A.D. 220)*. Seattle and London: University of Washington Press.

Hsu, Mei-Ling. 1978. "The Hàn Maps and Early Chinese Cartography". *Annals of the Association of American Geographers*. Washington DC: Taylor and Francis, LXVIII, 1, 45–60.

Hsu, Mei-Ling. 1993. "The Qín Maps: A Clue to Later Chinese Cartographic Development". *Imago Mundi*. London: Routledge, XLV, 90–100.

Hú, Píngshēng 胡平生. 2004. "Dú Lǐyē Qín jiǎn zhájì 讀里耶秦簡札記". *Jiǎndú xué yánjiū* 簡牘學研, 4.

Hú, Píngshēng 胡平生. 2009. "Lǐyē Qín jiǎn 8-455 hào mù fāng xìngzhì chúyì 里耶秦簡8-455號木方性質芻議". *Jiǎnbó* 簡帛, 4, 17–25.

Hú, Píngshēng 胡平生. 2010. "Lùn jiǎnbó biànwěi yǔ liúshī jiǎndú qiǎngjiù 論簡帛辨僞與流失簡牘搶救". *Zhōngguó wénhuà yíchǎn yánjiūyuàn* 中國文化遺產研究院, *Chūtǔ wénxiàn yánjiū* 出土文獻研究. Běijīng: *Zhōnghuá shūjú* 中華書局, 9, 76–108.

Hú, Píngshēng 胡平生. 2012a. "Dú 'Lǐyē Qín jiǎn (yī)' bǐjì 讀'里耶秦簡(壹)'筆記". *Jiǎnbó* 簡帛. http://www.bsm.org.cn/?qinjian/5856.html (last acc. 2024)

Hú, Píngshēng 胡平生. 2012b. "Lǐyē yōng gē fàn Wú Chǔ, diào gǔ gǎnhuái màn bēigē: Dú 'Lǐyē Qín jiǎn' yǔ 'Lǐyē Qín jiǎn jiàoshì' 里耶擁戈泛吳楚, 吊古感懷漫悲歌: 讀'里耶秦簡'與'里耶秦簡校釋'". *Jiǎnbó* 簡帛. http://www.bsm.org.cn/?qinjian/5912.html (last acc. May 31, 2024)

Hú, Píngshēng 胡平生, and Zhāng Défāng張德芳. 2001. *Dūnhuáng xuán quán Hàn jiǎn jiǎnshì* 敦煌懸泉漢簡釋粹. Shànghǎi: *Shànghǎi gǔjí chūbǎnshè* 上海古籍出版社.

Huá, Xuéchéng 華學誠. 2011. *Yāng Xióng 'Fāngyán' jiàoshì lùngǎo* 揚雄'方言'校釋論稿. Běijīng: *Zhōnghuá shūjú* 中華書局.

Huán, Tán 桓譚. 1977. *Xīn lùn* 新論. Shànghǎi: *Shànghǎi rénmín chūbǎnshè* 上海人民出版社.

Huáng, Àiméi 黃愛梅. 1997. "Shuìhǔdì Qín jiǎn yú Lónggǎng Qín jiǎn de bǐjiào 睡虎地秦簡與龍崗秦簡的比較". *Huádōng shīfàn dàxué xuébào* 華東師範大學學報, 4.

Huáng, Shèngzhāng 黃盛璋. 1979. "Yúnmèng Qín jiǎn biànzhèng 雲夢秦簡辨正". *Kǎogǔ xuébào* 考古學報, 1, 1–24.

Húběi shěng Jīngzhōushì Zhōuliángyùqiáo yízhǐ bówùguǎn 湖北省荊州市周梁玉橋遺址博物館. 2001. *Guānjú Qín Hàn mù jiǎndú* 關沮秦漢墓簡牘.

Hulsewé, Anthony F. P. 1961. "Hàn Measures". *T'oung Pao*. Leiden: Brill, 206–207.

Hulsewé, Anthony F. P. 1978. "The Ch'in Documents Discovered in Hupei". *T'oung Pao*. Leiden: Brill, 1, 175–217.

Hulsewé, Anthony F. P. 1985. *Remnants of Ch'in Law: An Annotated Translation of the Ch'in Legal and Administrative Rules of the 3rd century B.C. Discovered in Yün-meng Prefecture, Hu-pei Province, in 1975.* Leiden: Brill.

Húnán shěng wénwù kǎogǔ yánjiūsuǒ 湖南省文物考古研究所. 2003. "Húnán Lóngshān xiàn Lǐyē Zhànguó Qín Hàn chéngzhǐ jí Qín dài jiǎndú 湖南龍山縣里耶戰國秦漢城址及秦代簡牘". *Kǎogǔ* 考古, 15–19.

Húnán shěng wénwù kǎogǔ yánjiūsuǒ 湖南省文物考古研究所. 2007. *Lǐyē fājié bàogào* 里耶發掘報告. Chángshā: *Yuèlù chūbǎnshè* 嶽麓書出版社.

Húnán shěng wénwù kǎogǔ yánjiūsuǒ 湖南省文物考古研究所. 2012a. *Lǐyē Qín jiǎn (yī)* 里耶秦簡(壹). *Wénwù* 文物. Běijīng: *Wénwù chūbǎnshè* 文物出版社.

Húnán shěng wénwù kǎogǔ yánjiūsuǒ 湖南省文物考古研究所. 2012b. *'Lǐyē Qín jiǎn (yī)' jiǎn hào 8:155* '里耶秦簡(壹)'簡號8:155. *Wénwù* 文物. Běijīng: *Wénwù chūbǎnshè* 文物出版社.

Húnán shěng wénwù kǎogǔ yánjiūsuǒ 湖南省文物考古研究所. 2017. *Lǐyē Qín jiǎn (èr)* 里耶秦簡(貳). *Wénwù* 文物. Běijīng: *Wénwù chūbǎnshè* 文物出版社.

Húnán shěng wénwù kǎogǔ yánjiūsuǒ, Xiāngxī Tújiā miáozú zìzhìzhōu wénwù chǔ 湖南省文物考古研究所, 湘西圖佳苗族自治州文物處. 2003. "Xiāngxī Lǐyē Qín dài jiǎndú xuǎn shì 湘西里耶秦代簡牘選釋". *Zhōngguó lìshǐ wénwù* 中國歷史文物, 8–25.

Húnán shěng wénwù kǎogǔ yánjiūsuǒ, Xiāngxī Tǔjiā miáozú zìzhìzhōu wénwù chǔ, Lóngshān xiàn wénwù guǎnlǐsuǒ 湖南省文物考古研究所, 湘西土家族苗族自治州文物處, 龍山縣文物管理所. 2003. "Húnán Lóngshān Lǐyē Zhànguó: Qín dài gǔchéng yīhào jǐng fājué jiǎnbào 湖南龍山里耶戰國:秦代古城一號井發掘簡報". *Wénwù* 文物, 1, 4–35.

Huo, Wei. 2008. "Center and Periphery: The Expansion and Metamorphosis of Hàn Culture". *Frontiers of History in China*. Běijīng: Higher Education Press, III, 2, 293–322.

I-tien, Hsing. 2013. "Qín-Hàn Census and Tax and Corvée Administration: Notes on Newly Discovered Texts". *Birth of an Empire: The State of Qín Revisited*, by Yuri Pines et al. Berkeley: University of California Press, 155–186.

Itō, Michiharu 伊藤道治, and Takashima Ken-ichi 高嶋謙. 1996. *Studies in Early Chinese Civilization: Religion, Society, Language and Paleography*. Osaka: Kansai Gaidai University Publication.

Iversen, Paul A. 2017. "The Calendar on the Antikythera Mechanism and the Corinthian Family of Calendars". *Hesperia*. Athens: The Journal of the American School of Classical Studies at Athens, 1–3, 129–203.

James, Paul. 1996. *Nation Formation: Towards a Theory of Abstract Community*. London: Sage.

Jīn, Lí 金立. 1976. "Jiānglíng Fènghuáng 8 hào Hàn mù zhújiǎn shìshì 江陵鳳凰8號漢墓竹簡試釋". *Wénwù* 文物. Běijīng: *Wénwù chūbǎnshè* 文物出版社, 6.

Jīngzhōu dìqū bówùguǎn 荆州地區博物館. 1985. "Jiānglíng Zhāngjiāshān sān zuò Hàn mù chūtǔ dàpī zhújiǎn 江陵張家山三座漢墓出土大批竹簡." *Wénwù* 文物. Běijīng: *Wénwù chūbǎnshè* 文物出版社, 1–8.

Kanaya, Osamu 金谷治著. 1960. *Shin Kan Shisōshi Kenkyū* 秦漢思想史研究. Tōkyō: *Nihon Gakujutsu Shinkōkai* 日本学術振興会.
Kantorowicz, Ernst H. 1957. *The King's Two Bodies: A Study in Mediaeval Political Theology*. Princeton: Princeton University Press.
Keightley, David N. 1999. "The Shāng: China's First Historical Dynasty". *The Cambridge History of Ancient China*, by Michael Loewe and Edward L. Shaughnessy. Cambridge: Cambridge University Press, 232–291.
Keightley, David N. 2001. "The Diviners' Notebooks: Shāng Oracle-Bone Inscriptions as Secondary Sources". *Actes du Colloque International Commemorant le Centenaire de la Decouverte des Inscriptions sur os eat Carapaces (Proceedings of the International Symposium in Commemoration of the Oracle-Bone Inscriptions Discovery)*, by Yau Shun-chiu and Chrystelle Marechal. Paris: Editions Languages Croises.
Kern, Martin. 2000. *The Stele Inscriptions of Ch'in Shih-huang: Text and Ritual in Early Chinese Imperial Representation*. Ann Harbor: American Oriental Society.
Kern, Martin. 2009. "Bronze Inscriptions, the Shījīng and the Shāngshū: The Evolution of the Ancestral Sacrifice During the Western Zhōu". *Early Chinese Religion, Part One: Shāng through Hàn (1250 BC to 220 AD)*, by John Lagerwey and Marc Kalinowski. Leiden: Brill, 143–200.
Kesner, Ladislav. 1995. "Likeness of No One: (Re)presenting the First Emperor's Army". *The Art Bulletin*. New York: College Art Association, 115–132.
Kim, Moonsil Lee. 2016. "Discrepancy between Laws and their Implementation: An Analysis of Granaries, Statutes, and Rations during China's Qín and Hàn Periods". *Journal of the Economic and Social History of the Orient*. Leiden: Brill, 555–589.
Kiser, Edgar, and Cài Yǒng 蔡泳. 2003. "War and Bureaucratization in Qín China: Exploring an Anomalous Case". *American Sociological Review*. London: Sage, 8, 511–539.
Kominami, Ichirō 小南一郎. (1992). "Temmei to toku 天命と德". *Tōōhō gakuhō* 東方學 報, 64, 1–59.
Kudō, Motoo 工藤元男. 2010. *Shuìhǔdì Qín jiǎn suǒjiàn Qín dài guójiā yú shèhuì* 睡虎地秦簡所見秦代國家與社會. Shànghǎi: *Shànghǎi gǔjí chūbǎnshè* 上海古籍出版社.
Lander, Brian G. 2015. *Environmental Change and the Rise of the Qín Empire: A Political Ecology of Ancient North China*. Doctoral Dissertation. New York: Columbia University.
Langford, Heather. 2009. *The Textiles of the Hàn Dynasty and Their Relationship with Society*. Master's Thesis. Adelaide: University of Adelaide Press.
Lao, Kan. 1977. "The Unification of the States and the Fall of Ch'in". *Bulletin of the Institute of History and Philology*. Táiběi: Academia Sinica, 302–307.
Lau, Ulrich. 1999. "Hàn-zeitliche Rechtsentscheidungen aus Auskunftsquellen zur Stellung der Frau". *Frauenleben im Traditionellen China: Grenzen und Möglichkeiten einer Rekonstruktion*, by Monika Übelhör. Marburg: Universitätsbibliothek Marburg, 37–59.
Lau, Ulrich. 2002. *Die Rekonstruktion des Strafprozesses und die Prinzipien der Strafzumessung zu Beginn der Hàn-Zeit im Lichte des Zòuyànshū*. Vol. II. *Und folge nun dem, was mein Herz begehrt: Festschrift für Ulrich Unger zum 70. Geburtstag*, by Reinhard Emmerich and Hans Stumpfeldt. Hamburg: Hamburger Sinologische Schriften.
Lau, Ulrich. 2014. "Qín Criminal Case Records of the Collection 'Wéi yù děng zhuàng'". *Oriens Extremus*. Wiesbaden: Harrassowitz Verlag, 139–192.
Lau, Ulrich, and Thies Staack. 2016. *Legal Practice in the Formative Stages of the Chinese Empire: An Annotated Translation of the Exemplary Qín Criminal Cases from the Yuèlù Academy Collection*. Leiden: Brill.
Lewis, Marc E. 1990. *Sanctioned Violence in Early China*. Albany: State University of New York Press.
Lewis, Marc E. 1999. *Writing and Authority in Early China*. Albany: State University of New York Press.

Lewis, Marc E. 2006. *The Construction of Space in Early China*. Albany: State University of New York Press.

Lewis, Marc E. 2007. *The Early Chinese Empires: Qín and Hàn (History of Imperial China)*. Cambridge: Harvard University Press.

Lǐ, Fēng 李豐. 2008. *Bureaucracy and the State in Early China: Governing the Western Zhōu*. Cambridge: Cambridge University Press.

Lǐ, Héngquán 李恒全. 2018. "Lùn 'shuì tián' hé QÍn de tián shuì zhēngshōu fāngshì 論 '稅田'和秦的田稅徵收方式". *Jiǎnbó* 簡帛. http://www.bsm.org.cn/?qinjian/7713.html (last acc. May 31, 2024)

Lǐ Hóngcái 李洪財. 2016. "Qín jiǎndú 'cóngrén' kǎo 秦簡牘'從人'考". *Wénwù* 文物, 12, 65–71.

Lǐ, Jiànpíng 李建平 and Zhāng, Xiǎnchéng 張顯成. 2009. "Xiān Qín liǎng Hàn Wèi Jìn jiǎnbó liàngcí xī lùn 先秦兩漢魏晉簡帛量詞析論". *Zhōnghuá wénhuà lùntán* 中華文化論壇, 4, 69–75.

Lǐ, Lì 李力. 2007. *'Lìchén qiè' shēn fèn zài yánjiū* ' 隸臣妾'身份再研究. Běijīng: *Zhōngguó fǎzhì chūbǎnshè* 中國法制出版社.

Lǐ, Lì 李力. 2009. "Lùn 'túlì' de shēnfen: Cóng xīn chūtǔ Lǐyē Qín jiǎn rùshǒu 論'徒隸'的身份: 從新出土里耶秦簡入手". *Chūtǔ wénxiàn yánjiū* 出土文獻研究. Shànghǎi: *Shànghǎi gǔjí chūbǎnshè* 上海古籍出版社, 8, 33–42.

Lǐ, Líng 李零. 2011. "Qín jiǎn de dìngmíng yǔ fēnlèi 秦簡的定名與分類". *Jiǎnbó* 簡帛, 1–11.

Lǐ, Shìchí 李世持. 2017. *Qín jiǎn rénmíng zhěnglǐ yǔ mìngmíng yánjiū* 秦簡人名整理與命名研究. *Xīnán dàxué bóshì xuéwèi lùnwén* 西南大學博士學位論文. Chóngqìng: *Xīnán dàxué* 西南大學.

Lǐ, Shǒukuí 李守奎. 2003. *Chǔ wénzì biān* 楚文字編. Shànghǎi: *Huádōng shīfàn dàxué* 華東師範大學.

Lǐ, Xiùzhēn 李秀珍 et al. 2016. "Marking Practices and the Making of the Qín Terracotta Army". *Journal of Anthropological Archaeology*. Amsterdam: Elsevier, 169–183.

Lǐ, Xuéqín 李學勤. 1981. "Qín jiǎn de gǔ wénzìxué kǎochá 秦簡的古文字學考察". *Yúnmèng Qín jiǎn yánjiū* 雲夢秦簡研究. Běijīng: *Zhōnghuá shūjú* 中華書局.

Lǐ, Xuéqín 李學勤. 2002a. "Shì shuō zhāng jiāshān jiǎn 'shǐ lǜ' 試說張家山簡'史律'". *Wénwù* 文物, 69–72.

Lǐ, Xuéqín 李學勤. 2002b. "The Xià-Shāng-Zhōu Chronology Project: Methodology and results". *Journal of East Asian Archaeology*. Leiden: Brill, 321–333.

Lǐ, Xuéqín 李學勤. 2003. "Chū dú Lǐyē Qín jiǎn 初讀理耶秦簡". *Wénwù* 文物, 73–81.

Lǐ, Xuéqín 李学勤, and Wén Xíng 文邢. 2001. "New Light on the Early-Hàn Code: A Reappraisal of the Zhāngjiāshān Bamboo-slip Legal Texts". *Asia Major*. Táiběi: Academia Sinica, 125–146.

Lǐ, Yùfú 李玉福. 2002. *Qín Hàn zhìdù shǐlùn* 秦漢制度史論. Jǐnán: *Shāndōng dàxué* 山東大學.

Liáng, Wànrú 梁萬如, and Ráo Zōngyí 饒宗頤. 2017. *Shāngjūn shū* 商君書. Hong Kong: *Zhōnghuá shūjú* 中華書局.

Liáng, Wěijié 梁煒傑. 2013. "Lǐyē Qín jiǎn (yī) zhá jì – 'zuò tú bù' lèxíng fǎnyìng de Qín 'jù' yìyì 里耶秦簡(壹)札記–'作徒簿'類型反映的秦'冣'意義". *Jiǎnbó* 簡帛. http://www.bsm.org.cn/?qinjian/6127.html (las acc. May 31, 2024)

Liáng, Zhìpíng 梁治平. 1989. "Explicating 'Law': A Comparative Perspective of Chinese and Western Legal Cultures". *Columbia Journal of Chinese Law*. New York: Columbia University Press, III, 1, 55–92.

Liào, Bóyuán 廖伯源. 2003. "Hàn chū xiàn lì zhī zhì jiē jí qí rènmìng: Zhāngjiāshān Hàn jiǎn yánjiū zhī yī 漢初縣吏之秩階及其任命: 張家山漢簡研究之一". *Shèhuì kēxué zhànxiàn* 社會科學戰線, XIII, 2, 100–107.

Lín, Jiànmíng 林剑鸣. 1981. *Qín shǐ gǎo* 秦史稿. Shànghǎi: *Shànghǎi rénmín chūbǎnshè* 上海人民出版社.

Liú, Gāng 劉剛. 2012. *Hú Xiāng lìdài shūfǎ xuǎnjí* 湖湘历代書法選集. Vol. IV. Chángshā: *Húnán měishù chūbǎnshè* 湖南美术出版社.

Liú, Guózhōng 劉國忠. 2016. *Introduction to the Tsinghua Bamboo-Strip Manuscripts*. Leiden: Brill.

Liú, Hǎinián 劉海年. 1985. "Guānyú Zhōngguó suìxíng de qǐyuán 關於中國歲刑的起 源". *Fǎxué yánjiū* 法學研究, 5, 67–76.

Liú, Mǐn 劉敏. 2004. "Zhāngjiāshān Hàn jiǎn 'xiǎo jué' yìshì 張家山漢簡'小爵'臆釋". *Zhōngguó shǐ yánjiū* 中國史研究, 3, 94–104.

Liú, Mǐn 劉敏. 2009. "Qín Hàn shíqī de 'cì mín jué' jí 'xiǎo jué' 秦漢時期的'賜民爵'及'小爵'". *Shǐxué yuèkān* 史學月刊. Kāifēng: *Hénán dàxué* 河南大學.

Liú, Qìngzhù 劉慶柱, and Anthony Barbieri-Low. 2007. "Archeological Discovery and Research into the Layout of the Palaces and Ancestral Shrines of Hàn Dynasty Cháng'ān: A Comparative Essay on the Capital Cities of Ancient Chinese Kingdoms and Empires". *Early China*. Cambridge: Cambridge University Press, 113–143.

Liú, Ruì 劉瑞. 2003. "Lǐyē Qín dài mùdú língshí 里耶秦代木牘零拾". *Zhōngguó wénwù bào* 中國文物報.

Lǐyē gǔchéng Qín jiǎn yǔ Qín wénhuà yánjiū 里耶古城秦简与秦文化研究. 2009. *Zhōngguó shèhuì kēxuéyuàn* 中國社會科學院. Běijīng: *Kēxué chūbǎnshè* 科学出版社.

Lǐyē Qín jiǎndú jiàoshì xiǎozǔ 里耶秦簡牘校釋小組. 2014. "Xīn jiàn Lǐyē Qín jiǎndú zīliào xuǎnjiào (yī) 新見里耶秦簡牘資料選校(一)". *Jiǎnbó* 簡帛. http://m.bsm.org.cn/?qinjian/6246.html (last acc. May 31, 2024)

Lobell, Jarrett A. 2007. "The Antikythera Mechanism". *Archaeology*. Boston: Archaeological Institute of America, LX, 2, 42–45.

Loewe, Michael. 1977. "Manuscripts Found Recently in China: A Preliminary Survey". *T'oung Pao*. Leiden: Brill, 99–136.

Loewe, Michael. 1994. "China's Sense of Unity as Seen in the Early Empires". *T'oung Pao*. Leiden: Brill, 80, 6–26.

Loewe, Michael. 2011. *Dǒng Zhòngshū, a 'Confucian' Heritage and the Chūnqiū Fánlù*. Leiden: Brill.

Lóng, Jīngshā 龍京沙. 2001. "Xiāngxī Lǐyē Qín jiǎn dú xuǎn shì 揀湘西里耶秦簡牘選釋". *Zhōngguó lìshǐ wénwù* 中國歷史文物. Běijīng: *Zhōngguó guójiā bówùguǎn* 中國國家博物館.

Lóng, Jīngshā 龍京沙. 1999. *The Cambridge History of Ancient China: From the Origins of Civilization to 221 B.C.* Cambridge, New York: Cambridge University Press.

Lóng, Jīngshā 龍京沙. 2005. "On the Terms *bǎo zǐ*, *yín gōng*, *yín guān*, *huàn*, and *shōu*: Was Zhào Gāo a Eunuch?" *T'oung Pao*. Leiden: Brill, 301–319.

Lǔ, Jiāliàng 魯家亮. 2013. "Lǐyē chūtǔ Qīn 'bǔniǎo qiúyǔ' jiàn chūtàn 里耶出土秦'捕鳥求羽'簡初探". *Gǔdài Chángjiāng zhōngyóu shèhuì yánjiū* 古代長江中游社會研究, by Wèi Bīn 魏斌. Shànghǎi: *Shànghǎi gǔjí chūbǎnshè* 上海古籍出版社, 91–111.

Lǔ, Jiāliàng 魯家亮. 2014. "Lǐyē Qín jiǎn 'lìngshǐ xíng miào' wénshū zài tàn 里耶秦簡'令史行廟'文書再 探". *Jiǎnbó yánjiū* 簡帛研究. Guìlín: *Guǎngxī shīfàn dàxué chūbǎnshè* 廣西師範大學出版社, 45–47.

Lüdke, Michael. 2016. *Professional Practice: Law in Qín and Hàn China*. Doctoral Dissertation. Heidelberger Dokumentenserver der Universitätsbibliothek Heidelberg. http://archiv.ub.uniheidelberg.de/volltextserver/22321/1/Professional%20Practice.%20Law%20in%20Qin%20and%20Han%20China%20%28pdf-A.2%29.pdf.

Lycas, Alexis. 2019. "The Southern Mán People as a Political and Fiscal Problem in Hàn Times". *Monumenta Serica: Journal of Oriental Studies*. Sankt Augustin: Institut Monumenta Serica, 145–164.

Ma, Tsang Wing 馬增榮. 2017. "Scribes, Assistants, and the Materiality of Administrative Documents in Qín–Early Hàn China: Excavated Evidence from Lǐyē, Shuìhǔdì, and Zhāngjiāshān". *T'oung Pao*. Leiden: Brill, 297–333.

Ma, Tsang Wing 馬增榮. 2020a. “12.B Excavated Texts. Volume 1: Contexts”. *Handbook of Ancient Afro-Eurasian Economies*. Berlin: De Gruyter, 529–556.

Ma, Tsang Wing 馬增榮. 2020b. “Qín dài jiǎndú wénshū xué de gè’àn yánjiū: Lǐyē Qín jiǎn 9-2283, 16-5 hé 16-6 sān dú de wùzhì xíngtài, wénshū gòuchéng hé chuándì fāngshì 秦代簡牘文書學的個案研究: 里耶秦簡9-2283、16-5和16-6三牘的物質形態、文書構成和傳遞方式”. *Bulletin of the Institute of History and Philology*. Táiběi: Academia Sinica, 348–418.

Mǎ, Fēibǎi 馬非百. 1985. *Qín Shǐhuángdì zhuàn* 秦始皇帝傳. Nánjīng: *Nánjīng Jiāngsū gǔjí chūbǎnshè* 江蘇古籍出版社.

Mǎ, Yí 馬怡. 2005. “Lǐyē Qín jiǎn xuǎnjiào 里耶秦簡選校”. *Jiǎnbó* 簡帛. http://www.bsm.org.cn/show_article.php?id=113 (last acc. January 12, 2019).

Mǎ, Yí 馬怡. 2006. “Biǎnshū shìtàn 扁書試探”. *Jiǎnbó* 簡帛, 415–428.

Milburn, Olivia. 2020. “Featherwork in Early and Medieval China”. *Journal of the American Oriental Society*. Ann Harbor: American Oriental Society, CXL, 3, 549–564.

Mittag, Achim. 2003. “The Qín Bamboo Annals of Shuìhǔdì: A Random Note from the Perspective of Chinese Historiography”. *Monumenta Serica: Journal of Oriental Studies*. Sankt Augustin: Institut Monumenta Serica, 543–570.

Mizuma, Daisuke 水間大輔. 2012. “Qín Hàn shíqī chéngdān fùyù de jīguān yǔ guānlì 秦漢時期承擔覆獄的機關與官吏”. *Jiǎnbó* 簡帛, 277–295.

Murthy, Viren. 2006. “Modernity against Modernity: Wāng Huī’s Critical History of Chinese Thought”. *Modern Intellectual History*. Cambridge: Cambridge University Press, 137–165.Ne

Needham, Joseph. 1995. *Science and Civilisation in China. III: Mathematics and the Sciences of the Heavens and the Earth*. Cambridge: Cambridge University Press.

Needham, Joseph. 2000. *Science and Civilisation in China. IV: Mechanical Engineering*. Cambridge: Cambridge University Press.

Needham, Joseph, and Wáng Líng 王鈴. 1959. *Science and Civilisation in China: Mathematics and the Sciences of the Heavens and the Earth*. Vol. 3. Cambridge: Cambridge University Press.

Neininger, Ulrich. 1983. *Burning the Scholars Alive: On the Origin of a Confucian Martyrs’ Legend*. Vol. II, *East Asian Civilizations: New Attempts at Understanding Traditions*, by Wolfram Eberhard, Krysztof Gawlikowski and Carl-Albrecht Seyschab. Munich: Simon & Magiera, 121–136.

Nickel, Lukas. 2013. “The First Emperor and Sculpture in China”. *Bulletin of the School of Oriental and African Studies*. Cambridge: Cambridge University Press, LXXVI, 3, 413–447.

Nienhauser, William H. 1994. *The Grand Scribe’s Records. The Memoirs of Pre-Hàn China by Ssu-ma Ch’ien*. Vol. VII. Bloomington and Indianapolis: Indiana University Press.

Nylan, Michael. 1994. “The Chin Wen/Ku Wen (New Text/Old Text) Controversy in Hàn Times”. *T’oung Pao*. Leiden: Brill, 83–145.

Nylan, Michael. 2011. *Yáng Xióng and the Pleasures of Reading and Classical Learning in China*. Ann Harbor: American Oriental Society.

Nylan, Michael, and Michael Loewe. 2010. *China’s Early Empires: A Re-appraisal*. Cambridge: Cambridge University Press.

Ōba, Osamu 大庭脩. 1991. *Qín Hàn fǎzhì shǐ yánjiū* 秦漢法制史研究. Translated by Lín Jiànmíng 林剑鸣and others. Shànghǎi: *Shànghǎi rénmín chūbǎnshè* 上海人民出版社.

Onishi, Katsuya 大西克也. 1998.「毆」「也」の交替–六国統一前後に於ける書面言語の一側面 “Chūgoku shutsudo shiryō kenkyū 中國出土資料研究中国出土資料研究.” Osaka: *Chūgoku shutsudo shiryō kenkyūkai* 中国出土資料研究会, II, 4–24.

Onishi, Katsuya 大西克也. 2018. “An Investigation of Clerical Script in Chǔ Regions during the Qín and Hàn Periods, and its Relationship to ‘Scribal Writing’”. *Bamboo and Silk*. Leiden: Brill, I, 2, 359–402.

Pelliot, Paul. 1916. "Le Chou Kung en Caractères Anciens et le Chang Chou Che Wen". *Mémoires Concernant l'Asie Orientale*. Paris: Académie des Inscriptions et Belles-Lettres, II, 123–177.

Péng, Hào 彭浩. 2001. *Zhāngjiāshān Hàn jiǎn 'Suànshù shū' zhùshì* 張家山漢簡'算數書'註釋. Běijīng: *Kēxué chūbǎnshè* 科學出版社.

Péng, Hào 彭浩. 2002. "Dú Zhāngjiāshān Hàn jiǎn 'Xíng shū lǜ' 讀張家山漢簡'行書律'". *Wénwù* 文物. Běijīng: *Wénwù chūbǎnshè* 文物出版社.

Péng, Hào 彭浩. 2008. "Dú Lǐyē 'cí Xiān Nóng' jiǎn 讀里耶'祠先農'簡". *Jiǎnbó* 簡帛. http://m.bsm.org.cn/?qinjian/5054.html (last acc. May 31, 2024)

Péng, Hào 彭浩, Chén Wěi 陳偉, and Kudō Motoo 工藤元男. 2007. *Èrnián lǜlìng yǔ Zòuyànshū (Zhāngjiāshān èrsìqī hào Hàn mù chūtǔ fǎlǜ wénxiàn shìdú)* 二年律令與奏讞書(張家山二四七號漢墓出土法律文獻釋讀). Shànghǎi: *Shànghǎi gǔjí chūbǎnshè* 上海古籍出版社.

Perelomov, Leonard S. 1961. *Imperiia Tsin' – Pervoe Tsentralizovannoe Gosudarstvo v Kitae*. Moscow: Nauka.

Petersen, Jens Østergård. 1995. "Which Books did the First Emperor of Ch'in Burn? On the Meaning of Pai Chia". *Monumenta Serica: Journal of Oriental Studies*. Sankt Augustin: Institut Monumenta Serica, XLIII, 1–52.

Pīn, Yǔqiān 駢宇騫, and Duàn Shū'ān 段書安. 2006. *Èrshí shìjì chūtǔ jiǎnbó zòngshù* 二十世紀出土簡帛綜述. Běijīng: *Wénwù chūbǎnshè* 文物出版社.

Pines, Yuri. 2000. "'The One That Pervades the All' in Ancient Chinese Political Thought: The Origins of 'The Great Unity' Paradigm". *T'oung Pao*. Leiden: Brill, 280–324.

Pines, Yuri. 2004. "The Question of Interpretation: Qín history in Light of New Epigraphic Sources". *Early China*. Cambridge: Cambridge University Press, 1–44.

Pines, Yuri. 2005. "Biases and Their Sources: Qín History in the 'Shǐjì'". *Oriens Extremus*. Wiesbaden: Harrassowitz Verlag, 1–34.

Pines, Yuri. 2005/2006. "Subversion Unearthed: Criticism of Hereditary Succession in the Newly Discovered Manuscripts". *Oriens Extremus*. Wiesbaden: Harrassowitz Verlag, 159–178.

Pines, Yuri. 2008. "Imagining the Empire? Concepts of 'Primeval Unity' in Pre-imperial Historiographic Tradition". *Conceiving the Empire: China and Rome Compared*, by Fritz-Heiner Mutschler and Achim Mittag. Oxford: Oxford University Press.

Pines, Yuri. 2009. *Envisioning Eternal Empire: Chinese Political Thought of the Warring States Era*. Honolulu: University of Hawai'i Press.

Pines, Yuri. 2017a. "Contested Sovereignty: Heaven, the Monarch, the People, and the Intellectuals in Traditional China". *The Scaffolding of Sovereignty: Global and Aesthetic Perspectives on the History of a Concept*, by Zvi Ben-Dor Benite, Stefanos Geroulanos and Nicole Jerr. New York: Columbia University Press, 80–101.

Pines, Yuri. 2017b. *The Book of Lord Shāng: Apologetics of State Power in Early China*. New York: Columbia University Press.

Poo, Mu-chou. 1998. *In Search of Personal Welfare: A View of Ancient Chinese Religion*. Albany: State University of New York Press.

Pú, Mùzhōu 蒲慕州. 2005. *Enemies of Civilization: Attitudes toward Foreigners in Ancient Mesopotamia, Egypt, and China*. Albany: State University of New York Press.

Puett, Michael. 2001. *The Ambivalence of Creation*. Stanford: Stanford University Press.

Puett, Michael. 2005. "Following the Commands of Heaven: The Notion of Mìng in Early China". *The Magnitude of Ming: Command, Allotment, and Fate in Chinese Culture*, by Christopher Lupke. Honolulu: University of Hawai'i Press, 49–69.

Qiān, Sīmǎ 遷司馬. 1959. *Shǐjì: Qín Shǐhuáng běnjì* 史記: 秦始皇本紀. Běijīng: *Zhōnghuá shūjú* 中華書局.

Qiān, Sīmǎ 遷司馬. 1979. *Shǐjì xuǎn* 史記選. Běijīng: *Rénmín wénxué chūbǎnshè* 人民文學出版社.

Sīmǎ, Qiān 司馬遷, and Yáo Zǔ'ēn姚祖恩. 1988. *Shǐjì jīnghuá lù* 史記菁華錄. Shànghǎi: *Shànghǎi gǔjí chūbǎnshè* 上海古籍出版社.

Qiū, Guāngmíng 丘光明. 1992. *Zhōngguó lìdài dùliànghéng kǎo* 中國歷代度量衡考. Běijīng: *Kēxué jìshù wénxiàn chūbanshè* 科學技術文獻出版社.

Qiū, Guāngmíng 丘光明, and Lóng Yángpíng 隆楊平. 2001. *Zhōngguó kēxué jìshù shǐ: dùliànghéng juàn* 中國科學技術史: 度量衡卷. Běijīng: *Kēxué chūbǎnshè* 科學出版社.

Qiú, Xīguī 裘錫圭. 1981. *Zhànguó shídài shèhuì xìngzhì shìtàn* 戰國時代社會性質試探. Jílín: *Jílín rénmín chūbǎnshè* 吉林人民出版社.

Qiú, Xīguī 裘錫圭. 1992. "Sèfū chūtàn 嗇夫初探". *Gǔdài wén shǐ yánjiū xīntàn* 古代文史研究新探. Nánjīng: *Jiāngsū gǔjí chūbǎnshè* 江蘇古籍出版社, 430–523.

Richter, Matthias. 2005. *Guān rén: Texte der altchinesischen Literatur zur Charakterkunde und Beamtenrekrutierung*. Bern: Peter Lang.

Qiú, Xīguī 裘錫圭. 2006. "Tentative Criteria for Discenrning Individual Hands in the Guōdiàn Manuscripts". *Rethinking Confucianism: Selected Papers from the Third International Conference on Excavated Chinese Manuscripts, Mount Holyoke College, April 2004*, by Xíng Wén 邢文. San Antonio: Trinity University Press, 132–147.

Qiú, Xīguī 裘錫圭. 2016. "Punctuation, Premodern". *Encyclopedia of Chinese Language and Linguistics*, by Rint Sybesma. Leiden: Brill, III, 501–506.

Riya Shinkan kōdokukai 里耶秦簡講讀會. 2004. "Riya Shinkan yakuchū 里耶秦簡譯 註". *Chūgoku shutsudo shiryō kenkyū* 中國出土資料研究, 8, 88–137.

Robinson, Rebecca. 2016. *Cult and Calendars in the Ancient Empires of Qín, Hàn, and Rome*. Doctoral Dissertation. Montreal: McGill University.

Sanft, Charles. 2005. "Rituals that Don't Reach, Punishments that Don't Impugn: Jiǎ Yì on the Exclusions from Punishment and Ritual". *Journal of the American Oriental Society*. Ann Harbor: American Oriental Society, 31–44.

Sanft, Charles. 2008. "Progress and Publicity in Early China: Qín Shǐhuáng, Ritual, and Common Knowledge". *Journal of Ritual Studies*. Pittsburgh: University of Pittsburgh Press, 21–37.

Sanft, Charles. 2010. "Law and Communication in Qín and Western Hàn China". *Journal of the Economic and Social History of the Orient*. Leiden: Brill, 679–711.

Sanft, Charles. 2010–2011. "Dǒng Zhòngshū's 'Chūnqiū Juéyù' Reconsidered: On the Legal Interests in Subjective States and the Privilege of Hiding Family Members' Crimes as Developments from Earlier Practice". *Early China*. Cambridge: Cambridge University Press, 141–169.

Sanft, Charles. 2014a. *Communication and Cooperation in Early Imperial China: Publicizing the Qín dynasty*. New York: Suny Press.

Sanft, Charles. 2014b. "Paleographic Evidence of Qín Religious Practice From Lǐyē and Zhōujiātái". *Early China*. Cambridge: Cambridge University Press, 327–358.

Schafer, Edward H. 1990. "The Yeh chung chi". *T'oung Pao*. Leiden: Brill, 147–207.

Scheidel, Walter. 2015. *State Power in Ancient China and Rome*. Oxford: Oxford University Press.

Schuessler, Axel. 2009. *Minimal Old Chinese and Later Hàn Chinese: A Companion to Grammata Serica Recensa*. Honolulu: University of Hawai'i Press.

Selbitschka, Armin. 2018. "I Write Therefore I Am: Scribes, Literacy, and Identity in Early China". *Harvard Journal of Asiatic Studies*. Cambridge: Harvard University Press, 413–478.

Shāng, Qìngfū 商慶夫. 1985. "Qín xínglǜ de yuānyuán jí qí yǎnjìn 秦刑律的淵源及其演進". *Lìshǐ lùncóng* 歷史論叢. Qílǔ: *Qílǔ shūshè* 齊魯書社, 15–39.

Shaughnessy, Edward L. 2009. "Chronologies of Ancient China: A Critique of the 'Xià–Shāng–Zhōu Chronology Project'". *Windows on the Chinese World: Reflections by Five Historians*, by Ho Wing-chung. Lanham: Lexington Books, 15–28.

Shaughnessy, Edward L. 2015. "Unearthed Documents and the Question of the Oral Versus Written Nature of the *Classic of Poetry*". *Harvard Journal of Asiatic Studies*. Cambridge: Harvard University Press, LXXV, 2, 331–375.

Shelach, Gideon, and Yuri Pines. 2006. "Secondary State Formation and the Development of Local Identity: Change and Continuity in the State of Qín (770-221 B.C.)" *Archeology of Asia*, by Miriam T. Stark. Oxford: Blackwell Publishing Ltd., 202–230.

Shěn, Gāng 瀋剛. 2013. "Lǐyē Qín jiǎn [yī] zhōng de 'kè' yǔ jì–jiān tán Zhànguó Qín Hàn shíqī kǎojì zhìdù de liúbiàn 里耶秦簡(壹)中的'課'與'計'—兼談戰國秦漢時期考績制度的流變". *Jiǎnbó* 簡帛. http://www.bsm.org.cn/show_article.php?id=2077 (last acc. January 26, 2016).

Shěn, Gāng 瀋剛. 2015. "Lǐyē Qín jiǎn (yī) suǒ jiàn zuò tú guǎnlǐ wèntí tàntǎo 里耶秦簡(壹)所見作徒管理問題探討". *Shǐxué yuèkān* 史學月刊. Kāifēng: *Hénán dàxué* 河南大學.

Shěn, Sòngjīn 沈頌金. 2003. "Xiāngxī Lǐyē Qín jiǎn de jiàzhí jí qí yánjiū 湘西里耶秦簡的價值及其研究". *Zhōngguó shǐ yánjiū dòngtài* 中國史研究動態, 17–21.

Shǐ, Zhìlóng 史志龍. 2009. "Qín 'cí Xiān Nóng' jiǎn zài tàn 秦'祠先農'簡再探". *Jiǎnbó* 簡帛. http://m.bsm.org.cn/?qinjian/5283.html (last acc. May 31, 2024)

Shuìhúdì Qín mù zhújiǎn zhènglǐ xiǎo zǔbiān 睡虎地秦墓竹簡整理小組編. 1990. *Shuìhúdì Qín mù zhú jiǎn* 睡虎地秦墓竹簡. Běijīng: *Wénwù chūbǎnshè* 文物出版社.

Sirén, Osvald. 1956. *Chinese Paintings: Leading Masters and Principles*. Vol. I. London: Lund Humphries.

Škrabal. Ondřej. 2019. "Writing before Inscribing: On the Use of Manuscripts in the Production of Western Zhōu Bronze Inscriptions". *Early China*. Cambridge: Cambridge University Press, XLII, 273–332.

Smith, Adam. 2011a. "The Chinese Sexagenary Cycle and the Ritual Origins of the Calendar". *Calendars and Years II: Astronomy and time in the ancient and medieval world*, by John M. Steele. Oxford: Oxbow, 1–37.

Smith, Adam. 2011b. "The Evidence for Scribal Training at Ānyáng". *Writing and Literacy in Early China: Studies from the Columbia Early China Seminar*, by Lǐ Fēng 李豐 and David Prager Branner. Seattle and London: University of Washington Press, 173–205.

Smith, Adam. 2020. "The Understanding and Uses of the 'Day Court Diagram' in Qín-Hàn Hemerology and Calendrical Astrology". *Bamboo and Silk*, by Marc Kalinowski. Leiden: Brill, 173–205.

Smith, Anthony D. 1981. "War and Ethnicity: The Role of Warfare in the Formation, Self Images and Cohesion of Ethnic Communities". *Ethnic and Racial Studies*, IX, 4, 375–397.

Sòng, Shǎohuá 宋少華. 2008. "Chángshā chūtǔ de jiǎndú jí xiāngguān kǎochá 長沙出土的簡牘及相關考察". *Jiǎnbó yánjiū èrlínglínglliù* 簡帛研究二〇〇六, by Bǔ Xiànqún 卜憲群 and Yáng Zhènhóng 楊振紅. Guìlín: *Guǎngxī shífàn dàxué chūbǎnshè* 廣西師範大學出版社.

Sòng, Shǎohuá 宋少華, Zhāng Chūnlóng 張春龍, Zhèng Shǔbīn 鄭曙斌, and Huáng Pǔhuá 黃樸華. 2013. *Húnán chūtǔ jiǎndú xuǎnbiān* 湖南出土簡牘選編. Chángshā: *Yuèlù shūshè* 岳麓書社.

Sou, Daniel S. 2013. *In the Government's Service: A Study of the Role and Practice of Early China's Officials Based on Excavated Manuscripts*. Doctoral Dissertation. Philadelphia: University of Pennsylvania.

Sørensen, Ærenlund. 2010. "How the First Emperor Unified the Minds of Contemporary Historians: The Inadecuate Source Criticism in Modern Historical Works about the Chinese Bronze Age". *Monumenta Serica*. Sankt Augustin: Institut Monumenta Serica, 1–30.

Staack, Thiel. 2019. "'Drafting,' 'Copying,' and 'Adding Notes': On the Semantic Field of 'Writing' as Reflected by Qín and Early Hàn Legal and Administrative Documents". *Bamboo and Silk*. Leiden: Brill, II, 2, 290–318.

Sterckx, Roel. 2009. "The Economics of Religion in Warring States and Early Imperial China". *Early Chinese Religion*, by John Lagerwey and Marc Kalinowski. Leiden: Brill, 862–864.

Sterckx, Roel. 2010. "Religious Practice in Qín and Hàn". *China's Early Empires: A Re-appraisal*, by Michael Nylan and Michael Loewe. Cambridge: Cambridge University Press, 415–429.

Sterckx, Roel. 2020. "Agrarian and Mercantile Ideologies in Western Hàn". *Journal of the Ecomonic and Social History of the Orient*. Leiden: Brill, 63, 465–504.

Sūn Míng 孫銘. 2018. "'Kòu jiàng, yǐwéi lìchén' jiě '寇降, 以為隸臣'解". *Chūtǔ wénxiàn yǔ fǎlǜ shǐ yánjiū* 出土文獻與法律史研究, 7, 138–50.

Sūn, Wénbó 孫聞博. 2014a. "Lǐyē Qín jiǎn 'shǒu', 'shǒu chéng' xīn kǎo – jiān tán Qín Hàn de shǒuguān zhìdù 里耶秦簡'守'、'守丞'新考–兼談秦漢的守官制度". *Jiǎnbó* 簡帛. http://m.bsm.org.cn/?qinjian/6200.html (last acc. May 31, 2024)

Sūn, Wénbó 孫聞博. 2014b. "Qín xiàn de liè cáo yǔ zhū guān 秦縣的列曹與諸官". *Jiǎnbó* 簡帛. http://m.bsm.org.cn/?qinjian/6255.html (last acc. May 31, 2024)

Sūn, Wénbó 孫聞博. 2018. "Bureaus and Offices in Qín County-Level Administration: In Light of an Excerpt from the Lost Hóngfàn wǔxíng zhuàn (Great Plan Five Phases Commentary)". *Bamboo and Silk*. Leiden: Brill, 71–120.

Sūn, Zhānyǔ 孫占宇. 2013. "Tiānshuǐ Fàngmǎtān Qín jiǎn jíshì 天水放馬灘秦簡集釋". *Gānsù wénhuà* 甘肅文化. Lánzhōu: *Gānsù wénhuà chūbǎnshè* 甘肅文化出版社.

Tán, Qíxiāng 譚其驤. 1982. *Zhōngguó lìshǐ dìtújí dì èr cè: Qín, Xī Hàn, Dōng Hàn shíqī* 中國歷史地圖集第二冊: 秦、西漢、東漢時期. Běijīng: *Zhōngguó dìtú chūbǎnshè* 中國地圖出版社.

Takashima, Kenichi 伊藤道治. 1996. *Studies in Early Chinese Civilization: Religion, Society, Language, and Palaeography*. Osaka: Kansai Gaidai University Publication.

Thierry, François. 2017. Les Monnaies de la Chine Ancienne: Des origines à la fin de l'empire. Paris: Les Belles Lettres.

Thorp, Robert L. 1986. "Architectural Principles in Early Imperial China: Structural Problems and Their Solution". *The Art Bulletin*. New York: College Art Association, 360–378.

Tián, Rénlóng 田人隆. 1992. *Xiàng Yǔ* 項羽. Vol. III, *Zhōngguó dà báikè quánshū: Zhōngguó lìshǐ* 中國大百科全書: 中國歷史. Běijīng and Shànghǎi: *Zhōngguó dà báikè quánshū chūbǎnshè* 中國大百科全書出版社.

Tin-Bor, Hui, V. 2001. "The Emergence and Demise of Nascent Constitutional Rights: Comparing Ancient China and Early Modern Europe". Journal of Political Philosophy. Hoboken: Wiley-Blackwell, 373–403.

Tin-bor, Hui V. 2005. *War and State Formation in Ancient China and Early Modern Europe*. New York: Cambridge University Press.

Tomiya, Itaru 冨谷至. 2010. *Monjo gyōsei no Kan teikoku: mokkan, chikukan no jidai* 文書行政の漢帝国: 木簡, 竹簡の時代. Nagoya: Nagoya University Press.

Tong, Chun Fung. 2021. "The Construction of Territories in the Qín Empire". *T'oung Pao*, CVII, 5–6, 509–554.

Tsui, Chung-hui 崔中慧. 2020. *Chinese Calligraphy and Early Buddhist Manuscripts*. Oxford: Indica et Buddhica.

Twitchett, Denis. 1963. *Financial Administration under the T'ang Dynasty*. Cambridge: Cambridge University Press.

Unger, Ulrich. 1989. *Glossar des Klassischen Chinesisch*. Wiesbaden: Harrassowitz Verlag.

Valmisa, Mercedes. 2019. "The Reification of Fate in Early China". *Early China*. Cambridge: Cambridge University Press, 147–199.

van Evera, Stephen. 1999. *Causes of War: Power and the Roots of Conflict*. Ithaca and London: Cornell University Press.

Venture, Olivier (Fēng, Yíchéng 風儀誠). 2012. "Qín dài huìzì, guānfāng cíyǔ yǐjí Qíndài yòngzì xíguàn – cóng Lǐyē Qínjiǎn shuōqǐ 秦代諱字、官方詞語以及秦代用字習慣–從里耶秦簡說起". *Jiánbó* 簡帛, 147–157.

Vogelsang, Kai. 2007. *Geschichte als Problem: Entstehung, Formen und Funktionen von Geschichtsschreibung im Alten China*. Wiesbaden: Harrassowitz Verlag.

von Glahn, Richard. 2017. *The Economic History of China: From Antiquity to the Nineteenth Century*. Cambridge: Cambridge University Press.

Wang, Gungwu. 1974. "Burning Books and Burying Scholars Alive: Some Recent Interpretations Concerning Ch'in Shih Huang". *Papers on Far Eastern History*. Canberra: Australian National University, 144–153, 165–166.

Wáng, Guìhǎi 汪桂海. 1999. *Hàn dài guán wénshū zhìdù* 漢代官文書制度. Nánníng: *Guǎngxī jiàoyù chūbǎnshè* 廣西教育出版社.

Wáng, Guìyuán 王貴元. 2007. "Zhōujiātái Qín mù jiǎndú shìdú bǔzhèng 周家台秦墓簡牘釋讀補正". *Jiǎnbó* 簡帛. http://m.bsm.org.cn/?qinjian/4778.html (last acc. May 31, 2024)

Wáng, Huànlín 王煥林. 2007. *Lǐyē Qín jiǎn jiào gǔ* 里耶秦簡校詁. Táiběi: *Zhōngguó wénlián chūbǎnshè* 中國文聯出版社.

Wáng, Huànlín 王煥林. 2016. "Lǐyē Qín jiǎn shūfǎ tànlüè 里耶秦簡書法探略". *Shūfǎ* 書法. Shànghǎi: *Shànghǎi shūhuà chūbǎnshè* 上海书画出版社, 2, 67–73.

Wáng, Huī 王輝. 1990. *Qīngtóngqì míngwén biān nián jí shì* 青銅器銘文編年集釋. Xī'ān: *Sānqín chūbǎnshè* 三秦出版社.

Wáng, Huī 亡輝, and Chéng Xuéhuá 程學華. 1999. *Qín wénzì jízhèng* 秦文字集證. Táiběi: *Yìwén yìnshūguǎn* 藝文印書館.

Wáng, Lì 王力. 2003. *Wáng Lì gǔ hànyǔ zìdiǎn* 王力古漢語字典. Běijīng: *Zhōnghuá shūjú* 中華書局.

Wáng, Wěi 王偉, and Sūn Zhàohuá 孫兆華. 2014. *'Jī hù' yǔ 'jiàn hù': Lǐyē Qín jiǎn suǒ jiàn Qiānlíng biānhù shùliàng* '積戶'與'見戶': 里耶秦簡所見遷陵編戶數量. Chéngdū: *Sìchuān wénwù* 四川文物.

Wáng, Xiānqiān 王先謙. 1985. *Xúnzǐ jíjiě* 荀子集解. Běijīng: *Zhōnghuá shūjú* 中華書局.

Wáng, Yànhuī 王彥輝. 2012. "Lǐyē Qín jiǎn (yī) suǒ jiàn Qín dài xiàn xiāng jīgòu shèzhì wèntí lícè 里耶秦簡(壹)所見秦代縣鄉機構設置問題蠡測". *Gǔdài wénmíng* 古代文明, 4, 46–57.

Wáng, Yànhuī 王彥輝. 2016. *Qín Hàn hùjí guǎnlǐ yǔ fùyì zhìdù yánjiū* 秦漢户籍管理與賦役制度研究. Běijīng: *Zhōnghuá shūjú* 中華書局, 121–290.

Wáng, Zǐjīn 王子今. 1994. *Qín Hàn jiāotōng shǐ gǎo* 秦漢交通史稿. Běijīng: *Zhōng-Gòng Zhōngyāng dǎngxiào* 中共中央党校.

Wáng, Zǐjīn 王子今. 2016. "Lǐyē Qínjiǎn bǔyǔ de xiāofèi wèntí 里耶秦簡捕羽的消費問題". *Húnán dàxué xuébào* 湖南大學學報, XXX, 4, 27–31.

Watson, Burton. 1961. *Records of the Grand Historian of China: Translations from the Shih chi of Ssu-ma Ch'ien*. New York: Columbia University Press.

Wilkinson, Endymion. 2013. *Chinese History: A New Manual*. Cambridge: Harvard University Press.

Wu, Hung. 1988. "From Temple to Tomb: Ancient Chinese Art and Religion in Transition". *Early China*. Cambridge: Cambridge University Press, 78–115.

Wu, Hung. 2010. *The Art of the Yellow Springs: Understanding Chinese Tombs*. Honolulu: University of Hawai'i Press.

Wú, Jiànwěi 吳建偉. 2006. *Zhànguó Chǔ yīnxì jí Chǔ wénzì gòujiàn xìtǒng yánjiū* 戰國楚音系及楚文字構件系統研究. Jǐnán: *Qí Lǔ* 齐鲁.

Wǔhàn dàxué jiǎnbó yánjiū zhōngxīn 武漢大學簡帛研究中心. 2014. "Xīn jiàn Lǐyē Qín jiǎndú zīliào xuǎn jiào (yī) 新見里耶秦簡牘資料選校(一)". *Jiǎnbó* 簡帛. http://m.bsm.org.cn/?qinjian/6246.html (last acc. May 31, 2024)

Xiè, Kūn 謝坤. 2015. "Dú 'Lǐyē Qín jiǎn (yī)' zhájì (yī) 讀'里耶秦簡(壹)'札記(一)". *Jiǎnbó* 簡帛. http://m.bsm.org.cn/?qinjian/6436.html (last acc. May 31, 2024)

Xiè, Kūn 謝坤. 2016. "Dú 'Lǐyē Qín jiǎn (yī)' zhájì (sān) 讀'里耶秦簡(壹)'札記(三)". *Jiǎnbó* 簡帛. http://m.bsm.org.cn/?qinjian/7437.html (last acc. May 31, 2024)

Xiè, Kūn 謝坤. 2017. "Lǐyē Qín jiǎn suǒjiàn táowáng xiànxiàng: Cóng 'Liáokě táowáng' wénshū de fùyuán shuōqǐ 里耶秦簡所見逃亡現象: 從'繚可逃亡'文書的復原說 起". *Gǔdài wénmíng* 古代文明,1, 48–53.

Xīn, Déyǒng 辛德勇. 2009. "Qín Shǐhuáng sānshíliù jùn xīn kǎo 秦始皇三十六郡新考". *Qín Hàn zhèng qū yǔ biānjiè dìlǐ yánjiū* 秦漢政區與邊界地理研究. Běijīng: *Zhōnghuá shūjú* 中華書局.

Xíng, Wén 邢文. 2016. "Běidà jiǎn Lǎozǐ biànwěi 北大簡老子辨偽". *Guāngmíng rìbào* 光明日報. Běijīng: *Shāngwù yìn shūguǎn* 商務印書館.

Xíng, Yìtián 邢意田. 1983. "Qín Hàn de lǜlìng xué: jiān lùn Cáo Wèi lǜ bóshì de chūxiàn 秦漢的律令學: 兼論曹魏律博士的出現". *Zhōngyāng yánjiūyuàn lìshǐ yǔyán yánjiūsuǒ jíkān* 中央研究院歷史語言研究所集刊. Táiběi: *Guólì Táiwān dàxué túshū guǎn* 國立臺灣大學圖書館, 51–101.

Xíng, Yìtián 邢意田. 2008. "Qín huò Xī Hàn chū hé jiān àn zhōng suǒ jiàn de qīnshǔ lúnlǐ guānxi 秦或西漢初和姦案中所見的親屬倫理關係". *Jiǎnbó* 簡帛. http://m.bsm.org.cn/?hanjian/5052.html (last acc. May 31, 2024)

Xíng, Yìtián 邢意田. 2012a. "Hàn zhì Sānguó gōngwénshū zhōng de qiānshǔ 漢至三國公文書中的簽署". *Wén shǐ* 文史, 3, 166–188.

Xíng, Yìtián 邢意田. 2012b. "'Shǒu', 'bàn', 'yuē wǔ yuē jīng' yǔ 'Qiānlíng gōng' – Lǐyē Qín jiǎn chū dú zhīyī '手'、'半'、'曰啎曰荊'與'遷陵公'–里耶秦簡初讀之一". *Jiǎnbó* 簡帛. http://m.bsm.org.cn/?qinjian/5871.html (last acc. May 31, 2024)

Xióng, Gōngzhé 熊公哲. 1995. *Xúnzǐ jīnzhù jīnyì* 荀子今註今譯. Táiběi: *Táiwān shāngwù yìn shūguǎn* 臺灣商務印書館.

Xú, Fùchāng 徐富昌. 1992. *Shuìhǔdì Qín jiǎn yánjiū* 睡虎地秦簡研究. Vol. I. Táiběi: *Guólì Táiwān dàxué Zhōngguó wénxué yánjiūsuǒ bóshì lùnwén* 國立臺灣大學中國文學研究所博士論文.

Xǔ, Míngqiāng 許名瑲. 2013a. "'Lǐyē Qí jiǎn (yī)' lì rì jiào zhù bǔzhèng '里耶秦簡(壹)'曆日校注補正". *Jiǎnbó* 簡帛. http://m.bsm.org.cn/?qinjian/6069.html (last acc. May 31, 2024)

Xǔ, Míngqiāng 許名瑲. 2013b. "Qín lì shuò rì fùyuán – yǐ chūtǔ jiǎndú wéi xiànsuǒ 秦曆朔日復原–以出土簡牘為線索". *Jiǎnbó* 簡帛. http://m.bsm.org.cn/?qinjian/6052.html (last acc. May 31, 2024)

Xú, Shìhóng 徐世虹. 1997. "Hàn dài de lìfǎ xíngshì yǔ lìfǎ yǔyán 漢代的立法刑事與立法語言". *Nèiměnggǔ dàxué xuébào* 內蒙古大學學報, 66–72.

Xú, Shìhóng 徐世虹. 2004. "'Sanhuángzhī', 'xíng fù chéngdàn chōng', 'jì chéngdàn chōng mǒu suì' jiě: Dú 'Èrnián lǜlìng' zhájì '三環之'、'刑復城旦舂'、'繫城旦舂某歲'解: 讀'二年律令'札記". *Chūtǔ wénxiàn yánjiū* 出土文獻研究. Shànghǎi: *Zhōngxī shūjú* 中西書局.

Xu, Zhi, et al. 2008. "Mitochondrial DNA Evidence for a Diversified Origin of Workers Building Mausoleum for First Emperor of China". *PLoS ONE*, 10: e3275. www.plosone.org doi:10.1371/journal.pone.0003275.

Yamada, Katsuyoshi 山田勝芳. 1990. "Offices and Officials of Works, Markets and Lands in the Ch'in Dynasty". *Acta Asiatica: Bulletin of the Institute of Eastern Culture.* Tōkyō: Tōhō Gakkai, 58, 1–23.

Yán, Xiǎojūn 閆曉君. 2005. "Lüèlùn Qín lǜ duì Hàn lǜ de yǐngxiǎng 略論秦律對漢律的影響". *Gānsù zhèngfǎ xuéyuàn xuébào* 甘肅政法學院學報. Lánzhōu: *Gānsù zhèngfǎ dàxué* 甘肅政法大學, 5.

Yàn, Chāngguì 晏昌貴. 2006. "Èrnián lǜlìng – 'Zhì lǜ' yǔ Hàn chū zhèng qū dìlǐ 二年律令–'秩律'与漢初政區地理". *Lìshǐ dìlǐ* 歷史地理. Shànghǎi: *Shànghǎi rénmín chūbǎnshè* 上海人民出版社, 21, 41–51.

Yàn, Chāngguì 晏昌貴. 2013. "Lǐyē Qín dú 9-712+9-758 bǔshì 里耶秦牘9-712+9-758補釋". *Jiǎnbó* 簡帛. http://m.bsm.org.cn/?qinjian/6147.html (last acc. May 31, 2024)

Yàn, Chāngguì 晏昌貴. 2014. "Lǐyē Qín jiǎn dú suǒ jiàn jùn xiàn mínglù 里耶秦簡牘所見郡縣名錄". *Lìshǐ dìlǐ* 歷史地理. Shànghǎi: *Shànghǎi rénmín chūbǎnshè* 上海人民出版社.

Yáng, Xiānyún 楊先云. 2014. "Lǐyē Qín jiǎn shízì sān zé 里耶秦簡識字三則". *Jiǎnbó* 簡帛. http://m.bsm.org.cn/?qinjian/6171.html (last acc. May 31, 2024)

Yáng, Zhènhóng 楊振紅. 2014. "Cóng Qín 'bāng', 'nèishǐ' de yǎnbiàn kàn Zhànguó-Qín-Hàn shíqī jùnxiàn zhì de fāzhǎn 從秦'邦'、'内史'的演變看戰國秦漢時期郡縣制的發展". *Zhōngguó shǐ yánjiū* 中國史研究, 49–68.

Yáng, Zhènhóng 楊振紅, and Dān Yìnfēi 單印飛. 2014. "Lǐyē Qín jiǎn J1 (16)5, J1 (16)6 de shìdú yǔ wénshū de zhìzuò, chuándì 里耶秦簡J1(16)5、J1(16)6的釋讀與文書的製作、傳遞". *Zhèjiāng xuékān* 浙江學刊. Hángzhōu: *Shuāng yuè* 雙月.

Yáng, Zōngbīng 楊宗兵. 2004. "Lǐyē Qín jiǎn xiàn 'shǒu', 'chéng', 'shǒuchéng' tóngyì shuō 里耶秦簡縣'守'、'丞'、'守丞'同意說". *Běifāng lùncóng* 北方論叢. Hā'ěrbīn: *Shuāng yuè kān* 雙月刊, 6, 11–14.

Yao Alice. 2016. *The Ancient Highlands of Southwest China: From the Bronze Age to the Hàn Empire*. Oxford: Oxford University Press.

Yáo, Lěi 姚磊. 2015. "'Lǐyē Qín jiǎn (yī)' suǒ jiàn 'jiǎn' chūtàn '里耶秦簡(壹)'所見'檢'初探". *Jiǎnbó* 簡帛. http://m.bsm.org.cn/?qinjian/6568.html (last acc. May 31, 2024)

Yates, Robin D.S. 1985–1987. "Review: Some Notes on Ch'in Law (Reviewed Work: Article of Remnants of Ch'in Law A. F. P. Hulsewé)". *Early China*. Cambridge: Cambridge University Press, 243–275.

Yates, Robin D.S. 1987. "Social Status in The Ch'in: Evidence From The Yün-meng Legal Documents. Part One: Commoners". *Harvard Journal of Asiatic Studies*. Cambridge: Harvard University Press, 6, 197–237.

Yates, Robin D.S. 1990. "War, Food Shortages, and Relief Measures in Early China". *Hunger in History: Food Short-age, Poverty, and Deprivation*, by Lucile F. Newman, 147–188. New York: Blackwell.

Yates, Robin D.S. 1995. "State Control of Bureaucrats under the Qín: Techniques and Procedures". *Early China*. Cambridge: Cambridge University Press, 331–365.

Yates, Robin D.S. 2009. "Law and the Military in Early China". *Military Culture in Imperial China*, by Nicola Di Cosmo. Cambridge: Harvard University Press, 23–44.

Yates, Robin D.S. 2012–2013. "The Qín Slips and Boards from Well No. 1, Lǐyē, Húnán: A Brief Introduction". *Early China*. Cambridge: Cambridge University Press, 291–329.

Yates, Robin D.S. 2021. "The Economic Activities of a Qín Local Administration: Qiānlíng County, Modern Lǐyē, Húnán Province, 222–209 B.C.E". *Between Command and Market: Economic Thought and Practice in Early China*, by Elisa Sabattini and Christian Schwermann. Leiden: Brill, 244–317.

Yates, Robin D.S. 2022. "The Fate of the Defeated: Qín's Treatment of Their Enemies". *Bamboo and Silk*. Leiden: Brill, V, 1.

Yates, Robin D.S., and Anthony J. Barbieri-Low. 2015. *Law, State, and Society in Early Imperial China*. Leiden: Brill.

Yè Shūshān 葉書珊. 2021. *Lǐyē Qín jiǎn (yī) wénzì yánjiū* 里耶秦簡(壹)文字研究. Xīnběi: *Huā mùlán wénhuà shìyè yǒuxiàn gōngsī* 花木蘭文化事業有限公司, XI.

Yī, Qiáng 伊強. 2012. "Lǐyē Qín jiǎn (yī) zìcí kǎoshì sān zé 里耶秦簡(壹)字詞考釋三則". *Jiǎnbó* 簡帛. http://m.bsm.org.cn/?qinjian/5927.html (last acc. May 31, 2024)

Yóu, Yìfēi 游逸飛. 2013. "Píng Chén Wěi zhǔbiān 'Lǐyē Qín jiǎndú jiàoshì dì yī juàn' 評陳偉主編'里耶秦簡牘校釋第一卷'". *Xīn shǐxué* 新史學, 177–186.

Yóu, Yìfēi 游逸飛. 2015. "Lǐyē Qín jiǎn suǒjiàn de Dòngtíng jùn 里耶秦簡所見的洞庭 郡". *Zhōngguó wénhuà yánjiū suǒ xuébào* 中國文化研究所學報, 29–67.

Yú, Kūnqí 于琨奇. 2012. *Zhànguó Qín Hàn xiǎo nóng jīngjì yánjiū* 戰國秦漢小農經濟研究. Běijīng: *Shāngwù yìnshūguǎn* 商務印書館.

Yú, Xǐngwú 於省吾. 1962. "Huángdì chēnghào de yóulái hé Qín Shǐhuáng de zhèngshì chēnghào 皇帝稱號的由來和秦始皇的正式稱號". *Jílín dàxué shèhuì kēxué xuébào* 吉林大學社會科學學報, II.

Yú, Zhènbō 于振波. 2013. "Qín lǜ zhōng de jiǎ dùn bǐjià jí xiāngguān wèntí 秦律中的甲盾比價及相關問題". *Shǐxué jíkān* 史學集刊, V.

Zāng, Zhīfēi 臧知非. 2017. *Qín Hàn tǔdì fùyì zhìdù yánjiū* 秦漢土地賦役制度研究. Běijīng: *Zhōngyāng biānyì chūbǎnshè* 中央編譯出版社, 112–306.

Zēng, Zhāoyù 曾昭燏, Jiǎng Bǎogēng 蔣寶庚 Jiǎng, and Lí Zhōngyì 黎忠義. 1956. *Yínán gǔ huàxiàngshí mù fājué bàogào* 沂南古畫像石墓發掘報告. Běijīng: *Wénhuàbù wénwù guǎnlǐjú* 文化部文物管理局.

Zhāng, Chūnlóng 張春龍. 2007. "Lǐyē Qín jiǎn cí xiān nóng, cí [pronunciation unknown] hé cí jiào quàn 里耶秦簡祠先農, 祠窨和祠校券". *Jiǎnbó* 簡帛.

Zhāng, Chūnlóng 張春龍. 2012. *Gǔ wénzì yǔ gǔ lìshǐ* 古文字與古代史. Táiběi: *Zhōngyāng Yánjiūyuàn lìshǐ yǔyán yánjiūsuǒ* 中央研究院歷史語言研究所.

Zhāng, Chūnlóng 張春龍. 2019. "Lǐyē gǔchéng de chūtǔ Qín jiǎn chuándì le nǎxiē xìnxī – Zhāng Chūnlóng zài 'Guó bó jiǎngtáng' de yǎnjiǎng 里耶古城的出土秦簡傳遞了哪些信息–張春龍在'國博講堂'的演講". *Péng pài* 澎湃. https://www.thepaper.cn/newsDetail_forward_4475484 (last acc. May 31, 2024)

Zhāng, Chūnlóng 張春龍, and Lóng Jīngshā 龍京沙. 2003. "Xiāngxī Lǐyē Qín dài jiǎndú xuǎnshì 湘西里耶秦代簡牘選釋". *Zhōngguó lìshǐ wénwù* 中國歷史文物. Běijīng: *Zhōngguó guójiā bówùguǎn* 中國國家博物館.

Zhāng, Chūnlóng 張春龍, Ohkawa Toshitaka 大川俊隆, and Momiyama Akira 籾山明. 2015. "Lǐyē Qín jiǎn kèchǐ jiǎn yánjiū: Jiān lùn Yuèlù Qín jiǎn shù zhōng de wèi jiědú jiǎn 里耶秦簡刻齒簡研究: 兼論嶽麓秦簡數中的未解讀簡". *Wénwù* 文物. Běijīng: *Wénwù chūbǎnshè* 文物出版社.

Zhāng, Dàkě 張大可. 1990. *Shǐjì quánběn xīn zhù* 史記全本新注. Xī'ān: *Sānqín chūbǎnshè* 三秦出版社.

Zhāng, Jīn 張今. 2016. "Lǐyē Qín jiǎn zhōng de jié 里耶秦簡中的楬". *Jiǎnbó* 簡帛. http://m.bsm.org.cn/?qinjian/6766.html (last acc. May 31, 2024)

Zhāng, Jīnguāng 張金光. 2004. *Qín zhì yánjiū* 秦制研究. Shànghǎi: *Shànghǎi gǔjí chūbǎnshè* 上海古籍出版社.

Zhāng, Jùnmín 張俊民. 2003. "Lǐyē Qín jiǎn 'zú shǔ' biàn 里耶秦簡'卒署'辨". *Jiǎnbó* 簡帛.

Zhāng, Péiyú 張培瑜. 1987. *Zhōngguó xiān Qín lìshǐ biǎo* 中國先秦史曆表. Běijīng: *Qílǔ shūshè* 齊魯書社.

Zhāng, Róngmíng 張榮明. 2001. "Zhōngguó shànggǔ guójiā de chǎnshēng jí tèzhēng 中國上古國家的產生及特徵". *Shǐxué yuèkān* 史學月刊, 12–15.

Zhāng, Xīnchāo 張新超. 2014. "Qín dài 'chéngdàn chōng' kǎo biàn: Jiān lùn Qín lǜ de yīxiē tèdiǎn 秦代'城旦舂'考辯: 兼論秦律的一些特點". *Shǐxué yuèkān* 史學月刊, 10, 24–34.

Zhāng, Xiūguī 張修桂. 1981. "Dòngtínghú yǎnbiàn de lìshǐ guòchéng 洞庭湖演變的歷史過程". *Lìshǐ dìlǐ* 歷史地理. Shànghǎi: *Shànghǎi rénmín chūbǎnshè* 上海人民出版社.

Zhāng, Zhāoyáng 張朝陽. 2008. "A Note on Civil Cases in Early China". *Journal of the American Oriental Society*. Ann Harbor: American Oriental Society, 1–3, 121–130.

Zhào, Shǒuzhèng 趙守正. 1987. *Guǎnzǐ zhùyì* 管子註譯. Nánníng: *Guǎngxī rénmín chūbǎnshè* 廣西人民出版社.

Zhī, Qiáng 支強. 2012. "Qín jiǎn zhōng suǒ jiàn de 'bié shū': Dú Lǐyē Qín jiǎn zhájì 秦簡中所見的'別書': 讀里耶秦簡劄記". *Jiǎnbó* 簡帛. http://m.bsm.org.cn/?qinjian/5918.html (last acc. May 31, 2024)

Zhōng, Wěi 鐘煒. 2005. "Qín Dòngtíng, Cāngwú liǎng jùn yuánliú jí dì wàng xīn tàn 秦洞庭、蒼梧两郡源流及地望新探". *Jiǎnbó* 簡帛. http://www.bsm.org.cn/show_article.php?id=135.

Zhōng, Wěi 鍾煒, and Yàn Chāngguì 晏昌貴. 2008. "Chǔ Qín Dòngtíng Cāngwú jí yuánliú yǎnbiàn 楚秦洞庭蒼梧及源流演變". *Jiāng Hàn kǎogǔ* 江漢考古. Wǔhàn: *Húběi shěng wénwù kǎogǔ yánjiūsuǒ* 湖北省文物考古研究所, 92–100.

Zhōu Bèi 周蓓 and Gěng Xiāngxī 耿相新. 2011. "Qièsì kǎo – jiǎn bó shūjí de shōunà yǔ chéngjù 篋笥考–簡帛書籍的收納與盛具". *Zhōngyuán wénwù* 中原文物, 1.

Zhōu, Xiǎolù 周曉陸, and Lù Dōngzhī 路東之 Lù. 2000. *Qín fēngní jí* 秦封泥集. *Sānqín chūbǎnshè* 三秦出版社.

Zhōu, Zhènhè 周鎮鶴. 2005. "Qíndài Dòngtíng, Cāngwú liǎngjùn xuánxiǎng 秦代洞 庭、蒼梧兩郡懸想". *Fùdàn xuébào* 復旦學報, 5, 63–67.

Zhōu, Zhènhè 周振鶴. 2010. *Qín Hàn zōngjiào dìlǐ lüèshuō* 秦漢宗教地理略說. Shànghǎi: *Fùdàn dàxué chūbǎnshè* 復旦大學出版社.

Zhū, Hànmín 朱漢民, and Chén Sōngzhǎng 陳松長. 2010. *Yuèlù shūyuàn cáng Qín jiǎn (yī)* 嶽麓書院藏秦簡(壹). Shànghǎi: *Shànghǎi císhū* 上海辭書.

Zhū, Shèngmíng 朱聖明. 2014. "Lǐyē Qín jiǎn suǒjiàn Qín dài Qiānlíng xiàn gōng chuán xiāngguān wèntí yánjiū 里耶秦簡所見秦代遷陵縣公船相關問題研究". *Gǔdài wénmíng* 古代文明, 4, 47–59.

Zolo, Danilo. 2007. "Contemporary Uses of the Notion of 'Empire'". *The Monist*, XC, 1, 48–64.

Index

https://doi.org/10.1515/9783110791853-012